Communicated and Reviewed
by David Boyce

# Dynamic Travel Choice Models

A Variational Inequality Approach

Springer
*Berlin*
*Heidelberg*
*New York*
*Barcelona*
*Hongkong*
*London*
*Milan*
*Paris*
*Singapore*
*Tokyo*

Huey-Kuo Chen

# Dynamic Travel Choice Models

## A Variational Inequality Approach

With 42 Figures
and 95 Tables

Springer

Professor Huey-Kuo Chen
National Central University
Department of Civil Engineering
No. 38, Wu-Chuan Li
Chung-Li, Taiwan 32054 ROC

ISBN-13: 978-3-642-64207-4 Springer-Verlag Berlin Heidelberg New York

Library of Congress Cataloging-in-Publication Data
Die Deutsche Bibliothek - CIP-Einheitsaufnahme
**Chen, Huey-Kuo**
Dynamic travel choice models : a variational inequality approach;
with 95 tables / Huey-Kuo Chen. - Berlin; Heidelberg; New York :
Springer, 1999
ISBN-13: 978-3-642-64207-4



ISBN-13: 978-3-642-64207-4 e-ISBN-13: 978-3-642-59980-4
DOI: 10.1007/978-3-642-59980-4



Softcover reprint of the hardcover 1st edition 1999



Hardcover design: Erich Kirchner, Heidelberg
SPIN 10692566 42/2202-5 4 3 2 1 0 - Printed on acid-free paper

# Preface

This book is the result of several years of research into the modeling and algorithm of problems in dynamic travel choice and related areas. Three types of discrete-time dynamic travel choice models, along with numerical examples, are presented, i.e., deterministic, stochastic and fuzzy models. The notable features pertaining to these models are as follows:

1. The asymmetric property of the dynamic link travel time function is clearly verified, which leads to a variational inequality formulation.
2. The flow propagation constraint is implicitly defined in our model through the incidence relationship between the link inflow and route flow.
3. The inflow, exit flow, and number of vehicles on a link are identified as the three different *states* over time for the *same* vehicles under the flow propagation process. Therefore, only one link variable needs to be used in our dynamic travel choice models. This treatment can largely simplify the models' complexity, and possibly reduce the computation time.
4. The developed time-space network is consistent with scaling units both in temporal and spatial dimensions.
5. The proposed nested diagonalization method takes care of two types of link interactions, i.e., actual link travel times, and inflows other than that on the subject time-space link.
6. The equivalence analysis between a dynamic travel choice model and its corresponding equilibrium condition is performed by proving both the necessary and sufficient conditions under the presumption of equilibrated actual travel times.
7. The feasible region associated with any dynamic travel choice model is essentially nonconvex.

The aim of this book is to provide a unified account for the development of models and methods for the problem of estimating dynamic equilibrium traffic flows in urban areas. Also, the aim is to show the scope and–just as importantly–the limitations of present traffic models. The development is described and analyzed using the powerful instruments of variational inequalities and nonlinear programming within the field of operations research. Chapter 1 describes the progress of intelligent transportation systems and highlights the role that dynamic travel choice models could play. Chapter 2 provides the background knowledge of variational inequalities and some related mathematical programming concepts. Chapter 3 elaborates the network constraints that are necessary for the dynamic travel choice models and analyses the dynamic link travel time function. Chapter 4 presents the basic dynamic user-optimal (DUO) route choice model. Chapter 5 discusses the proposed dynamic nested diagonalization method in detail. Starting from Chapter 6, a repetitive format is used for various dynamic travel choice models, including the DUO departure time/route choice problem in Chapter 6, the DUO variable demand/route choice problem in Chapter 7, the DUO mode choice models in Chapter 8, the DUO O-D choice models in Chapter 9, and the dynamic system-optimal (DSO) route choice problem in Chapter 10. In Chapter 11, the inclusion of traffic control policies into the DUO route choice model is explored. In Chapters 12 and 13, the stochastic and fuzzy DUO route choice problems, are explored, respectively. Chapter 14 concludes the book with a look at possible applications and future research directions.

This book is intended for an advanced undergraduate or graduate transportation planning and network analysis course, as well as for transportation researchers who want a compilation of developments in the rapidly growing application of operations research. As much as possible, each chapter was organized to be self-contained for easier reference. An effort was made to provide flexibility regarding the order in which topics can be covered. With an appropriate selection of topics, the book can be used for a variety of one- or two-semester courses in dynamic transportation modeling and network analysis. For a one semester course, Chapters 1 through 9 are recommended, whereas for a two-semester course, the entire book can be covered. Since the models introduced in each chapter are repetitively described in a self-contained manner, the reader who is interested in a particular model can jump into the relevant chapter of the book without having to refer back.

Many researchers have contributed their ideas and comments to this work. I especially thank Professor D. E. Boyce for guiding me into the area of transportation planning while I was a Ph.D. student at the University of Illinois, Urbana-Champaign, and for providing valuable comments and editing assistance upon which parts of the book are based. I am also indebted to Professor Kan Chen, formerly at University of Michigan, Ann Arbor, who gave two valuable lectures about the general concept of intelligent vehicle/highway systems (IVHS) at National Central University, in May 1991. Those lectures stimulated me into dedicating myself to the research direction of dynamic modeling. I am very grateful to Professor Bin Ran, then at the Massachusetts Institute of Technology (now at the University of Wisconsin, Madison) for his most valuable insights presented during an intensive course, titled "Dynamic Transportation Networks and Intelligent Vehicle/Highway Systems", that was held at National Central University in the summer of 1994. That course had inspired much of our research activities. Special thanks go to Mr. David Ta-Wei Poo, then the Vice Chairman of China Engineering Consultants, Inc., R.O.C. (Now a Vice President of the Parsons Corporation), who had spent an immeasurable amount of time to read, edit the manuscript and discuss every detail of the contents with me. Some of the discussions invoked valuable rethinking on the book's contents and hence improved the presentation of the book.

I would also like to thank all the members of my "Intelligent Transportation Systems" research group, which was established at National Central University in 1992; Ms. Ya-Li Chiu completed the first group's ITS-related master thesis, titled "Instantaneous Traffic Assignment Models", Mr. Che-Fu Hsueh developed some deterministic dynamic travel choice models by which later research is largely based on, Mr. Ywe-Jeng Chen contributed his effort on comparing computational efficiency of projection-type solution algorithms, Ms. Li-Ly Tseng studied the applicability of the cell transmission model in developing the dynamic travel time function, Mr. Ming-Lin Tu began the study on the stochastic/dynamic user-optimal route choice model, Mr. Chia-Wei Chang accommodated route-based solution algorithms into the DUO route choice model, Ms. Mei-Shiang Chang adopted the concept of fuzzy traffic information into the DUO route choice model, Mr. Chung-Yung Wang extended the DUO route choice model into a capacitated network problem, and Mr. Chung-Jyi Chang dedicated his research on the adaptive vehicles separation control with neuro-fuzzy architecture. The work of all these students had helped me better understand the field of dynamic transportation models, and I am indebted to their contributions.

Appreciation is extended to Professor Nathan Gartner and Professor Chronis Stamatiadis at University of Massachusetts, Lowell, Dr. Der-Horng Lee at the University of California, Irvine, Mr. Owen Jianwen Chen at the Massachusetts Institute of Technology, Dr. Shaw-Pin Miaou at Oak Ridge National Laboratory and my colleagues, Professors Chaio-Fuei Ouyang, Jhy-Pyng Tang, Wei-Ling Chiang, Huei-Wen Chang and John Lee at National Central University for their constant encouragement and support.

Last but not the least, I would like to thank my wife, Huey-Wen, for initializing the idea of writing this book and taking excellent care of our family, and to my three sons, Jiann-Yeu, Jiann-Ann, and Jiann-Yeang for their sacrifice and the understanding they had shown to my obvious neglect during the period of manuscript preparation. Without them, this book would not have been possible.

Huey-Kuo Chen

Chungli, Taiwan (R.O.C)
June, 1998

# Contents

## 12 Stochastic/Dynamic User-Optimal Route Choice Model 229

# Chapter 1

# Analysis of Dynamic Transportation Systems

## 1.1 Introduction

Intelligent transportation systems (ITS) have attracted much interest among researchers and practitioners in the past decade. The main objectives of ITS are to improve the efficiency of transportation networks, enhance traffic safety, and reduce delays and negative environmental effects by utilizing real-time or predicted traffic information. To this end, six system functions have been identified: advanced traveler information systems (ATIS), advanced traffic management systems (ATMS), commercial vehicle operations (CVO), advanced vehicle control systems (AVCS), advanced public transportation systems (APTS) and advanced rural transportation systems (ARTS). (In 1996, US DOT re-classified 29 ITS user services into six function groups, i.e., travel and traffic management, commercial vehicle operations, public transportation management, electronic payment, emergency management, and advanced vehicle control and safety systems. However, use of the previous acronyms persist.) The application of ITS technologies is characterized by the exchange of vast amounts of data among transportation system users, vehicles, transportation operators, and transportation infrastructure, which makes possible the warning and avoidance of congestion or hazardous conditions, automatic collection of tolls, efficient dispatching of trucks and buses, dramatic improvements in traffic safety and many other benefits. In order to advance the development of such technologies, both hardware and software components must become more sophisticated; otherwise, the full benefit of ITS cannot be attained.

Today, hardware component innovations, such as communication, automation, and built-in image processing technologies, are commonly used in ITS; however, software component innovations, such as dynamic transportation network models and traffic incident detection software, are much less frequently applied. The lack of sophisticated software components has delayed the deployment of many large-scale ITS demonstration systems; the failure of some ongoing large-scale demonstration systems shows that the complexity of software components cannot be underestimated. Tackling all of the components of a large-scale system simultaneously may be too ambitious a project to be successful at this time. Decomposing a large-scale system into several smaller subsystems by means of modularization may be a better strategy for further exploration, since each small subsystem *per se* is more manageable.

One of the most challenging modules of software components for ITS concerns the dynamic transportation network models. Though their static counterparts were well developed and applied in transportation planning for several decades, dynamic transportation network models are still in their embryonic stage. In general, dynamic models may be classified into two categories, the vehicle-based approach model, and the flow-based approach model.

The vehicle-based approach emphasizes each individual driver's behavior, and usually adopts a simulation technique to describe the traffic system. This approach has the advantage of closely approximating the behavior of individual drivers in considering alternative traffic control systems; however, the number of attributes involved is normally quite large, and the properties of the solution remain uncertain. In contrast, the flow-based approach is essentially macroscopic, and usually can be formulated as an analytical model. The analytical model concerns the *average* driver's behavior and can by solved by analytical methods. The significant advantages of the analytical models are noted as follows:

1. The derived optimality conditions can be characterized by assumed driver behavior principles, such as utility maximization or equilibrium conditions.
2. Sensitivity analysis for different scenarios is easier to perform since the procedure is less time consuming than with a simulation approach. Also, different traffic policies or control measures can be effectively evaluated and compared in the framework of a model formulation.
3. The time-dependent O-D matrix can be estimated using a criterion, such as squared error minimization or maximum likelihood, subject to some network dispersion measure.

In the past, many analytical dynamic transportation network models have been presented (Merchant and Nemhauser, 1978a, 1978b; Friesz et al, 1989, 1993; Carey, 1986, 1987, 1992; Wie et al, 1990, 1994; Janson, 1991, 1995; Ran and Boyce, 1994, 1996; Boyce, Lee and Janson, 1996); however, no model has solved all of the critical issues. Moreover, acceptance of many models and algorithms have been slowed by the lack of meaningful numerical examples, or because they are too complicated to be readily grasped. Until critical properties have been clearly understood, the actual application of dynamic transportation network models in the real world will be severely restricted.

## 1.2 Significance of Dynamic Transportation Network Models

Dynamic transportation network models differ from their static counterparts in view of the fact that they represent traffic variations over time. Thus, the inclusion of temporal dynamics can better represent real situations. For example, in a large network where a trip could take longer than an hour to complete, a static traffic assignment of hourly travel demand analysis will misrepresent the real situation.

The importance of dynamic transportation network models is best described by the development of ITS and its subsystems.

1. In ATIS, dynamic transportation network models provide anticipatory traffic information to help travelers choose the best route from their origins toward their destinations, select the best departure and arrival times, determine which transportation mode to take, and/or reconsider whether the trip is worth making.
2. In ATMS, dynamic transportation network models are important for predicting the traffic flow patterns for which traffic controls can re-optimize the signal timing for intersections, yield the right-of-way for ambulances, police patrols, public transportation and other mission-oriented vehicles, and impose tolls to prevent the worsening of congestion and negative environmental effects. If dynamic transportation network models and incident detection systems are used in combination, route diversion, and consequently, trip rerouting, becomes possible.
3. In CVO, dynamic transportation network models are useful for fleet dispatching, vehicle routing, recommending alternative routes, etc. More sophisticated and comprehensive logistics operations, such as crew scheduling and stock and freight management systems, will also be possible when relevant technologies, such as global positioning systems, geographic information systems and telecommunication transceivers, are adopted.
4. In APTS, dynamic transportation network models provide relatively precise traffic information as a basis for more reliable traveler information displays. People tend to make more use of public transportation when the boarding time for the public transportation is foreseeable and the travel time difference between public and private vehicles is small.

## 1.3 Traffic Information and Transportation Network Models

Traffic information is the major factor on which travelers base their travel choice decisions. Thus, different types of traffic information have been grouped to yield different flow patterns over transportation networks. We may characterize the traffic information by the following criteria: 1) static versus dynamic; 2) user-optimal versus system-optimal; 3) deterministic, stochastic and/or fuzzy; 4) reactive versus predictive. Any combination of the above four criteria will result in a specific type of traffic information as shown in Table 1.1. For example, the shaded cell in Table 1.1 characterizes predictive deterministic and dynamic user-optimal traffic information.

Table 1.1: Taxonomy of Traffic Information Types

| Traffic Information Type | | Static | | Dynamic | |
|---|---|---|---|---|---|
| | | User-Optimal | System-Optimal | User-Optimal | System-Optimal |
| Deterministic | Reactive | NA | NA | -- | -- |
| | Predictive | -- | -- | 4,6,7,8,9 | 10,11 |
| Stochastic | Reactive | NA | NA | -- | -- |
| | Predictive | -- | -- | 12 | -- |
| Fuzzy | Reactive | NA | NA | -- | -- |
| | Predictive | -- | -- | 13 | -- |

NA means *not applicable*
-- implies *not emphasized*
Numbers refer to *Chapter numbers*

Each type of traffic information can be used to formulate various transportation network models based on different travel choice decisions, i.e., trip generation, origin-destination (O-D) choice, mode choice, and route choice. In addition, combined travel choice models may also be formulated by simultaneously taking two or more travel choice decisions into consideration. One of the most common combined models is *trip distribution and traffic assignment*, which considers origin-destination and route choice within a unified framework. Based on this classification, a multitude of transportation network models can be formulated and explored. However, to examine all possible transportation network models is too ambitious a task; therefore, only selected models are studied. In this book, the focus is mainly on the following *dynamic* travel choice models:

1. user-optimal route choice
2. user-optimal departure time/route choice
3. user-optimal variable demand/route choice
4. user-optimal variable demand/departure time/route choice
5. user-optimal mode choice/route choice
6. user-optimal mode choice/departure time/route choice
7. user-optimal O-D choice/route choice
8. user-optimal O-D choice/departure time/route choice
9. system-optimal route choice
10. network signal control system
11. traffic-responsive signal control system
12. stochastic user-optimal route choice
13. fuzzy user-optimal route choice

The first eleven travel choice models assume perfect traffic information, whereas the last two travel choice models allow for objective and subjective errors, respectively.

## 1.4 Equilibrium over Transportation Networks

Equilibrium plays a central role in many areas of activity, such as economics and engineering disciplines. In economics, for a competitive free market, the equilibrium is the point at which the amount of goods supplied is equivalent to the quantity of goods demanded. In the transportation arena, the equilibrium concept represents a stabilized transportation system for which no (net) forces try to push the system to some other state. In the past, the concept of traffic equilibrium was mainly adopted for *static* transportation network models, which are interpreted as a natural result of the day-to-day adjustments of drivers under the assumption of rational driving behavior. Given any type of traffic information, a reasonable behavioral assumption is that travelers use the shortest travel time route. Since each traveler independently makes his/her travel decision, a consequential route switching mechanism ultimately *stabilizes* the flow pattern over the network. This stabilized flow pattern can be characterized as the traffic ***equilibrium*** at which travelers have no incentive (or net force) for route switching.

When the transportation system is in disequilibrium, there are incentives (forces) that tend to move the system (via ripple effect) toward the equilibrium state by a route-switching mechanism. This ripple effect is governed by the interactions of two functions: 1) a demand function that describes how the flow of passengers increases with improved levels of service; and 2) a performance function that describes how the level of service deteriorates with increasing flow. While the traffic demand function may be analogous to other economic demand functions, the link performance function should not be interpreted as an economic supply function, because it does not take into account the prices of goods as arguments for the output of the quantities supplied.

Operationally, an equilibrium (denoted by superscript *) is in fact governed by the specified rule by which travelers choose a route. The most well-known rules in the field of transportation are Wardrop's principles (Wardrop, 1952). Wardrop's first principle states that the journey times on all routes actually used are equal, and less than (or equal to) those which would be experienced by a single vehicle on any unused route. Hence, the route travel times are equal; no traveler would be better off by unilaterally changing his/her route. Therefore, for each O-D pair *rs*, if the flow over route *p* is positive, i.e., $h_p^{rs*} > 0$, then the corresponding actual route travel time $c_p^{rs*}$ is minimal. However, if no flow occurs on route *p*, i.e., $h_p^{rs*} = 0$, then the corresponding actual route travel time $c_p^{rs*}$ is at least as great as the O-D travel time $\pi^{rs}$. These equilibrium conditions, also called the user-optimal state, can be mathematically expressed as follows:

$$c_p^{rs*} \begin{cases} = \pi^{rs} & \text{if } h_p^{rs*} > 0 \\ \geq \pi^{rs} & \text{if } h_p^{rs*} = 0 \end{cases} \quad \forall r,s,p \tag{1.1}$$

where

$$c_p^{rs} = \sum_a c_a \delta_{ap}^{rs} \quad \forall r,s,p \tag{1.2}$$

Symbol $c_a$ denotes travel time for link $a$, and symbol $\delta_{ap}^{rs}$ represents indicator variable to which either 0 or 1 is realized depending on whether link $a$ is in route $p$ between O-D pair $rs$.

Wardrop's second principle states that the average journey time is a minimum. It implies that for each origin-destination pair, every used route has the same marginal route travel time, that is the change in total travel time corresponding to a change in flow. Equally marginal route travel times for each O-D pair implies the total network travel time is minimal. Therefore, for each O-D pair $rs$, if the flow over route $p$ is positive, i.e., $h_p^{rs*} > 0$, then the corresponding marginal route travel time is minimal. However, if no flow occurs on route $p$, i.e., $h_p^{rs} = 0$, then the corresponding marginal route travel time is at least as great as the minimal marginal route travel time. These equilibrium conditions, also called the system-optimal state, can be mathematically expressed as follows:

$$\hat{c}_p^{rs*} \begin{cases} = \hat{\pi}^{rs} & \text{if } h_p^{rs*} > 0 \\ \geq \hat{\pi}^{rs} & \text{if } h_p^{rs*} = 0 \end{cases} \quad \forall r,s,p \tag{1.3}$$

where

$$\hat{c}_p^{rs} = \sum_a \hat{c}_a \delta_{ap}^{rs} \quad \forall r,s,p \tag{1.4}$$

$$\hat{c}_a = c_a + f_a \frac{\partial c_a}{\partial f_a} \quad \forall a \tag{1.5}$$

Symbol $f_a$ denotes flow on link $a$.

# 1.5 Dynamic Equilibrium Conditions

## 1.5.1 Warrants for Dynamic Extensions

The *static* equilibrium conditions, and hence Wardrop's principles, have long been adopted without much dispute by transportation planners. However, when extended to *dynamic* transportation network models, they do invoke much controversy. The most critical comment is that, in addition to some technical difficulties, dynamic equilibrium conditions do not even exist; therefore, the evolutionary process of the traffic flow pattern over time should be the focus of research. While these arguments may appear to be plausible, the fact is that dynamic extensions of traffic equilibrium models are indeed essential for the improvement of transportation network models for the following reasons:

1. Temporal evolution prevails everywhere in the real world. Therefore, more *accurate* traffic information can be captured only if the time dimension is appropriately taken into account. Indeed, day-to-day traffic information, as with static transportation network models, is not so useful for representing short term traffic variations.
2. Dynamic extensions of transportation network models include their static counterparts as special cases, but differ from their counterparts by giving additional consideration for the traffic variations over time. However, when this added time dimension is temporarily removed or fixed at a certain instant, the traditional static transportation network model should either result or be approximated. The possible advantages associated with static transportation network models, such as those depicted in Section 1.1, can be equally applied to dynamic transportation network models; from a modeling point of view, a general purpose model will give better performance and the possibility of broader applications.
3. The actual dynamic process can be represented by the model. Even though dynamic extensions of traffic equilibria may not be really achieved (as compared with the static traffic equilibria-- also not verified), because of ever-changing situations such as non-recurrent incidents, one cannot deny that the route switching mechanism, or travel choice mechanisms in a broader sense, based on the available traffic information is indeed inherent in travelers. If we treat the dynamic traffic equilibrium as a target or direction for travelers to approach under a certain scenario, and *jump* to another dynamic traffic equilibrium when the scenario changes, then the concept of a dynamic traffic equilibrium can be embedded as an intermediate *state* in a dynamic process. In other words, feedback adjustments can be accommodated for dynamic traffic equilibrium to better represent the dynamic process in the real world.
4. ITS development can be accelerated. Based on short-term, predictive route travel times, many on-line traffic controls, including signal optimization, arterial progression, right-of-way clearance for priority vehicles and real-time applications, such as vehicle route/guidance, emergency evacuation, fleet dispatching are possible.

To avoid possible ambiguity, the term *user-optimal* is preferred to *user-equilibrium* in the context of dynamic transportation network models. Occasionally, modelers may argue that *user-optimal* is not really *optimal* for travelers. Taking both arguments into account, a third term, *dynamic traffic assignment for users*, might be the best candidate for use. However, in consideration of the long tradition, we still adopt both terms, *user-optimal* and *user-equilibrium*, interchangeably for later use. In the following subsections, dynamic extensions of Wardrop's principles according to the types of available traffic information are discussed for route choice decisions.

### 1.5.2 Dynamic User-Optimal Conditions

The dynamic user-optimal conditions state that for each O-D pair, the actual route travel times experienced by travelers departing during the same interval are equal and minimal; no traveler would be better off by unilaterally changing his/her route. In other words, the actual route travel times of any unused route for each O-D pair is greater than or equal to the minimal actual route travel time. Therefore, for each O-D pair *rs*, if the flow over route *p* departing during interval *k* is positive, i.e., $h_p^{rs}(k) > 0$, then the corresponding actual route travel time is minimal. However, if no flow occurs on route *p*, i.e., $h_p^{rs}(k) = 0$, then the corresponding actual route travel time is at least as great as the minimal actual route travel time. These equilibrium conditions can be mathematically expressed as follows:

$$c_p^{rs*}(k)\begin{cases} = \pi^{rs}(k) & \text{if } h_p^{rs*}(k) > 0 \\ \geq \pi^{rs}(k) & \text{if } h_p^{rs*}(k) = 0 \end{cases} \quad \forall r,s,p,k \tag{1.6}$$

where

$$c_p^{rs}(k) = \sum_a \sum_t c_a(t)\delta_{apk}^{rs}(t) \quad \forall r,s,p,k \tag{1.7}$$

$$\pi^{rs}(k) = \min_{p,k}\left\{c_p^{rs*}(k)\right\} \quad \forall r,s \tag{1.8}$$

Symbol *t* denotes entering interval for a link, symbol *k* represents departure interval for a route, and symbol $\delta_{apk}^{rs}(t)$ represents indicator variable to which either 0 or 1 is realized depending on whether link *a* during interval *t* is in route *p* between O-D pair *rs* during interval *k*.

### 1.5.3 Dynamic System-Optimal Conditions

In a dynamic system-optimal problem, various objective functions can be formulated. Ran and Boyce (1994) listed five widely considered objective functions as follows:

1. minimize total travel time;
2. minimize total travel cost or disutility;
3. minimize total number of vehicles during analysis period;
4. minimize average congestion level during analysis period;
5. minimize the length of the congested time period.

For simplicity, only the first objective function is considered throughout this book. The dynamic system-optimal conditions state that for each O-D pair, the marginal route travel times experienced by travelers departing during the same interval are equal and minimal; no traveler would be better off by unilaterally changing his/her route. In other words, the marginal route travel time of any unused route for each

O-D pair is greater than or equal to the minimal marginal route travel time. Therefore, for each O-D pair *rs*, if the flow over route *p* departing during interval *k* is positive, i.e., $h_p^{rs}(k) > 0$, then the corresponding marginal route travel time is minimal. However, if no flow occurs on route *p*, i.e., $h_p^{rs}(k) = 0$, then the corresponding marginal route travel time is at least as great as the marginal route travel time. These equilibrium conditions can be mathematically expressed as follows:

$$\hat{c}_p^{rs*}(k)\begin{cases} = \hat{\pi}^{rs}(k) & \text{if } h_p^{rs*}(k) > 0 \\ \geq \hat{\pi}^{rs}(k) & \text{if } h_p^{rs*}(k) = 0 \end{cases} \quad \forall r,s,p,k \tag{1.9}$$

where

$$\hat{c}_p^{rs}(k) = \sum_a \sum_t \hat{c}_a(t)\delta_{apk}^{rs}(t) \quad \forall r,s,p,k \tag{1.10}$$

The dynamic marginal link travel time function $\hat{c}_a(t)$ can be obtained by taking the derivative of network total travel time with respect to link inflow, as follows:

$$\begin{aligned}\hat{c}_a(t) = \frac{\partial\left(\sum_{a'}\sum_{t'} c_{a'}(t')u_{a'}(t')\right)}{\partial u_a(t)} &= c_a(t)\frac{\partial u_a(t)}{\partial u_a(t)} + \sum_{a'}\sum_{t'} u_{a'}(t')\frac{\partial c_{a'}(t')}{\partial u_a(t)} \\ &= c_a(t) + \sum_{a'}\sum_{t'} u_{a'}(t')\frac{\partial c_{a'}(t')}{\partial u_a(t)}\end{aligned} \tag{1.11}$$

This dynamic marginal link travel time function can be interpreted as effect of an additional traveler on link *a* during interval *t* to the total travel time on all links during all intervals. It is the sum of two components, $c_a(t)$ which is the travel time experienced by that additional traveler when the total link inflow rate is $u_a(t)$, and $\sum_{a'}\sum_{t'} u_{a'}(t')\frac{\partial c_{a'}(t')}{\partial u_a(t)}$ which is the additional travel time that this traveler inflicts on each of the $u_{a'}(t')$ travelers using another link *a′* during another interval *t′*, where general link *a′* includes our specific link *a* and *t′* includes *t*. Strictly speaking, the effect is the result of increasing the flow $u_a(t)$ by one unit.

When the topological link interaction is not considered, equation (1.11) can be simplified to:

$$\hat{c}_a(t) = c_a(t) + \sum_{t'} u_a(t')\frac{\partial c_a(t')}{\partial u_a(t)} \tag{1.12}$$

### 1.5.4 Stochastic/Dynamic User-Optimal Conditions

Note that the basic assumption associated with Wardrop's principles is that the traffic information on which the travelers choose their shortest travel time route is *perfect*. However, in a practical situation, this basic assumption is generally not correct because traffic information is likely to be *imperfect* due to perception errors. When perception errors are associated with an objective probability distribution, then a stochastic model results. For example, if the Gumbel distribution is hypothesized, a *logit* model results. The corresponding equilibrium conditions can be mathematically expressed as follows:

$$\hat{c}_p^{rs*}(k)\begin{cases} = \hat{\pi}^{rs}(k) & \text{if } h_p^{rs*}(k) > 0 \\ \geq \hat{\pi}^{rs}(k) & \text{if } h_p^{rs*}(k) = 0 \end{cases} \quad \forall r,s,p,k \tag{1.13}$$

where

$$\hat{c}_p^{rs*}(k) = c_p^{rs*}(k) + \frac{1}{\theta}\ln h_p^{rs*}(k) \quad \forall r,s,p,k \tag{1.14}$$

The stochastic model reduces to the deterministic model as a special case if the associated error term is zero-valued. In addition, probability distributions other than the Gumbel distribution are also possible. If the normal distribution is adopted for the error term of the mean route travel time, the *probit* model results. The probit model can better represent human behavior, but its computational difficulty may make its application prohibitive.

### 1.5.5 Fuzzy/Dynamic User-Optimal Conditions

Instead of using an objective *probability* distribution in characterizing the error term of the mean route travel time, a subjective *possibility* distribution may also be employed. This introduces the so-called fuzzy concept. In a fuzzy environment, route travel times represented by fuzzy numbers are assumed to be incomplete and uncertain. Each fuzzy number $\tilde{c}$ is characterized by a real number $c$ and a *membership function* $\mu_{\tilde{c}}(c)$ as follows:

$$\tilde{c} = (c, \mu_{\tilde{c}}(c) | c \in R^n, \mu_{\tilde{c}}(c) \in [0,1]) \tag{1.15}$$

The membership function obeys a possibility distribution which is finitely bounded in both its higher and lower limits. The shape of membership functions is versatile; however, the most commonly used functions are either triangular or trapezoidal in form. The most likely possibility occurs at points where the degree of confidence, $\alpha$-cut, equals 1. If all fuzzy numbers were realized at $\alpha$-cut level equal to 1, then the *crisp* concept results. The least likely possibility occurs at points where the degree of confidence $\alpha$-cut equals 0. The entire range of a fuzzy variable with $\alpha$-cut level greater than zero is called *support*. The distance between the closest point (with $\alpha$-cut level equal to 1) and the bound is called *spread*. The length of spread reflects the level of fuzziness of the information.

The fuzzy/dynamic user-optimal conditions state that at a certain $\alpha$-cut level, for each O-D pair, the believed route travel times for travelers departing during the same interval are equal and minimal; no traveler would be better off in terms of believed route travel time by unilaterally changing his/her route. In other words, at a certain $\alpha$-cut level, the believed route travel times of any unused route for each O-D pair is greater than or equal to the minimal believed route travel time. Therefore, at a certain $\alpha$-cut level, for each O-D pair *rs*, if the flow over route $p$ departing during interval $k$ is positive, i.e., $h_p^{rs}(k) > 0$, then the corresponding believed route travel time is minimal. However, if no flow occurs on route $p$, i.e., $h_p^{rs}(k) = 0$, then the corresponding believed route travel time is at least as great as the minimal believed route travel time. These fuzzy traffic equilibrium conditions can be mathematically expressed as follows:

$$\overline{\overline{c}}_{p,\alpha}^{rs*}(k) \begin{cases} = \overline{\overline{\pi}}_{\alpha}^{rs}(k) & \text{if } h_p^{rs*}(k) > 0 \\ \geq \overline{\overline{\pi}}_{\alpha}^{rs}(k) & \text{if } h_p^{rs*}(k) = 0 \end{cases} \quad \forall r,s,p,k,\alpha \tag{1.16}$$

where

$$\overline{\overline{c}}_{p,\alpha}^{rs*}(k) = \sum_a \sum_t \overline{\overline{c}}_{a,\alpha}^{rs*}(t) \delta_{apk}^{rs,B*}(t) \quad \forall r,s,p,k,\alpha \tag{1.17}$$

The reader is referred to Chapter 13 for additional details.

## 1.6 Scope and Organization of the Book

This book is intended to provide a fresh look at dynamic transportation network models, or more precisely, a series of travel choice models classified according to different types of traffic information provided. All of the dynamic travel choice models are discrete, predictive, link-based (except for the stochastic/dynamic user-optimal route choice model) and formulated by the variational inequality (VI) approach. For convenience, we stratify the dynamic travel choice models in terms of the types of travel decisions, as follows.

1. Trip generation type: dynamic user-optimal (DUO) variable demand/route choice, DUO variable demand/departure time/route choice.
2. O-D choice type: DUO O-D choice/route choice, DUO O-D choice/departure time/route choice.
3. Mode choice type: DUO mode choice/route choice, DUO mode choice/departure time/route choice.
4. Route choice type: DUO route choice, DUO departure time/route choice.
5. Dynamic system optimal choices: dynamic system-optimal route choice, dynamic network signal control system, dynamic traffic-responsive signal control system.
6. Stochastic/dynamic user-optimal dynamic route choice.
7. Fuzzy/dynamic user-optimal dynamic route choice.

The first five types of models are deterministic in nature, while the last two types allow for imperfect traffic information. The presentation of the DUO travel choice models in this book is organized and classified by type of travel choice in Figure 1.1, and by hierarchical structure in Figure 1.2. Note that a similar taxonomy, though not depicted, can be adopted to structure the families of system-optimal (Chapter 10), stochastic (Chapter 12), and fuzzy (Chapter 13) travel choice models.

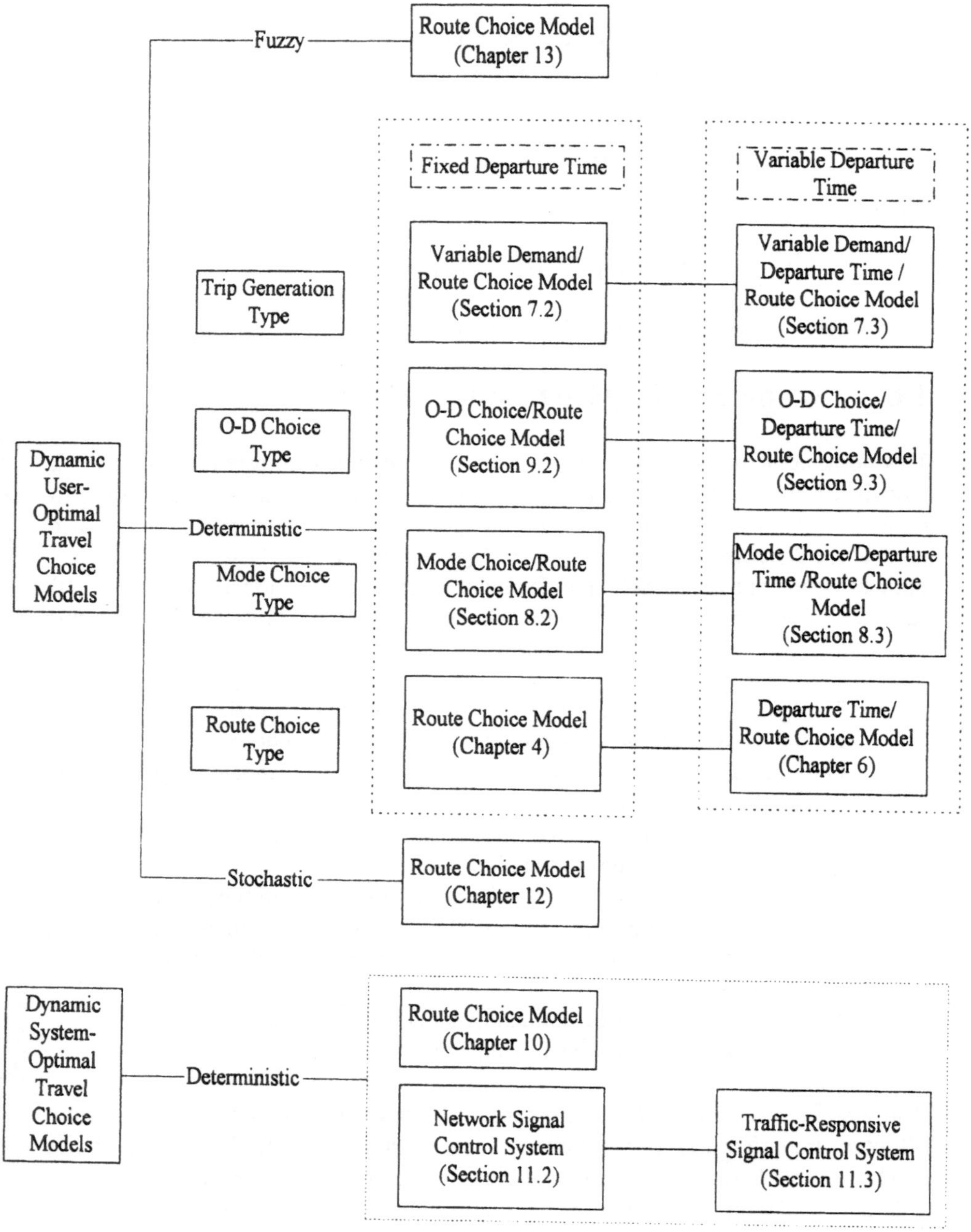


Figure 1.1: Organization of Dynamic Travel Choice Models
(By Type of Travel Choice)

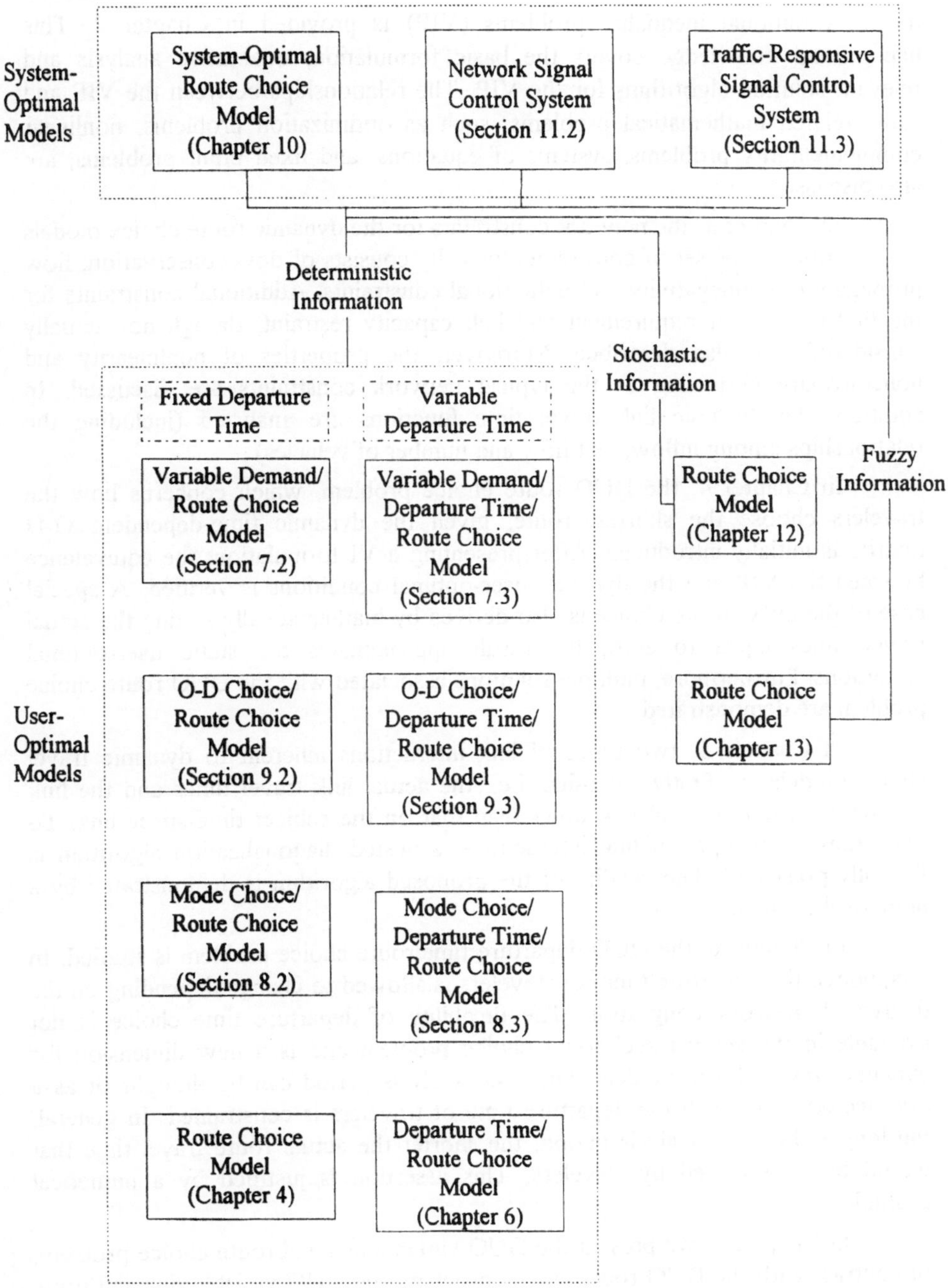


Figure 1.2: Organization of Dynamic Travel Choice Models (By Hierarchical Structure)

Before studying the above dynamic travel choice models, an introduction to the variational inequality problems (VIP) is provided in Chapter 2. This background knowledge covers the basic formulation, qualitative analysis and relevant solution algorithms for the VIP. The relationships between the VIP and some related mathematical problems, such as optimization problems, nonlinear complementarity problems, systems of equations, and fixed point problems, are also discussed.

In Chapter 3, the network constraints for the dynamic route choice models are described. This set of constraints typically consists of flow conservation, flow propagation, nonnegativity and definitional constraints. Additional constraints for the first-in-first-out requirement and link capacity restraint, though not actually considered, are also described. Moreover, the properties of nonlinearity and nonconvexity pertaining to the typical network constraints are discussed. In addition, the dynamic link travel time functions are analyzed (including the relationships among inflow, exit flow and number of vehicles).

In Chapter 4, the DUO route choice problem, which concerns how the travelers choose the shortest route, given the dynamic time-dependent O-D matrix, is initially introduced. After presenting a VI formulation, the equivalence between the VIP and the dynamic user-optimal conditions is verified. A special case of the DUO route choice is also derived by mathematically setting the actual travel times equal to zero; the result approximates the static user-optimal conditions. Furthermore, multiple solutions associated with the DUO route choice problem are demonstrated.

In Chapter 5, two types of link interactions inherent to dynamic travel choice models are firstly identified, i.e., the actual link travel times and the link inflows for each physical link other than that on the subject time-space link. To relax these two types of link interactions, a nested diagonalization algorithm is formally presented. The validity of the proposed algorithm is demonstrated by a numerical example.

In Chapter 6, the DUO departure time/route choice problem is studied. In this model, the departure time for travelers is allowed to change depending on the degree of network congestion. The flexibility of departure time choice is not available in the static travel route choice problem and is a new dimension for dynamic travel choice models. The total analysis period can be thought of as a *time window* to which the departure time of travelers is constrained. In general, the longer the total analysis period, the shorter the actual route travel time that would be experienced by travelers. This assertion is justified by a numerical example.

In Chapter 7, we present the DUO variable demand/route choice problem. In contrast with the DUO route choice problem, the DUO variable demand/route choice problem allows the time-dependent O-D demands to change. By appropriate network representations, both the DUO variable demand/route choice problem and the DUO variable demand/departure time/route choice problem can be transformed into the DUO route choice problem and the DUO departure time/route choice problem, respectively, and solved using the nested diagonalization solution algorithms presented in Chapter 5.

In Chapter 8, the mode choice dimension is introduced and added to the DUO models presented in Chapters 4 and 6. The resulting DUO travel choice models are named the DUO mode choice/route choice model and the mode choice/departure time/route choice model. To demonstrate the models, only binary mode choice, i.e., auto and bus, is considered and a logit type formula is accommodated for mode choice decision.

In Chapter 9, the DUO O-D choice models are presented. An emphasis is placed on the singly-constrained models, in which the number of trips either generated at origins or attracted to destinations is fixed. For the extension of these singly-constrained models to doubly-constrained models, the dynamic user-optimal equilibrium conditions may not be easily obtained since the number of trips both produced at origins and attracted to destinations may be difficult to determine when both the departure time and arrival time are fixed.

In Chapter 10, the dynamic system-optimal (DSO) route choice problem is treated. Except for the route choice criterion using minimal marginal travel time route, the DSO route choice model is exactly the same as the DUO route choice model. Braess's paradox and toll policies are also discussed in this chapter.

The inclusion of traffic control policies into the dynamic route choice model is introduced in Chapter 11. A conceptual network signal control system is first described without numerical examples provided because the whole model is complicated, time constraints prevented the coding tasks to test numerical examples in this presentation. In reality, an alternative traffic-responsive signal control system is applicable. The traffic-responsive signal control system can be formulated as a two-player game. For the *first player*, a signal optimization problem is formulated while for the *second player*, the DUO route choice problem is specified. This two-player game can be solved by the iterative optimization and assignment (IOA) approach in which the Frank-Wolfe method for the *first player* and the nested diagonalization method for the *second player* are accommodated sequentially until a prespecified convergence criterion is satisfied.

In Chapter 12, the stochastic DUO route choice problem is considered. While perfect traffic information is assumed for the deterministic DUO models, the stochastic DUO route choice model hypothesizes that the error term of an actual route travel time is randomly distributed. If the Gumbel distribution is hypothesized, a logit type model results for route choice behavior. A heuristic procedure, called the stochastic dynamic method (SADA), is developed for solutions.

In Chapter 13, the fuzzy DUO route choice problem is explored. In contrast with the stochastic DUO travel choice model, for each link, the fuzzy DUO route choice model adopts a subjective *possibilistic* distribution for a range of actual link travel times, which corresponds to a certain amount of link flows. Under a fuzzy mapping of link flows onto link travel times, the fuzzy DUO route choice problem can be formulated by a fuzzy variational inequality (FVI), which is difficult to solve directly. However, by varying the levels of $\alpha$-cut for actual link travel times, the fuzzy DUO route choice problem can be decomposed into a set of interval-valued variational inequality problems, in which a representative believed link travel time can be estimated and employed to solve the fuzzy DUO route choice problem. A numerical example is provided for demonstration.

Finally, possible applications of dynamic travel choice models are described in Chapter 14. In addition, some important research topics that should be explored in the near future such as link capacity constraint, route-based solution algorithm, and dynamic O-D estimation, are also discussed in hopes of stimulating further development in this enriched area of technology.

## 1.7 Notes

Dynamic transportation network models are superior to their static counterparts in predicting more accurate traffic information for travelers and operators; however, the development of dynamic transportation network models has unfortunately proceeded rather slowly in the past decade, and their increased complexity has hindered many researchers from entering this important field. Furthermore, some important issues are either unsolved or controversial, and therefore more research is needed for elaboration. In the meantime, we have tentatively classified all relevant issues into three categories: 1) conceptual issues; 2) formulation issues; 3) computational issues.

The conceptual issues raise four questions: (1) Which of the analytical-based models and the simulation-based models work best for ITS applications? Or, can they be combined together for ITS development? (2) Is the VI approach the best method for formulating analytical transportation network models? (3) Is temporal discretization necessary for the DUO route choice models? Or, to what extent can the time interval length be determined? (4) What advantages does the actual traffic information have over reactive traffic information for analytical network modeling? Or, more specifically, can reactive dynamic transportation network models be modeled appropriately to consider the flow propagation requirement?

Formulation issues concern the model's representation. Questions of concern include the following. (1) Can vehicles arrive earlier at the destination by leaving later from the origin? (2) Should the link capacity restraint be strictly imposed? (3) How can traffic spillover be considered in the model formulation? (4) Of variable-type and function-type approaches, which is more appropriate in representing the exit flow? (5) Is traffic deformation phenomenon allowed for flow propagation along a route? (6) How many link variables are required for the dynamic travel time function?

Computational issues include: (1) Which algorithm performs best in terms of computational efficiency in solving the DUO route choice model? (2) Which algorithm has a better chance of finding the global optimal solution? (3) How does the number of time intervals for the analysis period affect the computational time? (4) Is there any limitation on network size? (5) How should one calculate the actual route travel time in terms of actual link travel times? By performing the round-off operation for each actual link travel time before summing these times, or by accumulating the actual link travel times before performing the round-off operation for the actual route travel time? (6) How should one update actual link travel times from one iteration to the next? Will the cyclic (or oscillation) phenomenon occur between two (or more) sets of actual link travel times? (7) What kind of convergence criteria should be implemented?

Note that the issues in the different categories may be closely related to each other; for example, item (3) in the first category, item (1) in the second category and items (3), (4) and (6) in the third category all concern the trade-off between improving the real world representation and reducing the computational time requirement. Therefore, caution must be exercised if issues are studied independently. Most of above mentioned issues are discussed when appropriate in later chapters while the remaining issues still require more research before definite conclusions can be drawn.

In many circumstances, combined travel choice models can be transformed into an equivalent route choice problem by an appropriate network representation, termed a *supernetwork*. As a consequence, for those combined travel choice models only one model formulation and one solution algorithm will be needed if computational efficiency is not a primary concern.

# Chapter 2

# Mathematical Background

The dynamic travel choice problems introduced in this book are exclusively formulated with the variational inequality approach. Variational inequalities were originally developed as tools for the study of certain classes of partial differential equations, such as those that arise in mechanics, defined over infinite-dimensional spaces. In contrast, this book focuses on finite-dimensional variational inequality problems that are better suited for the analysis of dynamic travel choice models. To familiarize the reader with the basic background of this topic, we present the mathematical theory necessary for modeling and solving both static and dynamic travel choice problems.

In Section 2.1, the variational inequality problem (VIP) and some related mathematical problems, are introduced. In Section 2.2, the existence and uniqueness properties of the VIP are analyzed. Solution algorithms are presented in Section 2.3. Finally, some additional notes are provided in Section 2.4.

## 2.1 Variational Inequality Problem and Some Related Mathematical Problems

### 2.1.1 Variational Inequality Problem

Variational inequalities have been demonstrated to be quite useful and have been extensively employed in modeling various static and dynamic problems. The principal advantages of the VIP are fourfold:

1. The VIP is a general problem formulation that encompasses a plethora of

mathematical problems, including optimization problems, complementarity problems, systems of equations and fixed point problems.
2. The VIP is especially useful for its relation to an equivalent optimization formulation; an immediate benefit with the equivalent optimization formulation is that many solution algorithms are available.
3. The uniqueness proof for the VIP is simple.
4. The geometric interpretation for the VIP is meaningful.

A detailed definition of VIP is now presented.

***Definition 2.1*** The *static* finite-dimensional VIP is to find a decision vector $\mathbf{u}^* \in \Omega \subseteq R^n$ such that

$$\mathbf{c}^{*T}(\mathbf{u}-\mathbf{u}^*) \geq 0 \quad \forall \mathbf{u} \in \Omega \tag{2.1}$$

where $\mathbf{u}$ is a vector of decision variables $\mathbf{u} = (u_1, u_2, \cdots, u_a, \cdots, u_n)$, $\mathbf{c}$ is a vector of given continuous functions from $\Omega$ to $R^n$, $\mathbf{c} = (c_1(\mathbf{u}), c_2(\mathbf{u}), \cdots, c_a(\mathbf{u}), \cdots, c_n(\mathbf{u}))$, and $\Omega$ is a nonempty, closed and convex set.

Variational inequalities can be naturally extended to dynamic problems by associating all time intervals with decision *matrix* $\mathbf{u}$ and *matrix* of given continuous functions $\mathbf{c}$:

$$\mathbf{u} = [\mathbf{u}_1, \mathbf{u}_2, \cdots, \mathbf{u}_a, \cdots, \mathbf{u}_A]^T = [\mathbf{u}(1), \mathbf{u}(2), \cdots, \mathbf{u}(t), \cdots, \mathbf{u}(T)]$$
$$= \begin{bmatrix} u_1(1) & u_1(2) & \cdots & u_1(t) & \cdots & u_1(T) \\ u_2(1) & u_2(2) & \cdots & u_2(t) & \cdots & u_2(T) \\ \cdot & \cdot & \cdots & \cdot & \cdots & \cdot \\ u_a(1) & u_a(2) & \cdots & u_a(t) & \cdots & u_a(T) \\ \cdot & \cdot & \cdots & \cdot & \cdots & \cdot \\ u_A(1) & u_A(2) & \cdots & u_A(t) & \cdots & u_A(T) \end{bmatrix} \tag{2.2}$$

$$\mathbf{c} = \begin{bmatrix} c_1(\mathbf{u}(1)) & c_1(\mathbf{u}(1),\mathbf{u}(2)) & \cdots & c_1(\mathbf{u}(1),\mathbf{u}(2),\cdots,\mathbf{u}(t)) & \cdots & c_1(\mathbf{u}) \\ c_2(\mathbf{u}(1)) & c_2(\mathbf{u}(1),\mathbf{u}(2)) & \cdots & c_2(\mathbf{u}(1),\mathbf{u}(2),\cdots,\mathbf{u}(t)) & \cdots & c_2(\mathbf{u}) \\ \cdot & \cdot & \cdots & \cdot & \cdots & \cdot \\ c_a(\mathbf{u}(1)) & c_a(\mathbf{u}(1),\mathbf{u}(2)) & \cdots & c_a(\mathbf{u}(1),\mathbf{u}(2),\cdots,\mathbf{u}(t)) & \cdots & c_a(\mathbf{u}) \\ \cdot & \cdot & \cdots & \cdot & \cdots & \cdot \\ c_A(\mathbf{u}(1)) & c_A(\mathbf{u}(1),\mathbf{u}(2)) & \cdots & c_A(\mathbf{u}(1),\mathbf{u}(2),\cdots,\mathbf{u}(t)) & \cdots & c_A(\mathbf{u}) \end{bmatrix} \tag{2.3}$$

where $A$ denotes the maximal numbered link in the network and $T$ is the maximal numbered time interval in the analysis period. Note that dynamic link travel time function on link $a$ during interval $t$, $c_a(\mathbf{u}(1), \mathbf{u}(2), \cdots, \mathbf{u}(t))$, includes only the previously entered flows as its arguments because those following flows, $\mathbf{u}(t+1), \mathbf{u}(t+2), \cdots, \mathbf{u}(T)$, don't have any affect on its value. For simplicity, we hereafter use symbol $c_a(t)$ to denote $c_a(\mathbf{u}(1), \mathbf{u}(2), \cdots, \mathbf{u}(t))$.

***Definition* 2.2** The *dynamic* finite-dimensional VIP is to find a decision *matrix* $\mathbf{u}^* \in \Omega \subseteq R^{A \times T}$ such that

$$\mathbf{c}^{*T}(\mathbf{u} - \mathbf{u}^*) \geq 0 \quad \forall \mathbf{u} \in \Omega \tag{2.4}$$

where $\mathbf{u}$ is a *matrix* of decision variables, $\mathbf{c}$ is a *matrix* of given continuous functions from $\Omega$ to $R^{A \times T}$, and $\Omega$ is a nonempty, closed and convex set.

The VIP may be interpreted as the problem of finding a point $\mathbf{u}^* \in \Omega$ at which the vector field $\mathbf{c}^{*T}$ is an inward normal (or orthogonal) to the feasible set $\Omega$. This problem is also known as the *generalized equation* and *stationary point problem* (Patriksson, 1994). The variational inequality formulation is particularly convenient because it allows for a unified treatment of equilibrium problems and optimization problems. Therefore, to reformulate a set of equilibrium conditions as an equivalent optimization problem, the easiest procedure would be to go through two successive steps, i.e., transforming the equilibrium conditions into the VIP, and then further converting the VIP into the equivalent optimization problem.

We next discuss some mathematical problems that are closely related to the VIP. These problems include optimization problems, systems of equations, complementarity problems, and fixed point problems. Static cases are mainly considered; however, the analysis can be readily extended to their dynamic counterparts.

### 2.1.2 Optimization Problems

The optimization problem is in general characterized by a single objective function that is to be minimized or maximized, subject to a set of constraints. The optimization problem may be expressed as follows:

$$\min_{\mathbf{u} \in \Omega} z(\mathbf{u}) \tag{2.5}$$

where $z(\mathbf{u})$ is the specific objective represented as a function of decision variables $\mathbf{u} = (u_1, u_2, \cdots, u_n)$ and $\Omega$ defines the feasible set.

When the feasible set is not bounded, the optimization problem becomes unconstrained. In some circumstances, the decision maker may have more than one objective to be optimized, which results in the so-called multiple objective decision making (MODM) problem. The solution for the MODM problem is difficult to attain, especially when the problem is nonlinear. However, with known preferential information associated with each objective, the MODM problem can still be transformed into a single objective optimization problem. The MODM formulation is appropriate in representing human decision behavior and has great potential to be adopted by the fuzzy/dynamic user-optimal route choice problem.

The optimization problem has long been tackled by operations research professionals. Several solution algorithms are well developed for solving various types of optimization problems. In comparison with the optimization problem, the VIP is weak in solution algorithms, but strong in qualitative analysis. If we can delineate the domain on which the optimization problem and the VIP overlaps, then the advantages pertaining to both classes of mathematical problems can be fully utilized. The following two theorems describe the relationship between the single objective optimization problem and the VIP.

***Theorem 2.1*** (Optimality condition) Let z : $\Omega \mapsto R^1$ be continuously differentiable on a nonempty, closed and convex set $\Omega \subseteq R^n$. If $\mathbf{u}^* \in \Omega$ is an optimal solution to the optimization problem (2.5), then $\mathbf{u}^*$ is a solution to VIP (2.1), with $\mathbf{c}(\mathbf{u}) \equiv \nabla z(\mathbf{u})$. The converse holds whenever $z(\mathbf{u})$ is pseudoconvex.

***Proof of Necessity***: We first prove that if $\mathbf{u}^* \in \Omega$ is an optimal solution to the optimization problem (2.5), then $\mathbf{u}^*$ is a solution to VIP (2.1), with $\mathbf{c}(\mathbf{u}) \equiv \nabla z(\mathbf{u})$. Since $\mathbf{u}^*$ is a solution to the optimization problem (2.5), the objective value at the neighborhood $\sigma$ of the optimal solution must therefore be greater than or equal to that at the optimal solution.

$$\frac{z(\mathbf{u}^* + \sigma(\mathbf{u} - \mathbf{u}^*)) - z(\mathbf{u}^*)}{\sigma} \geq 0 \quad \forall \mathbf{u}, \mathbf{u}^* \in \Omega \tag{2.6}$$

Suppose $\sigma$ approaches zero, then the gradient in the direction of $(\mathbf{u} - \mathbf{u}^*)$ must be nonnegative

$$\nabla z^T(\mathbf{u}^*)(\mathbf{u} - \mathbf{u}^*) \geq 0 \tag{2.7}$$

***Proof of Sufficiency***: Next, we prove that the converse holds whenever $z(\mathbf{u})$ is pseudoconvex. Since $z(\mathbf{u})$ is pseudoconvex, then by definition, we have

$$\nabla z^T(\mathbf{u}^*)(\mathbf{u} - \mathbf{u}^*) \geq 0 \Rightarrow z(\mathbf{u}) \geq z(\mathbf{u}^*) \quad \forall \mathbf{u}, \mathbf{u}^* \in \Omega \tag{2.8}$$

In other words, $\mathbf{u}^*$ is a minimum of mathematical program (2.5). The proof is complete.

The results of the above theorem imply that whenever $\mathbf{c}(\mathbf{u})$ is the gradient of a real-valued function $z(\mathbf{u})$, a solution to (2.1) may be found through the solution of the optimization problem (2.5). Note that if the feasible set $\Omega = R^n$, then the unconstrained optimization problem is also the VIP.

***Theorem 2.2*** (Sufficient condition for $\mathbf{c}(\mathbf{u})$ to be a gradient) Let $\mathbf{c} : \Omega \mapsto R^n$ be continuously differentiable on an open convex set $\Omega_0 \subset \Omega \subseteq R^n$. Then $\mathbf{c}(\mathbf{u})$ is a gradient mapping on $\Omega_0$ if and only if $\nabla \mathbf{c}$ is symmetric for all $\mathbf{u} \in \Omega_0$.

***Proof***: See Ortega et al (1970), Theorem 4.1.6.

Under this symmetry condition, the line integral

$$z(\mathbf{u}) \overset{def}{=} \int_{\mathbf{0}}^{\mathbf{u}} \mathbf{c}(\mathbf{w})^T d\mathbf{w} \tag{2.9}$$

is path independent according to Green's theorem, and $\mathbf{c}(\mathbf{u})$ is integrable. The VIP problem (2.1) can hence be placed as an equivalent optimization program with an objective of the form equation (2.9). When both the symmetry condition and the positive semidefinite (p.s.d) condition hold, the VIP can be reformulated as a convex optimization problem.

The p.s.d. condition along with the positive definite condition (p.d) and strongly positive definite condition (s.p.d) are defined as follows:

***Definition 2.3*** An $n \times n$ matrix $\boldsymbol{M}(\mathbf{u})$, whose elements $m_{ij}(\mathbf{u})$; $i = 1, \cdots, n$; $j = 1, \cdots, n$ are functions defined on the set $\Omega \subset R^n$, is said to be p.s.d on $\Omega$ if

$$\mathbf{x}^T \boldsymbol{M}(\mathbf{u})\mathbf{x} \geq 0 \quad \forall \mathbf{x} \in R^n, \ \mathbf{u} \in \Omega$$

It is said to be p.d on $\Omega$ if

$$\mathbf{x}^T \boldsymbol{M}(\mathbf{u})\mathbf{x} > 0 \quad \forall \mathbf{x} \neq \mathbf{0}, \ \mathbf{x} \in R^n, \ \mathbf{u} \in \Omega$$

It is said to be s.p.d on $\Omega$ if

$$\mathbf{x}^T \boldsymbol{M}(\mathbf{u})\mathbf{x} \geq \sigma||\mathbf{x}|| \quad \text{for some } \sigma > 0, \quad \forall \mathbf{x} \in R^n, \ \mathbf{u} \in \Omega$$

It should be noted that reformulating the VIP as a convex optimization problem is possible only when the symmetry condition and the p.s.d condition hold true for the given function $\mathbf{c}(\mathbf{u})$. Unfortunately, the travel time function for dynamic travel choice models is usually asymmetric in nature, because the link travel time can be affected by the previously entered flow, while the converse is not true. Therefore, the equivalent optimization problem cannot be obtained. However, by temporarily fixing actual link travel times and flow interactions in different time units at current level, the original dynamic travel choice models can be reduced into an optimization subproblem which can be easily solved by prevailing solution algorithms. The Frank-Wolfe method is most commonly used in solving the quadratic subproblem of any travel choice model (see Section 2.3.5). In the following sections, symbol $\mathbf{c}$ may represent any vector, not necessarily the travel time function.

### 2.1.3 Complementarity Problems (CP)

***Definition 2.4*** Let $\mathbf{c}: R_{+}^{n} \mapsto R^{n}$ be continuous. The complementarity problem is to find an $\mathbf{u}^{*} \in R^{n}$ such that

$$\mathbf{c}(\mathbf{u}^{*})^{T}\mathbf{u}^{*} = 0 \tag{2.10}$$

$$\mathbf{c}(\mathbf{u}^{*})^{T} \geq \mathbf{0} \tag{2.11}$$

$$\mathbf{u}^{*} \geq \mathbf{0} \tag{2.12}$$

Complementarity problems are defined on the nonnegative orthant and consists of a system of equations and inequalities. Whenever, the mapping $\mathbf{c}$ is affine, that is, whenever $\mathbf{c}(\mathbf{u}) = \mathbf{Mu} + \mathbf{b}$, where $M$ is an $n \times n$ matrix and $\mathbf{b}$ an $n \times 1$ vector, program (2.10)~(2.12) is then known as the linear complementarity problem; otherwise, the nonlinear complementarity problem results.

***Theorem 2.3*** VIP (2.1) defined on $\Omega = R_{+}^{n}$ and the complementarity problem (2.10~2.12) have precisely the same solution, if any.

***Proof***: We first prove that complementarity problem (2.10~2.12) is equivalent to a variational inequality defined on $\Omega = R_{+}^{n}$. By inequalities (2.11) and (2.12)

$$\mathbf{c}(\mathbf{u}^{*})^{T}\mathbf{u} \geq 0 \quad \forall \mathbf{u} \in \Omega = R_{+}^{n} \tag{2.13}$$

Subtracting equation (2.10) from equation (2.13), we obtain

$$\mathbf{c}(\mathbf{u}^{*})^{T}(\mathbf{u} - \mathbf{u}^{*}) \geq 0 \quad \forall \mathbf{u}, \mathbf{u}^{*} \in \Omega = R_{+}^{n} \tag{2.14}$$

Conversely, if $\mathbf{u}^{*}$ is a solution to VIP (2.1) defined on $\Omega = R_{+}^{n}$, then it must also solve the complementary problem. Denote $\mathbf{e}_i$ as an $n$-dimensional vector with 1 in the $i$-th location and 0 elsewhere, i.e.,

$$\mathbf{e}_i = \begin{bmatrix} 0 & \cdots & 0 & 1 & 0 & \cdots & 0 \end{bmatrix}^{T}$$

Substituting $\mathbf{u} = \mathbf{u}^{*} + \mathbf{e}_i$ into VIP (2.1), we have $\mathbf{c}_i(\mathbf{u}^{*}) \geq \mathbf{0}$. We can choose any $\mathbf{e}_i$ with 1 at any $i$-th location so that each component $\mathbf{c}_i(\mathbf{u}^{*})$ is nonnegative. Thus, $\mathbf{c}(\mathbf{u}^{*}) \geq \mathbf{0}$.

Now, substituting $\mathbf{u} = 2\mathbf{u}^{*}$ into VIP (2.1), we obtain

$$\mathbf{c}(\mathbf{u}^{*})^{T}\mathbf{u}^{*} \geq 0 \tag{2.15}$$

Then, substituting $\mathbf{u} = \mathbf{0}$ into VIP (2.1), we obtain

$$\mathbf{c}(\mathbf{u}^{*})^{T}(-\mathbf{u}^{*}) \geq 0 \tag{2.16}$$

Inequalities (2.15) and (2.16) together imply that equation (2.10). This completes the proof.

### 2.1.4 Systems of Equations

***Definition 2.5*** Let $\mathbf{c} : \Omega \mapsto R^n$ on $\Omega = R^n$. The system of equation is to find an $\mathbf{u}^* \in \Omega$ such that

$$\mathbf{c}(\mathbf{u}^*) = \mathbf{0} \tag{2.17}$$

***Theorem 2.4*** A vector $\mathbf{u}^* \in \Omega$ solves VIP (2.1), if and only if $\mathbf{c}(\mathbf{u}^*) = \mathbf{0}$.

***Proof:*** If $\mathbf{c}(\mathbf{u}^*) = \mathbf{0}$, then VIP (2.1) holds with equality. Conversely, if $\mathbf{u}^*$ satisfies VIP (2.1), let $\mathbf{u} = \mathbf{u}^* - \mathbf{c}(\mathbf{u}^*)$, which implies that

$$\mathbf{c}(\mathbf{u}^*)^T(-\mathbf{c}(\mathbf{u}^*)) \geq 0, \text{ or } -\|\mathbf{c}(\mathbf{u}^*)\|^2 \geq 0 \tag{2.18}$$

and, therefore, $\mathbf{c}(\mathbf{u}^*) = \mathbf{0}$.

### 2.1.5 Fixed Point Problems

***Definition 2.6*** Let $\mathbf{c} : \Omega \mapsto R^n$ be continuous. The fixed point problem is to find an $\mathbf{u}^* \in \Omega$ such that

$$\mathbf{c}(\mathbf{u}^*) = \mathbf{u}^* \tag{2.19}$$

In the field of traffic equilibrium, fixed point problems have mostly been applied as instruments for establishing the existence of solutions to variational inequality or complementarity models. The general proof is based on the definition of an appropriate continuous mapping which transforms the original model into an equivalent fixed point problem, for which existence is then established by imposing strong enough properties onto the original problem data. There are several classical existence results for fixed point problems.

***Theorem 2.5*** (Existence of a fixed point, Brouwer's fixed point theorem) Let $\Omega$ be bounded, and $\mathbf{c}$ be a mapping from $\Omega$ to $\Omega$, with $\mathbf{c}$ continuous. Then there exists a solution to the fixed point problem.

***Theorem 2.6*** (Existence of a unique fixed point) Let $\mathbf{c}$ be contractive on $\Omega$. Then there exists a unique solution to the fixed point problem. Furthermore, the sequence $\{\mathbf{u}^k\}$, defined by $\mathbf{u}^0 \in \Omega$,

$$\mathbf{u}^{k+1} = \mathbf{c}(\mathbf{u}^k) \quad k = 0, 1, \cdots, \cdots$$

converges to the unique fixed point.

To prove the equivalence between the fixed point problem and the VIP, we first state without proofs the following lemma and theorem (Nagurney, 1993).

***Lemma 2.1*** Let $\Omega$ be a closed convex set in $R^n$. Then for each $\mathbf{x} \in R^n$, there is a unique point $\mathbf{u} \in \Omega$, such that

$$\|\mathbf{x} - \mathbf{u}\| \leq \|\mathbf{x} - \mathbf{y}\| \quad \forall \mathbf{y} \in \Omega \tag{2.20}$$

and $\mathbf{u}$ is known as the orthogonal projection of $\mathbf{x}$ on the set $\Omega$ with respect to the Euclidean norm, that is,

$$\mathbf{u} = P_\Omega \mathbf{x} = \arg\min_{\mathbf{y} \in \Omega} \|\mathbf{x} - \mathbf{y}\| \tag{2.21}$$

***Theorem 2.7*** Let $\Omega$ be a closed convex set in $R^n$. Then $\mathbf{u} = P_\Omega \mathbf{x}$ if and only if

$$\mathbf{u}^T(\mathbf{y} - \mathbf{u}) \geq \mathbf{x}^T(\mathbf{y} - \mathbf{u}) \quad \forall \mathbf{y} \in \Omega \tag{2.22}$$

or

$$(\mathbf{u} - \mathbf{x})^T(\mathbf{y} - \mathbf{u}) \geq 0 \quad \forall \mathbf{y} \in \Omega \tag{2.23}$$

The relationship between the VIP and a fixed point problem can be made through the use of a projection operator.

***Theorem 2.8*** A vector $\mathbf{u}^* \in \Omega$ is a solution of VIP (2.1), if and only if for any $\rho > 0$, $\mathbf{u}^*$ is a fixed point of the map

$$P_\Omega(\mathbf{I} - \gamma\mathbf{c}) \colon \Omega \mapsto \Omega,$$

that is,

$$\mathbf{u}^* = p_\Omega(\mathbf{u}^* - \rho\mathbf{c}(\mathbf{u}^*)) \tag{2.24}$$

***Proof***: Suppose that $\mathbf{u}^*$ is a solution of the VIP, i.e.,

$$\mathbf{c}(\mathbf{u}^*)^T(\mathbf{u} - \mathbf{u}^*) \geq 0 \quad \forall \mathbf{u} \in \Omega$$

By multiplying the above inequality by $-\rho < 0$, and adding $\mathbf{u}^{*T}(\mathbf{u} - \mathbf{u}^*)$ to both sides of the resulting inequality, one obtains

$$\mathbf{u}^{*T}(\mathbf{u} - \mathbf{u}^*) \geq [\mathbf{u}^* - \rho\mathbf{c}(\mathbf{u}^*)]^T(\mathbf{u} - \mathbf{u}^*) \quad \forall \mathbf{u} \in \Omega \tag{2.25}$$

From Theorem (2.7), one can conclude that $\mathbf{u}^*$ is the orthogonal projection of $\mathbf{u}^* - \rho\mathbf{c}(\mathbf{u}^*)$ on the set $\Omega$ with respect to the Euclidean norm, that is

$$\mathbf{u}^* = P_\Omega(\mathbf{u}^* - \rho\mathbf{c}(\mathbf{u}^*))$$

Conversely, if $\mathbf{u}^* = P_\Omega(\mathbf{u}^* - \rho\mathbf{c}(\mathbf{u}^*))$, for $\rho > 0$, then

$$\mathbf{u}^{*T}(\mathbf{u}-\mathbf{u}^*) \geq [\mathbf{u}^* - \rho\mathbf{c}(\mathbf{u}^*)]^T(\mathbf{u}-\mathbf{u}^*) \quad \forall \mathbf{u} \in \Omega$$

and, therefore,

$$\mathbf{c}(\mathbf{u}^*)^T(\mathbf{u}-\mathbf{u}^*) \geq 0 \quad \forall \mathbf{u} \in \Omega$$

In other words, the solution $\mathbf{u}^*$ of VIP (2.1) is the orthogonal projection on $\Omega$ of the vector $\mathbf{u}^* - \rho\mathbf{c}(\mathbf{u}^*)$ for any $\rho > 0$. This completes the proof.

## 2.2 Existence and Uniqueness

Qualitative properties of existence and uniqueness are easily obtained under certain monotonic conditions. To help the reader understand the meanings, we present the following definitions and theorems.

***Definition 2.7*** A vector of functions $\mathbf{c}(\mathbf{u})$ is monotone on $\Omega$ if

$$[\mathbf{c}(\mathbf{u}_1) - \mathbf{c}(\mathbf{u}_2)]^T(\mathbf{u}_1 - \mathbf{u}_2) \geq 0 \quad \forall \mathbf{u}_1, \mathbf{u}_2 \in \Omega \tag{2.26}$$

where $\mathbf{u}_1$ and $\mathbf{u}_2$ are any points on $\Omega$.

***Definition 2.8*** A vector of functions $\mathbf{c}(\mathbf{u})$ is strictly monotone on $\Omega$ if

$$[\mathbf{c}(\mathbf{u}_1) - \mathbf{c}(\mathbf{u}_2)]^T(\mathbf{u}_1 - \mathbf{u}_2) > 0 \quad \forall \mathbf{u}_1, \mathbf{u}_2 \in \Omega,\ \mathbf{u}_1 \neq \mathbf{u}_2 \tag{2.27}$$

***Definition 2.9*** A vector of functions $\mathbf{c}(\mathbf{u})$ is strongly monotone on $\Omega$ if for some $\sigma > 0$

$$[\mathbf{c}(\mathbf{u}_1) - \mathbf{c}(\mathbf{u}_2)]^T(\mathbf{u}_1 - \mathbf{u}_2) \geq \sigma\|\mathbf{u}_1 - \mathbf{u}_2\|^2 \quad \forall \mathbf{u}_1, \mathbf{u}_2 \in \Omega \tag{2.28}$$

***Definition 2.10*** A vector of functions $\mathbf{c}(\mathbf{u})$ is *Lipschitz* continuous on $\Omega$ if there exists an $L$>0 such that

$$\|\mathbf{c}(\mathbf{u}_1) - \mathbf{c}(\mathbf{u}_2)\| \leq L\|\mathbf{u}_1 - \mathbf{u}_2\| \quad \forall \mathbf{u}_1, \mathbf{u}_2 \in \Omega \tag{2.29}$$

Monotonicity is closely related to positive definiteness.

***Theorem 2.9*** Suppose that $\mathbf{c}(\mathbf{u})$ is continuously differentiable on $\Omega$ and the Jacobian matrix

$$\nabla \mathbf{c}(\mathbf{u}) = \begin{bmatrix} \frac{\partial c_1}{\partial u_1} & \frac{\partial c_1}{\partial u_2} & \cdots & \frac{\partial c_1}{\partial u_n} \\ \frac{\partial c_2}{\partial u_1} & \frac{\partial c_2}{\partial u_2} & \cdots & \frac{\partial c_2}{\partial u_n} \\ \vdots & \vdots & & \vdots \\ \frac{\partial c_n}{\partial u_1} & \frac{\partial c_n}{\partial u_2} & \cdots & \frac{\partial c_n}{\partial u_n} \end{bmatrix} \tag{2.30}$$

is p.s.d (or p.d), then $\mathbf{c}(\mathbf{u})$ is monotone (or strictly monotone).

***Theorem 2.10*** Assume that $\mathbf{c}$ is continuously differentiable at some $\overline{\mathbf{u}}$. Then $\mathbf{c}(\mathbf{u})$ is locally strictly (or strongly) monotone at $\overline{\mathbf{u}}$ if $\nabla\mathbf{c}(\overline{\mathbf{u}})$ is positive definite (or strongly positive definite); that is,

$$\mathbf{x}^T \mathbf{c}(\overline{\mathbf{u}})\mathbf{x} > 0 \quad \mathbf{x} \in R^n,\ \mathbf{x} \neq \mathbf{0} \tag{2.31}$$

$$\mathbf{x}^T \mathbf{c}(\overline{\mathbf{u}})\mathbf{x} \geq \sigma||\mathbf{x}||^2 \quad \text{for some } \sigma > 0,\ \forall\ \mathbf{x} \in R^n \tag{2.32}$$

where $\mathbf{x}$ is an arbitrary vector with components of real values.

***Theorem 2.11*** Assume that $\mathbf{c}(\mathbf{u})$ is continuously differentiable on $\Omega$ and that $\nabla\mathbf{c}(\mathbf{u})$ is strongly positive definite, then $\mathbf{c}(\mathbf{u})$ is strongly monotone.

Conditions under which the set of solutions to the variational inequality problem is guaranteed to be nonempty are given below.

***Theorem 2.12*** (Existence of solutions to the VIP) If $\Omega$ is a compact convex set and $\mathbf{c}(\mathbf{u})$ is continuous on $\Omega$, then the VIP admits at least one solution $\mathbf{u}^*$.

***Proof***: Under the condition that $\Omega$ is a closed convex set, the VIP is equivalent to the fixed point problem. With the additional property that $\Omega$ is also bounded, the fixed point problem has at least one solution according to Brouwer's fixed point theorem.

The uniqueness of the solution to the VIP is easily verified.

***Theorem 2.13*** (Uniqueness of solutions to the VIP) The solution set of the VIP, if nonempty, is a singleton if $\mathbf{c}$ is strictly monotone on $\Omega$.

***Proof***: We prove by showing that if there are two optimal solutions $\mathbf{u}_1, \mathbf{u}_2$, then $\mathbf{c}$ is not strictly monotone on $\Omega$. Since both $\mathbf{u}_1$, $\mathbf{u}_2$ are optimal to the VIP, the following must hold.

$$\mathbf{c}(\mathbf{u}_1)(\mathbf{u}-\mathbf{u}_1) \geq 0 \quad \forall \mathbf{u} \in \Omega \tag{2.33}$$

$$\mathbf{c}(\mathbf{u}_2)(\mathbf{u}-\mathbf{u}_2) \geq 0 \quad \forall \mathbf{u} \in \Omega \tag{2.34}$$

We substitute **u** in inequality (2.33) by $\mathbf{u}_2$, and inequality (2.34) by $\mathbf{u}_1$, the following two VIPs result:

$$\mathbf{c}(\mathbf{u}_1)(\mathbf{u}_2-\mathbf{u}_1) \geq 0 \quad \forall \mathbf{u}_1, \mathbf{u}_2 \in \Omega \tag{2.35}$$

$$\mathbf{c}(\mathbf{u}_2)(\mathbf{u}_1-\mathbf{u}_2) \geq 0 \quad \forall \mathbf{u}_1, \mathbf{u}_2 \in \Omega \tag{2.36}$$

Adding inequality (2.36) to inequality (2.35) yields

$$[\mathbf{c}(\mathbf{u}_1)-\mathbf{c}(\mathbf{u}_2)][\mathbf{u}_2-\mathbf{u}_1] \geq 0 \quad \forall \mathbf{u}_1, \mathbf{u}_2 \in \Omega \tag{2.37}$$

It is clear that inequality (2.37) violates the assumption that **c** is strictly monotone on $\Omega$, therefore, there is only one solution for the VIP. Note also that the separability property is not needed in the proof.

If $\mathbf{c}(\mathbf{u})$ is strongly monotone, then there exists precisely one solution $\mathbf{u}^*$ to the VIP. The proof of existence follows from the fact that strong monotonicity implies coercivity, whereas uniqueness follows from the fact that strong monotonicity implies strictly monotonicity. In conclusion, in the case of an unbounded feasible set $\Omega$, strong monotonicity of the function **c** guarantees both existence and uniqueness. If $\Omega$ is compact, then existence is guaranteed if **c** is continuous, and only the strict monotonicity condition is needed for uniqueness to be guaranteed.

## 2.3 Solution Algorithms

### 2.3.1 The General Iterative Scheme

In this section, a general iterative scheme for the solution of VIP (2.1) is presented. The iterative scheme induces, in special cases, such well-known algorithms as the projection method, linearization algorithms, and the relaxation (or diagonalization) method, and also induces new algorithms.

In particular, we seek to determine $\mathbf{u}^* \in \Omega \subseteq R^n$, such that

$$\mathbf{c}(\mathbf{u}^*)^T(\mathbf{u}-\mathbf{u}^*) \geq 0 \quad \forall \mathbf{u} \in \Omega \tag{2.1}$$

where **c** is a given continuous function from $\Omega$ to $R^n$ and $\Omega$ is a given closed, convex set. $\Omega$ is also assumed to be compact and **c**(**u**) continuously differentiable.

Assume that there exists a smooth function

$$F(\mathbf{u}, \mathbf{u}^{m-1}): \Omega \times \Omega \mapsto R^n \tag{2.38}$$

with the following properties:

(i) $F(\mathbf{u},\mathbf{u}) = \mathbf{c}(\mathbf{u}) \quad \forall \mathbf{u} \in \Omega,$

(ii) for every fixed $\mathbf{u}, \mathbf{u}^{m-1} \in \Omega$, the $n \times n$ matrix $\nabla_{\mathbf{u}} F(\mathbf{u}, \mathbf{u}^{m-1})$ is symmetric and positive definite.

Any function $F(\mathbf{u}, \mathbf{u}^{m-1})$ with the above properties generates the following.

***The General Iterative Algorithm***

**Step 0: Initialization**

Start with an $\mathbf{u}^0 \in \Omega$. Set $m = 1$.

**Step 1: Construction and Computation**

Compute $\mathbf{u}^m$ by solving the variational inequality subproblem:

$$F(\mathbf{u}^m, \mathbf{u}^{m-1})(\mathbf{u} - \mathbf{u}^m) \geq 0 \quad \forall \mathbf{u} \in \Omega \tag{2.39}$$

**Step 2: Convergence Verification**

If $|\mathbf{u}^m - \mathbf{u}^{m-1}| \leq \varepsilon$, for some $\varepsilon > 0$, a prespecified tolerance, then stop; otherwise, set $m = m + 1$ and go to Step 1.

Since $\nabla_{\mathbf{u}} F(\mathbf{u}, \mathbf{u}^{m-1})$ is assumed to be symmetric and positive definite, the line integral $\oint F(\mathbf{u}, \mathbf{u}^{m-1}) d\mathbf{x}$ defines a function $z(\mathbf{u}, \mathbf{u}^{m-1})$: $\Omega \times \Omega \mapsto R$ such that, for fixed $\mathbf{u}^{m-1} \in \Omega$, $z(\bullet, \mathbf{u}^{m-1})$ is strictly convex and

$$F(\mathbf{u}, \mathbf{p}) = \nabla_{\mathbf{u}} z(\mathbf{u}, \mathbf{p}) \tag{2.40}$$

Hence, VIP (2.1) is equivalent to the strictly convex mathematical programming problem

$$\min_{\mathbf{u} \in \Omega} z(\mathbf{u}, \mathbf{u}^{m-1}) \tag{2.41}$$

for which a unique solution $\mathbf{u}^m$ exists. The solution to expression (2.41) can be computed using any appropriate mathematical programming algorithm. If there is, however, a special-purpose algorithm that takes advantage of the problem's structure, then such an algorithm is usually preferable from an efficiency point of view. Of course, inequality (2.39) should be constructed in such a manner so that, at each iteration *m*, this subproblem is easy to solve. Note that if the sequence $\{\mathbf{u}^m\}$ is convergent, i.e., $\mathbf{u}^m \to \mathbf{u}^*$, as $m \to \infty$, then because of the continuity of $F(\mathbf{u}, \mathbf{u}^{m-1})$, inequality (2.39) yields

$$c(\mathbf{u}^*)^T (\mathbf{u} - \mathbf{u}^*) = F(\mathbf{u}^*, \mathbf{u}^*)(\mathbf{u} - \mathbf{u}^*) \geq 0 \quad \forall \mathbf{u} \in \Omega \tag{2.42}$$

and, consequently, $\mathbf{u}^*$ is a solution to VIP (2.1). A condition on $F(\mathbf{u}, \mathbf{u}^{m-1})$, which guarantees that the sequence $\{\mathbf{u}^m\}$ is convergent is found in Nagurney (1993).

### 2.3.2 The Diagonalization Method

The diagonalization (sometimes also called relaxation) method resolves VIP (2.1) into a sequence of subproblems (2.39) which are, in general, nonlinear programming problems. In the framework of the general iterative scheme, the diagonalization method corresponds to the choice:

$$F_i\left(\mathbf{u},\mathbf{u}^{m-1}\right)=\mathbf{c}_i\left(u_1^{m-1},\cdots,u_{i-1}^{m-1},u_i,u_{i+1}^{m-1},\cdots,u_n^{m-1}\right)\quad i=1,\cdots,n \tag{2.43}$$

The diagonalization method is stated as follows:

***The Diagonalization Method***

**Step 0: Initialization**

Find a set of feasible decision variables $\mathbf{u}^m$. Set $m=0$.

**Step 1: Diagonalization**

Solve the mathematical programming subproblem:

$$\min_{\mathbf{u}\in\Omega}\ z\left(\mathbf{u},\mathbf{u}^{m-1}\right) \tag{2.41}$$

obtaining solution $\mathbf{u}^m$.

**Step 2: Convergence Test**

If $|\mathbf{u}^m-\mathbf{u}^{m-1}|\le\varepsilon$, for some $\varepsilon>0$, a prespecified tolerance, then stop; otherwise, set $m:=m+1$ and go to Step 1.

The assumptions under which the diagonalization method converges are given in Nagurney (1993).

### 2.3.3 The Projection Method

The projection method resolves VIP (2.1) into a sequence of subproblem (2.39) (c.f. also (2.41)), which is equivalent to a quadratic programming problem. Quadratic programming problems are usually easier to solve than more highly nonlinear optimization problems, and effective algorithms have been developed for such problems. In the framework of the general iterative scheme, the projection method corresponds to the choice:

$$F\left(\mathbf{u},\mathbf{u}^{m-1}\right)=\mathbf{c}\left(\mathbf{u}^{m-1}\right)+\frac{1}{\rho}\mathbf{G}\left(\mathbf{u}-\mathbf{u}^{m-1}\right)\quad \rho>0 \tag{2.44}$$

where $\mathbf{G}$ is a fixed symmetric positive definite matrix. At each step $m$ of the projection method, the subproblem that must be solved is given by:

$$\min_{\mathbf{u}\in\Omega}\frac{1}{2}\mathbf{u}^T\mathbf{G}\mathbf{u}+\left(\rho\mathbf{c}\left(\mathbf{u}^{m-1}\right)-\mathbf{G}\mathbf{u}^{m-1}\right)^T\mathbf{u} \tag{2.45}$$

In particular, if $\mathbf{G}$ is selected to be a diagonal matrix, then expression (2.44) is a separable quadratic programming problem. Convergence conditions for the projection method are given in Nagurney (1993).

### 2.3.4 The Modified Projection Method

Note that a necessary condition for convergence of the general iterative scheme is that $\mathbf{c}(\mathbf{u})$ is strictly monotone. In the case that such a condition is not met by the application under consideration, a modified projection method may still be appropriate. This algorithm requires, instead, only monotonicity of $\mathbf{c}$, but with the *Lipschitz* continuity condition holding with constant $L$. The $\mathbf{G}$ matrix (cf. projection method) is now the identity matrix $\mathbf{I}$. The algorithm is now stated.

***The modified projection method***

**Step 0: Initialization**

Start with an $\mathbf{u}^0 \in \Omega$. Set $m$=1 and select $\rho$, such that $0 < \rho < \frac{1}{L}$, where $L$ is the *Lipschitz* constant for function $\mathbf{c}$ in the VIP.

**Step 1: Construction and Computation**

Compute $\bar{\mathbf{u}}^{m-1}$ by solving the variational inequality subproblem:

$$\left[\bar{\mathbf{u}}^{m-1} + \left(\rho\mathbf{c}\left(\mathbf{u}^{m-1}\right) - \mathbf{u}^{m-1}\right)\right]^T \left(\mathbf{u}' - \bar{\mathbf{u}}^{m-1}\right) \geq 0 \quad \forall \mathbf{u}' \in \Omega \tag{2.46}$$

**Step 2: Adaptation**

Compute $\mathbf{u}^m$ by solving the variational inequality subproblem:

$$\left[\mathbf{u}^{m} + \left(\rho\mathbf{c}\left(\bar{\mathbf{u}}^{m-1}\right) - \mathbf{u}^{m-1}\right)\right]^T \left(\mathbf{u}' - \mathbf{u}^{m}\right) \geq 0 \quad \forall \mathbf{u}' \in \Omega \tag{2.47}$$

**Step 3: Convergence Verification**

If $|\mathbf{u}^m - \mathbf{u}^{m-1}| \leq \varepsilon$, for $\varepsilon > 0$, a prespecified tolerance, then stop; otherwise, set $m$=$m$+1 and go to Step 1.

The algorithm converges to a solution of VIP (2.1).

### 2.3.5 The Frank-Wolfe Method

The Frank-Wolfe (FW) algorithm was originally suggested by Frank and Wolfe (1956) as a procedure for solving quadratic programming problems with linear constraints. It is also known as the convex combination algorithm. This method is extensively used in determining equilibrium flows in static transportation network problems. In this book, it is extended to solve the dynamic transportation network equilibrium subproblems.

We consider a convex minimization program with linear constraints:

$$\min_{\mathbf{u}} \ z(\mathbf{u}) \tag{2.48}$$

$$\text{s.t.} \ \sum_i a_{ij} u_i \geq b_j \quad \forall j \tag{2.49}$$

where $a_{ij}$ and $b_j$ are constant coefficients $(i = 1, \cdots, I;\ j = 1, \cdots, J)$. The algorithm is basically a feasible descent direction method. At iteration ($n$+1), it generates a point $\mathbf{u}^{n+1} = \left(u_1^{n+1}, \cdots, u_I^{n+1}\right)$ from $\mathbf{u}^n = \left(u_1^n, \cdots, u_I^n\right)$ such that $z\left(\mathbf{u}^{n+1}\right) < z\left(\mathbf{u}^n\right)$. Thus, the essence of this algorithm is the calculation of $\mathbf{u}^{n+1}$ from $\mathbf{u}^n$. The algorithmic step can be written in a standard form as

$$\mathbf{u}^{n+1} = \mathbf{u}^n + \lambda^n \mathbf{d}^n \tag{2.50}$$

where $\mathbf{d}^n = \left(\mathbf{d}_1^{n+1}, \cdots, \mathbf{d}_I^{n+1}\right)$ is a descent direction vector and $\lambda^n$ is a nonnegative scalar known as the step size or move size. This equation means that at each point, $\mathbf{u}^n$, a direction $\mathbf{d}^n$ is identified along which the objective function is decreasing. Then the step size $\lambda^n$ determines how far the next point $\mathbf{u}^{n+1}$ will be located along the direction $\mathbf{d}^n$.

The FW method selects the feasible descent direction based not only on how steep each candidate direction is in the vicinity of $\mathbf{u}^n$, but also according to how far it is possible to move along this direction. It chooses a direction based on the product of the rate of descent in the vicinity of $\mathbf{u}^n$ in a given direction and the length of the feasible region in that direction. This product is the *drop*, or the possible reduction in the objective function value, which can be achieved by moving in this direction. Thus, the algorithm uses the direction that maximizes the *drop*.

To find a descent direction, the algorithm checks the entire feasible region for an auxiliary feasible solution, $\mathbf{p}^n = \left[\mathbf{p}_1^n, \cdots, \mathbf{p}_I^n\right]$, such that the direction from $\mathbf{u}^n$ to $\mathbf{p}^n$ provides the maximum drop. In seeking the feasible direction, the bounding of the move size does not require a separate step of the algorithm. The bounding is accomplished as an integral part of choosing the descent direction. The direction from $\mathbf{u}^n$ to any feasible solution, $\mathbf{p}$, is the vector $\left(\mathbf{p} - \mathbf{u}^n\right)$ (or the unit vector $\left(\mathbf{p} - \mathbf{u}^n\right) / \| \mathbf{p} - \mathbf{u}^n \|$). The slope of $z\left(\mathbf{u}^n\right)$ in the direction of $\left(\mathbf{p} - \mathbf{u}^n\right)$ is given by the projection of the opposite gradient $\left[-\nabla z^T\left(\mathbf{u}^n\right)\right]$ in this direction, i.e.,

$$-\nabla z^T\left(\mathbf{u}^n\right) \frac{\left(\mathbf{p} - \mathbf{u}^n\right)}{\| \mathbf{p} - \mathbf{u}^n \|} \tag{2.51}$$

The drop in the objective function in the direction of $\left(\mathbf{p} - \mathbf{u}^n\right)$ is obtained by multiplying this slope by the distance from $\mathbf{u}^n$ to $\mathbf{p}^n$, $\| \mathbf{p} - \mathbf{u}^n \|$, i.e.,

$$-\nabla z^T\left(\mathbf{u}^n\right)\left(\mathbf{p} - \mathbf{u}^n\right) \tag{2.52}$$

This term has to be maximized (in $\mathbf{p}$) subject to the feasibility of $\mathbf{p}$. Alternatively, the term can be multiplied by (-1) and minimized. It follows that:

$$\min \nabla z^T(\mathbf{u}^n)(\mathbf{p}-\mathbf{u}^n) = \sum_i \frac{\partial z(\mathbf{u}^n)}{\partial u_i}(p_i - u_i^n) \tag{2.53}$$

$$\text{s.t. } \sum_i a_{ij} p_i \geq b_j \quad \forall j \tag{2.54}$$

where constraint set (2.54) is equivalent to the original constraint set (2.49) by replacing **u** with **p**. Thus, finding the descent direction amounts to solving a linear program, in which $p_i$ is the decision variable. Note that $\nabla z^T(\mathbf{u}^n)$ is constant at $\mathbf{u}^n$, and the term $\nabla z^T(\mathbf{u}^n)\cdot(\mathbf{u}^n)$ can be discarded from the objective function. Thus, the linearized problem can be simplified as:

$$\min \nabla z^T(\mathbf{u}^n)(\mathbf{p}) = \sum_i \frac{\partial z(\mathbf{u}^n)}{\partial u_i}(p_i) \tag{2.55}$$

s.t. Equation (2.54)

The objective function coefficients are $\frac{\partial z(\mathbf{u}^n)}{\partial u_1}, \frac{\partial z(\mathbf{u}^n)}{\partial u_2}, \ldots\ldots, \frac{\partial z(\mathbf{u}^n)}{\partial u_I}$. These coefficients are the derivatives of the original objective function at $\mathbf{u}^n$, which are known at iteration $n$. The decision variables of the optimization problem (2.53)~(2.54) are $\mathbf{p}^n = (p_1^n, p_2^n, \cdots, p_I^n)$ and the descent direction is the vector pointing from $\mathbf{u}^n$ to $\mathbf{p}^n$, i.e., $\mathbf{d}^n = (\mathbf{p}^n - \mathbf{u}^n)$, or in an expanded form, $d_i^n = (p_i^n - u_i^n), \forall i$. Once the descent direction is known, other algorithmic steps involve the determination of the move size and a convergence test.

As in many other descent methods, the move size in the direction of $\mathbf{d}^n$ equals the distance to the point along which $\mathbf{d}^n$ minimizes $z(\mathbf{u})$. The FW method does not require a special step to bracket the search for an optimal move size in order to maintain feasibility, but the new solution, $\mathbf{u}^{n+1}$, must lie between $\mathbf{u}^n$ and $\mathbf{p}^n$. Because $\mathbf{p}^n$ is a solution of the linearized problem, it naturally lies at the boundary of the feasible region. In other words, the search for a descent direction automatically generates a bound for a line search by accounting for all constraints when the descent direction is determined. Since the search interval is bracketed, the bisection or golden section method can be used to find the step size $\lambda$ by solving the minimization of $z(\mathbf{u})$ along $\mathbf{d}^n = (\mathbf{p}^n - \mathbf{u}^n)$. It follows that:

$$\min_{0\leq\lambda\leq 1} z\left[\mathbf{u}^n + \lambda(\mathbf{p}^n - \mathbf{u}^n)\right] \tag{2.56}$$

Once the optimal solution of this line search, $\lambda^n$, is found, the next point can be generated using the following formula.

$$\mathbf{u}^{n+1} = \mathbf{u}^n + \lambda^n(\mathbf{p}^n - \mathbf{u}^n) \tag{2.57}$$

Note that equation (2.57) can be written as:

$$\mathbf{u}^{n+1} = (1-\lambda^n)\mathbf{u}^n + \lambda^n \mathbf{p}^n \tag{2.58}$$

The new solution is thus a convex combination (or a weighted average) of $\mathbf{u}^n$ and $\mathbf{p}^n$. An appropriate convergence criterion is to check the lower bound of the objective function at each iteration. By convexity,

$$z(\mathbf{u}^*) \geq z((\mathbf{u}^n) + \nabla z(\mathbf{u}^n)(\mathbf{u}^* - \mathbf{u}^n) \tag{2.59}$$

Thus, the value of the linearized objective function yields a lower bound at $z(\mathbf{u}^n)$,

$$\mathrm{LB}(\mathbf{u}^n) \geq z(\mathbf{u}^n) + \nabla z(\mathbf{u}^n)(\mathbf{p} - \mathbf{u}^n) \tag{2.60}$$

An appropriate convergence criterion is:

$$\nabla z(\mathbf{u}^n)(\mathbf{p} - \mathbf{u}^n) / \mathrm{LB}(\mathbf{u}^n) < \varepsilon \tag{2.61}$$

The numerator of equation (2.61) is sometimes called the *gap*.

***FW Algorithm***

**Step 0: Initialization.**

Find a feasible solution $\mathbf{u}^0$. Set iteration counter $n$=0.

**Step 1: Direction Finding.**

Find $\mathbf{p}^n$ that solves the linear program of expressions (2.54) and (2.55).

**Step 2: Step Size Determination.**

Find $\lambda^n$ that solves

$$\min_{0 \leq \lambda \leq 1} z\left[\mathbf{u}^n + \lambda(\mathbf{p}^n - \mathbf{u}^n)\right] \tag{2.62}$$

**Step 3: Move.**

Set

$$\mathbf{u}^{n+1} = (1 - \lambda^n)\mathbf{u}^n + \lambda^n \mathbf{p}^n \tag{2.63}$$

**Step 4: Convergence Check.**

If $\nabla z(\mathbf{u}^n)(\mathbf{p} - \mathbf{u}^n) / \mathrm{LB}(\mathbf{u}^n) < \varepsilon$, stop. Otherwise, let $n$=$n$+1 and go to Step 1.

The algorithm converges in a finite number of iterations. Since the FW algorithm involves a minimization of a linear program as part of the direction-finding step, it is useful only in cases in which this linear program can be solved relatively easily. When applied to route choice problems, an advantage of mild memory requirements for the FW algorithm can be exploited because the route information is not stored. Therefore, the FW algorithm can be efficiently applied to large-sized route choice problems. Since dynamic travel choice problems usually involve a large number of variables, and hence are large-scale problems, the FW method is a good choice for solving their linearized subproblems in the direction finding step.

### 2.3.6 The Simplicial Decomposition Method

The FW method discussed in Section 2.3.5 only requires link flows to be stored, and thus can be applied to large-scale transportation network problems without too much difficulty. However, the troublesome zigzagging behavior at the vicinity of the optimal solution makes its convergence speed rather slow; numerous approaches have been attempted for making improvements to the FW method. One intuitive formalization is the simplicial decomposition method.

The simplicial decomposition is a special case of column generation in which the column generating subproblem is the same as the direction-finding subproblem of the FW algorithm, and the original objective is used in the solution of the restricted master problem. The basic theory and algorithm of the simplicial decomposition method may be found in Patriksson (1994). Another potential application of the simplicial decomposition method might be in vehicle route guidance systems, since the route information is produced and stored during the solution procedure.

## 2.4 Notes

Variational inequalities were first introduced to formulate transportation equilibrium problems by Dafermos (1980). Smith (1979) was the first to state the equilibrium conditions as a set of inequalities involving link flows only. Since then, many researchers have been involved in this approach (Ran and Boyce, 1996; Nagurney, 1993; Friesz et al., 1993; Fisk and Boyce, 1983). The advantage of variational inequalities is their capability in representing various real situations, such as interactions among link flows on link travel times, and in proving the uniqueness for the solution. Under such a framework, sensitivity analysis for different scenarios becomes possible, and the advantage becomes more pronounced when applied to dynamic travel choice problems.

The text by Kinderlehrer and Stampacchia (1980) provides an introduction to some variational inequality problems; a comprehensive summary of variational inequalities is provided by Nagurney (1993). Patriksson (1994) also discussed variational inequalities both in theoretical and algorithmic aspects, along with other mathematical models for the static traffic assignment problem. Ran and Boyce (1996) presented a brief mathematical background, mostly variational inequalities and optimal control theory, for dynamic transportation network models. The materials covered in this chapter are based on those publications.

# Chapter 3

# Network Flow Constraints and Link Travel Time Function Analysis

The network flow constraints construct the feasible region for dynamic travel choice models. Although the feasible region associated with each dynamic travel choice model may be different, the basic components of the network flow constraints are essentially the same. The complete requirements of dynamic travel choice models include the constraints of flow conservation, flow propagation, nonnegativity, definition, first-in-first-out (FIFO) and oversaturation.

In this chapter, we elaborate in depth the flow propagation requirement in both the discrete and continuous senses, whereas the continuous model may be closely approximated by its discrete counterpart using smaller time intervals. As for the other categories of network flow constraints, except for the flow conservation constraint, they are the same for all predictive dynamic travel choice models. Properties of the dynamic travel time function, including the number of independent decision variables, and asymmetric link interactions, are also discussed.

## 3.1 Flow Conservation Constraints

A traffic flow pattern which ensures flow conservation is required for a dynamic transportation network when considering flow propagation. For the sake of convenience, we discuss these flow conservation requirements with respect to four types of travel decisions, which are analogous to the four steps of traditional transportation planning.

### 3.1.1 Route Choice Type

For the DUO route choice model, each O-D pair must conserve the fixed time-dependent O-D demand $\bar{q}^{rs}(k)$ in terms of route flow as follows:

$$\sum_{p} h_p^{rs}(k) = \bar{q}^{rs}(k) \quad \forall r,s,k \tag{3.1}$$

Equation (3.1) states that for each O-D pair *rs*, summing up the flow departing during interval *k* over all routes must equal the departure flow during interval *k*.

For the DUO departure time/route choice model with the variable time-dependent O-D demand $q^{rs}(k)$ instead of $\bar{q}^{rs}(k)$, one more flow conservation constraint is needed.

$$\sum_{p} h_p^{rs}(k) = q^{rs}(k) \quad \forall r,s,k \tag{3.2}$$

$$\sum_{k} q^{rs}(k) = \bar{q}^{rs} \quad \forall r,s \tag{3.3}$$

Equation (3.2) conserves time-dependent O-D demand with respect to corresponding route flows, while equation (3.3) expresses the fixed time-independent O-D demand $\bar{q}^{rs}$ in terms of time-dependent O-D variable demands $q^{rs}(k)$. Hence, the time-dependent O-D demand $q^{rs}(k)$ is allowed to vary, but the sum over *k* is fixed and equal to the time-independent O-D demands $\bar{q}^{rs}$.

### 3.1.2 Trip Generation (Variable Demand) Type

In a general sense, the trip generation procedure concerns how trips attributed to origins are produced. The number of trips is usually related to the intensities and types of land use activities. For the sake of simplicity, we restrict our problem domain to a special case where trip generation relates to each O-D pair, i.e., variable demand travel choice. For the DUO variable demand/route choice model, each O-D pair must conserve the variable time-dependent O-D demand $q^{rs}(k)$ in terms of the route flow as follows:

$$\sum_{p} h_p^{rs}(k) = q^{rs}(k) \quad \forall r,s,k \tag{3.4}$$

The variable time-dependent O-D demand $q^{rs}(k)$ is a function of the O-D minimum actual route travel time during interval *k*. The variable time-dependent O-D demand function is normally defined by an additional term, in addition to the one for route choice, in the variational inequality expression (not in the constraint set) to preserve the nice structure of network flow constraints; see Section 7.1.1.2. From the computational point of view, an upper bound on the trips for each O-D pair is usually imposed. This implies the maximum O-D trips is equated to the sum of variable time-dependent O-D demand and the corresponding excess demand, as follows:

$$\sum_p h_p^{rs}(k) + e^{rs}(k) = \bar{q}_{\max}^{rs}(k) \quad \forall r,s,k \tag{3.5}$$

For the DUO variable demand/departure time/route choice model, in addition to equation (3.4), one more flow conservation constraint is needed.

$$\sum_p h_p^{rs}(k) = q^{rs}(k) \quad \forall r,s,k \tag{3.6}$$

$$\sum_k q^{rs}(k) = q^{rs} \quad \forall r,s \tag{3.7}$$

Equation (3.6) expresses the time-dependent O-D demand $q^{rs}(k)$ in terms of the route flow and equation (3.7) conserves the time-independent O-D demand $q^{rs}$. The variable time-independent O-D demand $q^{rs}$ is a function of the O-D minimum actual route travel time regardless of the departure time, and as previously noted, this variable time-independent O-D demand $q^{rs}$ is determined by an additional term, other than the one for route choice, in the variational inequality expression; see Section 7.2.1.2. Again, from the computational point of view, an upper bound on the trips for each O-D pair is usually imposed. This implies the maximum number of O-D trips is equated to the sum of the variable time-dependent O-D demand and the corresponding excess demand $e^{rs}$, as follows.

$$\sum_k q^{rs}(k) + e^{rs} = \bar{q}_{\max}^{rs} \quad \forall r,s \tag{3.8}$$

### 3.1.3 Mode Choice Type

In the real world, a traveler may choose from numerous transportation modes. However, in this section, we only consider the two most common modes for selection, i.e., the binary choice between private automobile, denoted by $m_1$, and public transit, denoted by $m_2$. For the DUO mode choice/route choice model, the flow conservation constraints are as follows:

$$\sum_p h_{mp}^{rs}(k) = q_m^{rs}(k) \quad \forall r,s,m \in (m_1, m_2), k \tag{3.9}$$

$$q_{m_1}^{rs}(k) + q_{m_2}^{rs}(k) = \bar{q}^{rs}(k) \quad \forall r,s,k \tag{3.10}$$

Equation (3.9) expresses for each mode the time-dependent O-D demand in terms of the route flow, and equation (3.10) indicates that each O-D pair must conserve the fixed time-dependent O-D demand $\bar{q}^{rs}(k)$ in terms of O-D demands for all modes.

For the mode choice/departure time/route choice model, the corresponding flow conservation constraints are described without explanation as follows.

$$\sum_p h_{mp}^{rs}(k) = q_m^{rs}(k) \quad \forall r,s,m \in (m_1, m_2), k \tag{3.11}$$

$$\sum_k q_m^{rs}(k) = q_m^{rs} \quad \forall r,s,m \in (m_1, m_2) \tag{3.12}$$

$$q_{m_1}^{rs} + q_{m_2}^{rs} = \bar{q}^{rs} \quad \forall r,s \tag{3.13}$$

### 3.1.4 O-D Choice Type

If the total number of trips originating from each origin node is given, then the singly constrained O-D choice determines the travelers' choice of destination. When both the total flow generated at each origin node and the total flow attracted to each destination node are fixed and known, the resulting models are doubly constrained. Since doubly-constrained models sometimes yield no feasible solution and are quite restricted in the dynamic sense, this section only deals with singly-constrained DUO O-D choice models. For the DUO O-D choice/route choice model, the flow conservation constraints are as follows:

$$\sum_p h_p^{rs}(k) = q^{rs}(k) \quad \forall r,s,k \tag{3.14}$$

$$\sum_s q^{rs}(k) = \bar{q}^{r}(k) \quad \forall r,k \tag{3.15}$$

Equation (3.14) expresses the time-dependent O-D demand in terms of corresponding route flows, while equation (3.15) conserves the number of trips originating from each origin node during interval $k$, and is sometimes called the trip production constraint. One of the simplest ways to ensure that the trip production constraint is an integral part of the demand function is to specify this function as a share model. A multinomial logit formula is commonly used in the following destination demand model:

$$q^{rs}(k) = \bar{q}^{r}(k) \frac{e^{-\theta\left(\pi^{rs}(k) - M^{s}\right)}}{\sum_{s'} e^{-\theta\left(\pi^{rs'}(k) - M^{s'}\right)}} \quad \forall r,s,k \tag{3.16}$$

where $M^s \geq 0$ denotes attractiveness associated with destination $s$, and $\theta > 0$ is a parameter of the logit model.

For the DUO O-D choice/departure time/route choice model, the flow conservation constraints include:

$$\sum_p h_p^{rs}(k) = q^{rs}(k) \quad \forall r,s,k \tag{3.17}$$

$$\sum_k q^{rs}(k) = q^{rs} \quad \forall r,s \tag{3.18}$$

$$\sum_s q^{rs} = \bar{q}^{r} \quad \forall r \tag{3.19}$$

Equation (3.17) expresses time-dependent O-D demand in terms of route flows, equation

(3.18) expresses time-independent O-D demand in terms of time-dependent O-D demands and equation (3.19) conserves the number of trips originating from each origin node, and is sometimes called the trip production constraint. To ensure that the trip production constraint is an integral part of the demand function, a multinomial logit formula is commonly used in the following destination demand model.

$$q^{rs} = \bar{q}^{r} \frac{e^{-\theta\left(\pi^{rs} - M^{s}\right)}}{\sum_{s'} e^{-\theta\left(\pi^{rs'} - M^{s'}\right)}} \qquad \forall r, s \tag{3.20}$$

## 3.2 Flow Propagation Constraints

Flow propagation is essential for differentiating dynamic travel choice models from their static counterparts, and depicts how vehicles advance along a route over time. Flow propagation can be better described by a continuous-time formulation. For instance, consider the inflow on link $a$ in route $p$ departing from origin $r$ during instant $k$ toward destination $s$; if the inflow experiences an actual link travel time $\tau_a(t)$, the corresponding exit flow will appear at time $t + \tau_a(t)$ as follows:

$$u_{apk}^{rs}(t) = v_{apk}^{rs}\left(t + \tau_a(t)\right) \qquad \forall r, s, a, p, k, t \tag{3.21}$$

Note that the above equation does not necessarily guarantee the first-in-first-out condition, as shown by the following example. Suppose that the actual link travel times at time instants $t$=1 and $t$=2 are 4 and 2 time units, respectively. Then, the exit time will be $t$=5 for the former, and $t$=4 for the latter, which implies a first-in-last-out phenomenon. Under the situation that the exit flow function $v_{apk}^{rs}(\bullet)$ is invertible, the actual link travel time may be analytically denoted as follows:

$$\tau_a(t) = \left(v_{apk}^{rs}\right)^{-1}\left(u_{apk}^{rs}(t)\right) - t \qquad \forall r, s, a, p, k, t \tag{3.22}$$

The actual link travel time plays a key role in exerting flow propagation of dynamic traffic assignment, and the rate of change of the actual link travel time $\tau_a(t)$ reflects the variation of degree of congestion over time on links. Three cases may be identified:

1. $\dot{\tau}_a(t) < 0$: A decreasing rate of actual travel time for link $a$ at time $t$ less than 0 implies that traffic congestion is gradually alleviated over time. Flow deformation may be visible. Furthermore, under the situation of $\dot{\tau}_a(t) \le -1$, the first-in-first-out condition cannot be preserved.
2. $\dot{\tau}_a(t) = 0$: The changing rate of actual travel time for link $a$ at time $t$ equal to 0 means the number of vehicles on link $a$ is unchanged. Since the inflow over route $p$ at time $t$ is exactly equal to the exit flow over route $p$ at time $\left(t + \tau_a(t)\right)$, a uniform flow pattern results.
3. $\dot{\tau}_a(t) > 0$: An increasing rate of actual travel time for link $a$ at time $t$

greater than 0 means the number of vehicles on link $a$ is increasing over time. The link becomes congested and a queue may form.

The above discussion pertains to continuous-time dynamic travel choice models. When a discrete-time formulation is desired, modifications are necessary. In such a case, we redefine the time instant $t$ as a time interval within the analysis time period. For each time interval, only the *average* traveler behavior is considered. The adopted actual link travel time can be interpreted as the average actual link travel time or by other reasonable measures of impedance. In any case, a uniform or single actual link travel time is assumed within each time interval. In other words, no flow deformation is allowed within each time interval, yielding a *packet* flow. With this *packet* movement assumption, equation (3.21) can still be adopted for discrete dynamic models. Except for *time instant* being replaced by *time interval*, the interpretation of equation (3.21) is the same as before. That is, inflow entering link $a$ over route $p$ during interval $t$ is required to exit the link during interval $t$ plus the *average* actual link travel time $\tau_a(t)$.

In discrete dynamic travel choice models, the actual link travel times $\tau_a(t)$ have to be approximated as integers, because only integer-valued time intervals are used as the argument of the link flow related function. Here, we take the exit flow $v_{apk}^{rs}(t+\tau_a(t))$ as an example. The argument $t+\tau_a(t)$ must be integer-valued. This means that round-off arithmetic operations are necessary for computing the actual link travel times during the solution procedure. This assumption, though essential, may cause some computational problems, such as non-convergence, for the discrete dynamic travel choice models. However, in consideration of the prohibitive computational burden associated with the continuous-time dynamic travel choice models, these problems or drawbacks cannot be avoided.

To continue the discussion, consider a sequence of links comprising route $p$ between O-D pair $rs$. For any two consecutive links in route $p$, the exit flow of the previous link $a$ can also be represented by the inflow associated with the succeeding link $b$, that is:

$$v_{apk}^{rs}(t+\tau_a(t)) = u_{bpk}^{rs}(t+\tau_a(t)) \quad \forall r,s,a,b \in A(a),p,k,t \tag{3.23}$$

By the above equation, the flow propagation constraint (3.21) can alternatively be denoted as:

$$u_{apk}^{rs}(t) = u_{bpk}^{rs}(t+\tau_a(t)) \quad \forall r,s,a,b \in A(a),p,k,t \tag{3.24}$$

For an urban network, and by using indicator variables, the above equation can be decomposed into the following two equations:

$$u_{apk}^{rs}(t) = h_p^{rs}(k)\delta_{apk}^{rs}(t) \quad \forall r,s,a \in A(r),p,k,t \tag{3.25}$$

$$u_{apk}^{rs}(t) = u_{bpk}^{rs}(t+\tau_a(t))\delta_{apk}^{rs}(t) \\ \forall r,s,p,k,t,a \in p,b \in p,a \in B(j),b \in A(j) \tag{3.26}$$

Equation (3.25) expresses the link inflow in terms of the route departure flow through the incidence relationship, and states that for each O-D pair $rs$ the flow departing during interval $k$ over route $p$ must enter onto the first link $a$ in route $p$ during interval $t$. Equation

(3.26) concerns flow propagation for any two successive links in route $p$, and the inflow on link $a$ propagates to its succeeding link $b$ in route $p$ after the actual link travel time has passed. In detail, this equation denotes for each O-D pair $rs$ that the flow departing during interval $k$ and entering link $a$ during interval $t$ cannot get on its succeeding link $b$ along route $p$ until the actual link travel time $\tau_a(t)$ elapses. Note that actual travel times have to be rounded because time intervals can only be counted by integers. Such a treatment is specific to discrete dynamic models. A compact form for equations (3.25) and (3.26) is as follows:

$$u_{apk}^{rs}(t) = h_p^{rs}(k)\delta_{apk}^{rs}(t) \quad \forall r,s,a,p,k,t \tag{3.27}$$

The resulting equation (3.27) describes for each O-D pair $rs$, how flows propagate throughout a route.

The indicator variables $\delta_{apk}^{rs}(t)$ defined above are 0-1 integer-valued because flow deformation is not possible in our models. If the indicator variable $\delta_{apk}^{rs}(t)$ is equal to 1, then for O-D pair $rs$, the flow departing during interval $k$ over route $p$ enters link $a$ during interval $t$. On the contrary, if the indicator variable $\delta_{apk}^{rs}(t)$ is equal to zero, it implies that link $a$ during interval $t$ is not in the route $p$ that is associated with O-D pair $rs$ and departure interval $k$. Consequently, the following two constraints must also be satisfied.

$$\sum_t \delta_{apk}^{rs}(t) = 1 \quad \forall r,s,p,a \in p,k \tag{3.28}$$

$$\delta_{apk}^{rs}(t) = \{0,1\} \quad \forall r,s,a,p,k,t \tag{3.29}$$

Equation (3.28) indicates for each O-D pair $rs$ that the flow departing during interval $k$ over route $p$ can be incident to link $a$ at most once during a specific time interval $t$. If the route flow is not present on link $a$, then it must get on one of other links in the network, unless the destination has been reached. Equation (3.29) designates that the indicator variables are integer-valued.

## 3.3 Nonnegativity Constraints

Nonnegativity constraints are established for all route flows.

$$h_p^{rs}(k) \geq 0 \quad \forall r,s,p,k \tag{3.30}$$

Since link flows can be represented by the summation of related route flows, equation (3.30) also implies nonnegative link flows.

## 3.4 Definitional Constraints

Definitional constraints include inflow, exit flow, number of vehicles and route travel time as follows:

$$u_{ap}^{rs}(t)=\sum_{k}u_{apk}^{rs}(t) \quad \forall r,s,a,p,t \tag{3.31}$$

$$u_{a}^{rs}(t)=\sum_{p}u_{ap}^{rs}(t) \quad \forall r,s,a,t \tag{3.32}$$

$$u_{a}(t)=\sum_{rs}u_{a}^{rs}(t) \quad \forall a,t \tag{3.33}$$

$$v_{ap}^{rs}(t)=\sum_{k}v_{apk}^{rs}(t) \quad \forall r,s,a,p,t \tag{3.34}$$

$$v_{a}^{rs}(t)=\sum_{p}v_{ap}^{rs}(t) \quad \forall r,s,a,t \tag{3.35}$$

$$v_{a}(t)=\sum_{rs}v_{a}^{rs}(t) \quad \forall a,t \tag{3.36}$$

$$x_{ap}^{rs}(t)=\sum_{k}x_{apk}^{rs}(t) \quad \forall r,s,a,p,t \tag{3.37}$$

$$x_{a}^{rs}(t)=\sum_{p}x_{ap}^{rs}(t) \quad \forall r,s,a,t \tag{3.38}$$

$$x_{a}(t)=\sum_{rs}x_{a}^{rs}(t) \quad \forall a,t \tag{3.39}$$

$$u_{a}(t)=\sum_{rs}\sum_{p}\sum_{k}u_{apk}^{rs}(t)=\sum_{rs}\sum_{p}\sum_{k}h_{p}^{rs}(k)\delta_{apk}^{rs}(t) \quad \forall a,t \tag{3.40}$$

$$c_{p}^{rs}(k)=\sum_{a}\sum_{t}c_{a}(t)\delta_{apk}^{rs}(t) \quad \forall r,s,p,k \tag{3.41}$$

Equations (3.31)~(3.39) are self-evident while equation (3.40) expresses the link flows in terms of the route flows through the use of the indicator variables, and implies that if flows from any used route $p$ arrive at link $a$ during interval $t$, then link $a$ is being used during interval $t$. The first equality of equation (3.40) expresses the inflows on link $a$ during interval $t$ in terms that route flows enter link $a$ during interval $t$. The second equality of equation (3.40) is simply a restatement of equation (3.27), where summations over $r,s,p,k$ are presented. Equation (3.41) expresses the actual route travel time in terms of the actual link travel times and computes the actual route travel time by adding up the actual travel times on those links along that route.

The indicator variable $\delta_{apk}^{rs}(t)$ defined above is a 0-1 integer *variable*, and not just an *index* as in the static models. Consequently, equations (3.25), (3.26), (3.40) and (3.41) are essentially nonlinear and nonconvex. The nonlinearity property comes from the multiplication of two variables (yielding a quadratic function), and the nonconvexity is due to the inclusion of 0-1 integer variables. Therefore, multiple local solutions could exist for the proposed dynamic travel choice models. However, once actual link travel times $\tau_{a}(t)$ are estimated, the indicator variables $\delta_{apk}^{rs}(t)$ would be accordingly simplified as indices.

Under the estimated actual travel times, the resulting feasible region of the subproblem is delineated by linear constraints and can be easily solved.

## 3.5 First-In-First-Out Constraints

The first-in-first-out condition (FIFO) follows the requirement that vehicles can arrive at the destination earlier only by leaving earlier. To preserve the FIFO condition, the changing rate of actual link travel times between two time intervals must be greater than or equal to -1. Ran and Boyce (1994) derived this condition as follows: Consider a simple network with one origin and one destination joined by one link. Suppose the actual travel times for the vehicles entering the link during time intervals $t$ and $t + \Delta t$ are $\tau(t)$ and $\tau(t + \Delta t)$, respectively .Then, the FIFO condition can be expressed mathematically as:

$$t + \tau(t) \leq t + \Delta t + \tau(t + \Delta t) \tag{3.42}$$

Rearranging the terms on both sides yields

$$\frac{\tau(t + \Delta t) - \tau(t)}{\Delta t} \geq -1 \tag{3.43}$$

Assuming that the link travel time function $\tau(t)$ is continuous and differentiable at time $t$, the left side of equation (3.43) becomes the derivative of the link travel time function with respect to time as the difference of arrival times between two consecutive vehicles approaches zero (i.e., $\Delta t \to 0$).

$$\dot{\tau}(t) \geq -1 \tag{3.44}$$

To conserve the FIFO condition, equation (3.44) denotes that the derivative of $\tau(t)$ at any time $t$ should never be less than -1.

The FIFO condition is crucial from the traffic engineer's point of view. For instance, consider a one-way one-lane street of a physical layout. There is no way to overtake on this street; however, one-way one-lane streets are rare and usually account for a very small percentage of passageways in any urban area. These types of streets can therefore deemed as less influential for inclusion in the urban network representation. For those (wider and longer) streets included in the transportation network, the FIFO condition is also not so striking. As mentioned in Ran and Boyce (1996, p. 80):

> In general, overtaking will not occur for most definitions of link lengths (over 100 feet) and time intervals (shorter than 2 minutes) if the free flow speed is assumed to be 50 miles/hour in the experiment. However, as we note in discrete models, no matter how accurate the link traffic dynamics model is, overtaking or a jump may still occur when the time interval is too large. On the other hand, for most problems, we are only interested in the aggregate behavior of flows and the FIFO is not so important in those situations. Furthermore, we should note that the FIFO assumption itself is also an approximation of reality.

From the mathematical point of view, there is an additional difficulty associated with the inclusion of the FIFO constraint in the dynamic user-optimal travel choice models. This added side constraint would make the problem much more difficult to solve because the advantages of network flow constraints have vanished. In his seminal paper, Carey (1987) constructed the FIFO constraint as follows:

$$\frac{v_a^s(t)}{v_a^{s'}(t)} = \frac{x_a^s(t)}{x_a^{s'}(t)} \tag{3.45}$$

With this added constraint, unfortunately, the feasible set of the subproblem (with the interactions of actual travel times, $\tau$, being relaxed in the dynamic user-optimal travel choice models) is also nonconvex. This nonconvex property of the subproblem is indeed troublesome and there is no way, to the best of the author's knowledge, to further relax it. In fact, there is no efficient solution algorithm available to solve a nonconvex mathematical problem. Moreover, according to an experiment performed at National Central University in Taiwan, both the FIFO requirement, and the dynamic user-optimal equilibrium conditions defined in this book, cannot be satisfied for a simple network with only one origin and destination joined by a single link. In other words, the feasible set is empty in this special case. From the above discussion, it may be concluded that, without including the FIFO constraint, the dynamic user-optimal equilibrium conditions defined in this book can be fulfilled, provided an optimal solution is obtained. On the other hand, including the FIFO constraint in the dynamic route choice model could make the feasible region nil.

## 3.6 Link Capacity and Oversaturation

There are two basic constraints for link capacity. The first constraint is the maximum number of vehicles on the link, while the second constraint is the maximum exit flow rate from the link.

### 3.6.1 Maximum Number of Vehicles on a Link

Let $l_a$ denote the length of link $a$ and $e_{a\max}(t)$ denote the maximum traffic density (vehicle/mile) of link $a$ during interval $t$. The maximum number of vehicles that link $a$ can accommodate is $l_a e_{a\max}(t)$. The number of vehicles on link $a$ must be less than or equal to the maximum number of vehicles on the link. It follows that:

$$x_a(t) \le l_a e_{a\max}(t) \quad \forall a, t \tag{3.46}$$

### 3.6.2 Maximum Exit Flow from a Link

Another constraint concerns the exit flow capacity $v_{a\max}(t)$ at the exit of link $a$ during interval $t$. It follows that:

$$v_a(t) \le v_{a\max}(t) \quad \forall a, t \tag{3.47}$$

In a network, the exit capacity constraint for an upstream link is also an inflow capacity constraint for downstream links. This constraint can be either added directly in the formulation or combined in the link travel time functions. If it is directly added in the variational inequality formulation, more analysis of the optimality conditions of dynamic network models is necessary when the exit flow capacity is reached; moreover, the computational algorithm needs to be revised. In most of our dynamic travel choice models, we implicitly consider this exit capacity constraint in the travel time functions. Thus, it is not necessary to define an explicit constraint in these travel choice models.

### 3.6.3 Constraints for Spillback

Oversaturation may occur anywhere and during any time interval when traffic demand exceeds capacity. When queues at critical intersections develop upstream, then they cause the so-called spillback problem. In an oversaturated situation, continuing excess demand relative to supply could transform local oversaturation to regional oversaturation. Thus, in dynamic travel choice models, corresponding constraints should be formulated to reflect this phenomenon. In an advanced control/assignment framework, those constraints should be consistent with each type of traffic control strategy. The two main types of traffic control strategies are: 1) minimize delay and stops; 2) keep traffic moving or maximize productivity (Lieberman, 1993). We leave the control policy for further study in the context of dynamic signal control schemes.

## 3.7 Link Travel Time Function Analysis

As always in transportation network models, dynamic link travel time functions, also named dynamic link performance functions, play a central role in the DUO route choice model. A good dynamic link travel time function should not only closely reflect the real situation, but also as much as possible satisfy ideal mathematical properties. However, these requirements can hardly ever be achieved at the same time, and therefore we have to settle for trade-offs in the practical applications. In this section, we are not attempting to develop the definitive function, but rather we only study the general properties pertaining to the dynamic link travel time function without going into details.

Up to now, there has been no consensus reached for a suitable form of dynamic link travel time functions. However, without going into detail, the dynamic link travel time function has been generally developed to represent the aggregate effect of the free flow travel time plus delay time due to congestion in the downstream of the link. Ran et al (1997) adopt the following function to calculate the dynamic link travel time, where $c_{1a}(t)$ and $c_{2a}(t)$ denote the free flow travel time and delay, respectively, for link $a$ during time interval $t$.

$$c_a(t) = c_{1a}(t) + c_{2a}(t) \tag{3.48}$$

Since both $c_{1a}(t)$ and $c_{2a}(t)$ can be expressed as a function of inflow $u_a(t)$, exit flow $v_a(t)$, and number of vehicles $x_a(t)$ on link *a*, equation (3.48) can thus be rewritten, without loss of generality, as follows:

$$c_a(t) = c_a(u_a(t), v_a(t), x_a(t)) \tag{3.49}$$

### 3.7.1 Treating Exit Flow as a Variable

Different treatments on exit flow can be found in research literature. Many researchers claim that exit flow is a function of the number of vehicles, $v_a(x_a(t))$, and this representation might describe the real situations to some extent; however, it does add some problems from both the computational and physical representation points of view:

1. The link exit flow becomes positive immediately whenever flows enter this link, i.e., $x_a(t) > 0$. This phenomenon is certainly not valid if the link flow initializes with zero, because the trips must take some time to traverse the entire link no matter how short the link length is. However, Wie et al (1994) defended with the argument that the initial flows with zeros may not come true in practical situations.
2. It is extremely hard to decompose the link exit flows by destination if the link exit flow function is nonlinear. Therefore, the dynamic network model with multiple destinations, which is required for practical applications, is difficult to be formulated and solved.
3. There is no idea or picture on what the nonlinear link exit flow function should look like, and hence there is no ideal way to calibrate it.

The problem caused by the nonlinear link exit flow function can be resolved by simply treating the exit flow as a variable rather than as a function (Ran and Shimazaki, 1989b). With this modification, many-to-many dynamic network models can be easily formulated, and in addition, a reduction of computational effort required for the dynamic network model can be expected. We adopt this type of treatment for exit flow throughout this book.

### 3.7.2 Relationships among Inflow, Exit Flow and Number of Vehicles

In dynamic networks, the inflow, exit flow and number of vehicles can be deemed as three different *states* for the *same* vehicles. This concept can be easily illustrated by a simple network with two nodes connected by a link. Suppose the actual link travel time for vehicles arriving during interval $t$ is $\tau_a(t)$. We can then easily identify and term these vehicles as: *inflow rate* during interval *t*, *number of vehicles* between time intervals (not included) $t$ and $t + \tau_a(t)$, and *exit flow rate* during interval $t + \tau_a(t)$. In other words, given actual link travel times, the relationships among inflow, exit flow (or outflow), and number of vehicles can be accordingly determined. Also, no variable can be claimed to be independent of the other two variables. The implication of these relationships is that we can accommodate one variable, rather than three variables, in our discrete dynamic travel choice models. We verify this premise by carrying out rigorous mathematical derivations as follows.

We first verify that inflows and exit flows can be substituted for each other. For a specific route $p$ between O-D pair $rs$ departing during interval $k$, the flow propagation constraint can be written as:

$$u_{apk}^{rs}(i) = v_{apk}^{rs}\left(i + \tau_a(i)\right) \quad \forall r, s, a, p, k, i \tag{3.50}$$

Equation (3.50) expresses that the inflows on link $a$ during interval $i$ over route $p$ can only exit the link exactly during interval $i + \tau_a(i)$. The exit flows on link $a$ during interval $i + \tau_a(i)$ may be comprised of several inflows from various intervals $i$, i.e.,

$$\sum_i u_a(i) = v_a\left(i + \tau_a(i)\right) \quad \forall a, \left(i + \tau_a(i)\right) \tag{3.51}$$

By taking equation (3.51) into account, and summing equation (3.50) with respect to $r,s,p,k$, it follows that:

$$\sum_{rs}\sum_{p}\sum_{k=1}^{i}\sum_{i} u_{apk}^{rs}(i) = \sum_{rs}\sum_{p}\sum_{k=1}^{i} v_{apk}^{rs}\left(i + \tau_a(i)\right) \quad \forall a, \left(i + \tau_a(i)\right) \tag{3.52}$$

By equations (3.31)~(3.36) and letting $t = i + \tau_a(i)$, the above equation can be simplified as:

$$\sum_{i=t-\tau_a(i)} u_a(i) = v_a(t) \quad \forall a, t \tag{3.53}$$

By introducing the indicator variable into equation (3.53), one obtains

$$v_a(t) = \sum_i u_a(i)\delta_{a_1}^i(t) \quad \forall a, t \tag{3.54}$$

where

$$\delta_{a_1}^i(t) = \begin{cases} 1 & if\ i + \tau_a(i) = t \\ 0 & otherwise \end{cases} \quad \forall a,i,t \tag{3.55}$$

Now, we can see that equations (3.54)~(3.55) indicate that the exit flows are equal to the sum of inflows entering the link $a$ no later than interval $t$, $\left(u_a(i), i \le t\right)$, and exiting the link exactly during interval $t$. The exit flows $v_a(t)$ relate to inflows $u_a(i)$ through the indicator variable $\delta_{a_1}^i(t)$; implying that link exit flows can be at least theoretically replaced by link inflows. Note that equations (3.54)~(3.55) result from equation (3.50) by summing over route $p$, interval $i$, O-D pair $rs$, and departure interval $k$. Clearly, the inflow and exit flow on link $a$ are mutually convertible under the estimated actual link travel time.

A similar approach can be used to deduce the relationship between inflows and number of vehicles. For a specific route $p$ between O-D pair $rs$ departing during interval $k$, the number of vehicles on link $a$ at the beginning of interval $t$, $x_a(t)$, can be denoted by the following expression.

$$x_{apk}^{rs}(t) = x_{apk}^{rs}(t-1) + u_{apk}^{rs}(t-1) - v_{apk}^{rs}(t-1) \tag{3.56}$$

By summing up equation (3.56) with respect to $r,s,p,k$, it follows:

$$x_a(t) = x_a(t-1) + u_a(t-1) - v_a(t-1) \tag{3.57}$$

Analogously, the number of vehicles on link $a$ from interval 1 to interval $t$-1 can be expressed as follows:

$$\begin{aligned} &x_a(t-1) = x_a(t-2) + u_a(t-2) - v_a(t-2) \\ &\quad\vdots \\ &x_a(2) = x_a(1) + u_a(1) - v_a(1) \\ &x_a(1) = 0 \end{aligned} \tag{3.58}$$

The sum of equations (3.57) and (3.58) would yield the number of vehicles at the beginning of interval $t$ equal to the accumulated number of the net increased vehicles on link $a$ before interval $t$ as follows:

$$x_a(t) = \sum_{i=1}^{t-1} u_a(i) - \sum_{i=1}^{t-1} v_a(i) \quad \forall a,t \tag{3.59}$$

Substituting equation (3.54) into equation (3.59) results in the following equation.

$$x_a(t) = \sum_{i=1}^{t-1} u_a(i) - \sum_{i=1}^{t-1}\left[\sum_j u_a(j)\delta_{a_1}^j(i)\right] \quad \forall a,t \tag{3.60}$$

Equation (3.60) states that $x_a(t)$ can be computed by adding up those inflows entering link $a$ before interval $t$ and exiting link $a$ after interval $t$ (included). Note that the number of vehicles on link $a$ under the estimated actual link travel time can be represented as a function of inflows. By canceling out those inflows appearing in both the first and second terms of equation (3.60), the following expressions are obtained:

$$x_a(t) = \sum_i u_a(i)\delta_{a_2}^i(t) \tag{3.61}$$

where

$$\delta_{a_2}^i(t) = \begin{cases} 1 & \text{if } i < t \text{ and } i + \tau_a(i) \geq t \\ 0 & \text{otherwise} \end{cases} \tag{3.62}$$

Since the three variables of inflow, exit flow and number of vehicles can be substituted for each other, the dynamic link travel time function shown in equation (3.49) can be readily represented as a function of inflow without loss of generality. The reduced form of the dynamic link travel time function can thus be expressed as follows:

$$c_a(t) = c(u_a(1), u_a(2), \cdots, u_a(t)) \tag{3.63}$$

Note that when the FIFO restraint is imposed, then the inflow entering the subject link before a specific interval $t$ would have exited the link already, this means that the inflows that need to be considered in the link travel time function are only those entering the subject link between intervals $t$ and $t'$, where $t'$ is the current interval, i.e.,

$$\begin{aligned} & c_a(t') = c\big(u_a(t), u_a(t+1), \cdots, u_a(t')\big) \\ & \qquad t \le t', t + \tau_a(t) \ge t', t - 1 + \tau_a(t-1) < t' \end{aligned} \tag{3.64}$$

In a recent paper, Daganzo (1995) showed by means of an example that inflow $u_a(t)$ and exit flows $v_a(t)$ should be excluded from the dynamic link travel time function represented by equation (3.49), in order to prevent possible violation of the FIFO condition. Consequently, the link travel time function (with capacity constraints for exit flows) is in the form of :

$$c_a(t) = c\big(x_a(t)\big) \tag{3.65}$$

Equation (3.65) states that the link travel time is only affected by the number of vehicles on that link. Even though this function form is acceptable, according to our previous discussion about the relationship of inflows and number of vehicles, we note that equation (3.65) essentially constructs a special case of equation (3.64) and can be rewritten as follows:

$$\begin{aligned} & c_a(t') = c\big(u_a(t), u_a(t+1), \cdots, u_a(t'-1)\big) \\ & \qquad t \le t', t + \tau_a(t) \ge t', t - 1 + \tau_a(t-1) < t' \end{aligned} \tag{3.66}$$

For further discussion on the dynamic link travel time function, see Chen and Tu (1996).

### 3.7.3 Asymmetric Link Interactions

Optimization problems are especially useful for obtaining solutions because many efficient solution algorithms are associated with them. Therefore, formulating a dynamic travel choice problem as an optimization problem naturally becomes the first step to solve the problem. However, in many circumstances, the equivalent optimization problem does not exist. Therefore, without knowing this fact, research in this direction becomes ineffective. In the following analysis, we first prove that a DUO travel choice problem cannot be formulated as an optimization problem due to temporal link interactions. Then, by simple observation we confirm that the asymmetric property of the dynamic link travel time functions still remains when topological interactions exist, implying an equivalent optimization formulation does not exist as well. Consequently, the variational inequality approach emerges.

#### 3.7.3.1 Temporal interactions

For each physical link, inflows may be blocked by those previously entered vehicles, yielding temporal link interactions.

***Theorem 3.1***: Any DUO travel choice problem with the dynamic travel time function $c_a(t)$ relating to inflows $(u_a(\omega),\ \omega \le t)$ as shown in equation (3.63) does not have an equivalent optimization formulation.

***Proof***: Consider a specific physical link $a$. The corresponding part of the Jacobian matrix of the dynamic travel time function as shown in (3.63) can be expressed as follows:

$$\mathbf{A}_a = \begin{bmatrix} \frac{\partial c_a(1)}{\partial u_a(1)} & \frac{\partial c_a(2)}{\partial u_a(1)} & \cdots & \frac{\partial c_a(T)}{\partial u_a(1)} \\ \frac{\partial c_a(1)}{\partial u_a(2)} & \frac{\partial c_a(2)}{\partial u_a(2)} & \cdots & \frac{\partial c_a(T)}{\partial u_a(2)} \\ \vdots & \vdots & \ddots & \vdots \\ \frac{\partial c_a(1)}{\partial u_a(T)} & \frac{\partial c_a(2)}{\partial u_a(T)} & \cdots & \frac{\partial c_a(T)}{\partial u_a(T)} \end{bmatrix} \tag{3.67}$$

Since the time-space link travel time of any vehicle can only be affected by those previously entered vehicles on that link, the partial derivatives of the dynamic travel time function with respect to the inflow at later time intervals would therefore be zero. The corresponding part of the Jacobian matrix of the dynamic travel time function for time-space link $a$ can thus be simplified as:

$$\mathbf{A}_a = \begin{bmatrix} \frac{\partial c_a(1)}{\partial u_a(1)} & \frac{\partial c_a(2)}{\partial u_a(1)} & \cdots & \frac{\partial c_a(T)}{\partial u_a(1)} \\ 0 & \frac{\partial c_a(2)}{\partial u_a(2)} & \cdots & \frac{\partial c_a(T)}{\partial u_a(2)} \\ \vdots & \vdots & \ddots & \vdots \\ 0 & 0 & \cdots & \frac{\partial c_a(T)}{\partial u_a(T)} \end{bmatrix} \tag{3.68}$$

Obviously, matrix $\mathbf{A}_a$ is asymmetric. Since every link on the network has a similar Jacobian form of equation (3.68), the full Jacobian matrix of dynamic travel time functions can be written as follows:

$$\nabla_u \mathbf{c} = \begin{bmatrix} \mathbf{A}_1 & \mathbf{0} & \cdots & \mathbf{0} \\ \mathbf{0} & \mathbf{A}_2 & \cdots & \mathbf{0} \\ \vdots & \vdots & \ddots & \vdots \\ \mathbf{0} & \mathbf{0} & \cdots & \mathbf{A}_n \end{bmatrix} \tag{3.69}$$

Note that the value of off-diagonal elements is a $\mathbf{0}$ matrix because at present the time-space link travel time is assumed not to be influenced by flows other than the subject time-space link. Since matrix $\mathbf{A}_a$ is asymmetric, the Jacobian matrix of the dynamic travel time function $\nabla_{\mathbf{u}} \mathbf{c}$ must therefore also be asymmetric. According Green's Theorem (Nagurney, 1993; Patriksson, 1994), an equivalent optimization problem can exist only when the Jacobian matrix of the travel time function is symmetric; see also Section 2.1.2. Since the Jacobian matrix of the dynamic travel time function is asymmetric, there exists no corresponding optimization model. This completes the proof.

#### 3.7.3.2 Topological interactions

In general, the actual link travel time is also affected by those links other than the subject physical link. In many circumstances, the link interaction is asymmetric. The asymmetric property means that the interaction effect of the current link on another link is not equivalent to that link on the current one. Examples include: overtaking on a two-lane-two-way street, turning movements at a congested intersection, merging/diverging at a freeway interchange, and movements at an unsignalized intersection. These situations will result in many off-diagonal elements in the matrix $\nabla_{\mathbf{u}} \mathbf{c}$ shown in expression (3.69) not being zero-valued. However, the asymmetric property of dynamic travel time functions still remain; implying that an equivalent optimization problem does not exist as well.

## 3.8 Notes

In this chapter, we have discussed sets of constraints including flow conservation, flow propagation, nonnegative, definitional, FIFO, and link capacity and oversaturation, which are necessary for dynamic travel choice models. However, not all constraints are considered in later chapters. The FIFO requirement, as mentioned in Section 3.5, may cause some difficulties in computation of solutions. This is especially true when a nonlinear FIFO condition is imposed. In lieu of imposing the difficult FIFO requirement in the constraint set, two other strategies may be adopted to alleviate the possible violation of the FIFO condition. One is to develop a flow movement mechanism that implicitly satisfies this requirement. Examples are using six link-based variables to represent the dynamics of traffic on a link (Ran and Boyce, 1994) and accommodating a cell transmission model (Daganzo, 1993). However, there is insufficient evidence that their results can be readily applied to satisfy the FIFO condition for the dynamic travel choice models. The other strategy is to accommodate a very short time interval length. By using smaller time intervals, the difference of time-dependent O-D demands associated with any two time intervals can be narrowed significantly or even smoothed out. This implies that the rate of change of actual link travel times for any two time intervals will be small enough such that the violation of the FIFO condition is unlikely to occur. Unfortunately, increasing the number of time intervals for the entire analysis period means that larger computer memory and more execution time are required, which may be prohibitive for large transportation networks.

Link capacity constraints and oversaturation are also troublesome. When the side constraint of link capacity is imposed, the attractive property of the network flow constraints is destroyed, and many efficient algorithms, such as the diagonalization method, are no longer applicable. Recently, Larsson and Patriksson (1995) proposed the augmented Lagrangian dual (ALD) method that embeds the disaggregate simplicial decomposition (DSD) method to solve the static capacitated route choice model where at equilibrium, a modified Wardrop's principle can be defined. Their computational experience has showed that the ALD method is quite promising in solving the static capacitated route choice model; likewise, this research direction is currently being adopted by the author for capacitated dynamic travel choice models. Note that under the estimated link travel times, inflows and exit flows can be substituted for each other as shown in equation (3.50); hence the link capacity constraint expressed by the maximum number of exit flow from a link in equation (3.47) is essentially equivalent to the following expression:

$$u_a(t) \leq v_{a\max}\big(t + \tau(t)\big) \quad \forall a, t \tag{3.70}$$

If the maximum number of exit flow from a link is invariant over time and is denoted by $CAP_a$, then the above equation can be readily replaced by:

$$u_a(t) \leq CAP_a \quad \forall a, t \tag{3.71}$$

It is also worth re-emphasizing that the dynamic travel time function may originally appear as a function of inflows, exit flows, and number of vehicles on a link; however, it can eventually be rewritten as only a function of inflows, without any difficulty. This treatment can simplify the model's formulation and possibly reduce the computation time.

As a final remark, it is conjectured that the link capacity constraint and the FIFO requirement are closely related to each other; however, more effort is needed to elaborate this hypothesis.

# Chapter 4

# Dynamic User-Optimal Route Choice Model

The dynamic user-optimal route choice problem explores how travelers make their route choice decision based on predicative traffic information. The fixed time-dependent O-D demand is assumed to be known and may be regarded as a natural result of day-to-day adjustments of commuters. In this chapter, the discrete-time dynamic user-optimal (DUO) route choice problem is formulated using the variational inequality approach. The dynamic user-optimal conditions and its equivalent problem, along with the concept of time-space network, are presented in Section 4.1. A numerical example is given in Section 4.2. The relationship between dynamic and static user-optimal route choice models is depicted in Section 4.3. The occurrence of multiple solutions is justified by a numerical example in Section 4.4, and finally, concluding notes are given in Section 4.5.

## 4.1 Equilibrium Conditions and Model Formulation

In this section, the dynamic user-optimal conditions are first defined to characterize travelers' driving behavior in using routes with the minimal actual travel time. The route choice decision for travelers is based on *actual* route travel times rather than *reactive* (or *instantaneous*) route travel times. The actual route travel times can be computed by adding up the actual link travel times, under the flow propagation procedure along that route, whereas the reactive route travel times are obtained by summing up reactive link travel times along that route based on the prevailing traffic conditions. The terms - *actual* travel time, *idealized* travel time, *realized* travel time and *experienced* travel time - may be used interchangeably. The variational inequality formulation is then presented for the DUO route choice problem. After that, the equivalence between the dynamic user-optimal conditions and the variational inequality formulation is verified by a proof.

### 4.1.1 Dynamic User-Optimal Conditions

Assuming O-D demands are fixed and time-dependent, the dynamic user-optimal conditions state for each O-D pair that the actual route travel times experienced by travelers departing during the same interval are equal and minimal; that is, no traveler would be better off by unilaterally changing his/her route. In contrast, the actual route travel time of any unused route for each O-D pair is greater than or equal to the actual used route travel times. In other words, at equilibrium, if the flow departing from origin $r$ during interval $k$ over route $p$ toward destination $s$ is positive, i.e., $h_p^{rs*}(k)>0$, then the corresponding actual route travel time is minimal. On the contrary, if no flow occurs on route $p$, i.e., $h_p^{rs*}(k)=0$, then the corresponding actual route travel time is at least as great as the minimal actual route travel time. This equilibrium conditions can be mathematically expressed as follows:

$$c_p^{rs*}(k)\begin{cases} = \pi^{rs}(k) & \text{if } h_p^{rs*}(k)>0 \\ \geq \pi^{rs}(k) & \text{if } h_p^{rs*}(k)=0 \end{cases} \quad \forall r,s,p,k \tag{4.1}$$

where

$$c_p^{rs*}(k)=\sum_a \sum_t c_a^*(t)\delta_{apk}^{rs}(t) \quad \forall r,s,p,k \tag{4.2}$$

$$\pi^{rs}(k)=\min_{p,k}\left\{c_p^{rs*}(k)\right\} \quad \forall r,s \tag{4.3}$$

### 4.1.2 Variational Inequality Formulation

The DUO route choice problem is formulated using variational inequalities. As is traditional in the model formulation, the assumptions and limitations are given first as follows:

1. Traffic information available to travelers is perfect.
2. Travelers are homogeneous with respect to driving behavior regardless of their income, age and other socio-economic characteristics.
3. Route choice criterion for travelers is the pre-trip minimal actual route travel time. No route switching decision is permitted en route based on *reactive* (or *instantaneous*) traffic information.
4. The signal timing plan for each signalized intersection is fixed (or simply ignored), which implies that link capacities remain the same for the entire analysis period.
5. A hard link capacity restraint is not imposed. Instead, link travel time functions that smoothly approach infinity as flow approaches infinity are adopted.
6. The model is discrete in time and, indeed, relatively independent of its continuous counterpart development. More specifically, flow propagation is exerted in terms of

discrete time intervals.

7. The first-in-first-out condition which prohibits vehicles from arriving at a destination earlier by leaving later is not considered.

The first five assumptions also apply to static route choice problems, whereas the last two limitations are specific to dynamic route choice problems. Items 5 and 7 are of great concern to traffic engineers.

The DUO route choice problem may be regarded as a generalization of the static user-optimal route choice problem. The associated feasible region is usually delineated by a set of constraints including flow conservation, flow propagation and nonnegativity requirements. Since the continuous DUO route choice problem is extremely difficult to model and solve for real-world network problems, only the discrete DUO route choice model is addressed; however, the discrete DUO route choice model may be reasonably regarded an approximation of its continuous counterpart, if it exists. The difference between the two types of dynamic models becomes negligible as the length of the time interval for the discrete DUO route choice model approaches zero.

The following theorem is stated for the DUO route choice model.

***Theorem 4.1***: The DUO route choice problem is equivalent to finding a solution $\mathbf{u}^* \in \Omega$ such that the following VIP holds.

$$\mathbf{c}^*[\mathbf{u}-\mathbf{u}^*] \geq 0 \quad \forall \mathbf{u} \in \Omega^* \tag{4.4}$$

Or, alternatively, in expanded form:

$$\sum_a \sum_t c_a^*(t)\left[u_a(t)-u_a^*(t)\right] \geq 0 \quad \forall \mathbf{u} \in \Omega^* \tag{4.5}$$

where $\Omega^*$ is a subset of $\Omega$ with $\delta_{apk}^{rs}(t)$ being realized at equilibrium, i.e., $\left(\delta_{apk}^{rs}(t)=\delta_{apk}^{rs*}(t)\right), \forall r,s,a,p,k,t$. The symbol $\Omega$ denotes the feasible region that is delineated below by flow conservation, flow propagation, nonnegativity, and definitional constraints.

Flow conservation constraint:

$$\sum_p h_p^{rs}(k)=\bar{q}^{rs}(k) \quad \forall r,s,k \tag{4.6}$$

Flow propagation constraints:

$$u_{apk}^{rs}(t)=h_p^{rs}(k)\delta_{apk}^{rs}(t) \quad \forall r,s,a,p,k,t \tag{4.7}$$

$$\sum_t \delta_{apk}^{rs}(t)=1 \quad \forall r,s,p,a \in p,k \tag{4.8}$$

$$\delta_{apk}^{rs}(t)=\{0,1\} \quad \forall r,s,a,p,k,t \tag{4.9}$$

Nonnegativity constraint:

$$h_p^{rs}(k) \geq 0 \quad \forall r,s,p,k \tag{4.10}$$

Definitional constraints:

$$u_a(t) = \sum_{rs}\sum_{p}\sum_{k} h_p^{rs}(k)\delta_{apk}^{rs}(t) \quad \forall a,t \tag{4.11}$$

$$c_p^{rs}(k) = \sum_{a}\sum_{t} c_a(t)\delta_{apk}^{rs}(t) \quad \forall r,s,p,k \tag{4.12}$$

For the meaning of each constraint, the reader is referred to Chapter 3.

### 4.1.3 Equivalence Analysis

The following theorem verifies the equivalence between equilibrium conditions (4.1) and VIP (4.5).

***Theorem 4.2***: Under a certain flow propagation relationship $\left(\delta_{apk}^{rs}(t) = \delta_{apk}^{rs*}(t)\right)$, equilibrium conditions (4.1) imply VIP (4.5) and vice versa.

***Proof of necessity***: We need to prove that under a certain propagation relationship $\left(\delta_{apk}^{rs}(t) = \delta_{apk}^{rs*}(t)\right)$, dynamic user-optimal equilibrium conditions (4.1) can be reformulated as VIP (4.5). We first rewrite equilibrium conditions (4.1) as follows:

$$\left[c_p^{rs*}(k) - \pi^{rs}(k)\right]h_p^{rs*}(k) = 0 \quad \forall r,s,p,k \tag{4.13}$$

Since $c_p^{rs*}(k) - \pi^{rs}(k) \geq 0, \forall r,s,p,k$, and $h_p^{rs}(k) \geq 0, \forall r,s,p,k$, it implies:

$$\left[c_p^{rs*}(k) - \pi^{rs}(k)\right]h_p^{rs}(k) \geq 0 \quad \forall r,s,p,k \tag{4.14}$$

Subtracting equation (4.13) from equation (4.14) results:

$$\left[c_p^{rs*}(k) - \pi^{rs}(k)\right]\left[h_p^{rs}(k) - h_p^{rs*}(k)\right] \geq 0 \quad \forall r,s,p,k \tag{4.15}$$

Summing over *r,s,p,k* yields:

$$\begin{aligned}&\sum_{rs}\sum_{k}\sum_{p} c_p^{rs*}(k)\left[h_p^{rs}(k) - h_p^{rs*}(k)\right] \\ &\qquad -\sum_{rs}\sum_{k}\pi^{rs}(k)\sum_{p}\left[h_p^{rs}(k) - h_p^{rs*}(k)\right] \geq 0\end{aligned} \tag{4.16}$$

By making the substitution of $\sum_{p} h_p^{rs}(k) = \sum_{p} h_p^{rs*}(k) = \bar{q}^{rs}(k)$, the second term vanishes; the remaining term results in the following VIP:

$$\sum_{rs}\sum_{k}\sum_{p} c_p^{rs*}(k)\left[h_p^{rs}(k)-h_p^{rs*}(k)\right]\geq 0 \qquad (4.17)$$

Equation (4.17), accompanied with the relevant constraints, is in fact a route-based DUO route choice model. This route-based DUO route choice model can be converted into a link-based DUO route choice model by using the following two definitional constraints.

$$u_a(t)=\sum_{rs}\sum_{k}\sum_{p} h_p^{rs}(k)\delta_{apk}^{rs*}(t) \quad \forall a,t \qquad (4.18)$$

$$c_p^{rs}(k)=\sum_{a}\sum_{t} c_a(t)\delta_{apk}^{rs*}(t) \quad \forall r,s,p,k \qquad (4.19)$$

By applying equation (4.19) to equation (4.17), one obtains:

$$\sum_{rs}\sum_{k}\sum_{p}\left[\sum_{a}\sum_{t} c_a^*(t)\delta_{apk}^{rs*}(t)\right]\left[h_p^{rs}(k)-h_p^{rs*}(k)\right]\geq 0 \qquad (4.20)$$

By changing the order of the summation, it follows that:

$$\sum_{a}\sum_{t} c_a^*(t)\sum_{rs}\sum_{k}\sum_{p}\delta_{apk}^{rs*}(t)\left[h_p^{rs}(k)-h_p^{rs*}(k)\right]\geq 0 \qquad (4.21)$$

By using equation (4.11), we have:

$$\sum_{a}\sum_{t} c_a^*(t)\left[u_a(t)-u_a^*(t)\right]\geq 0 \qquad (4.22)$$

Equation (4.22) is identical to VIP (4.5).

***Proof of sufficiency***: We next prove that VIP (4.5) can induce dynamic user-optimal conditions (4.1). Since equations (4.17)~(4.22) are essentially reversible, the remaining steps of the proof are to show that route-based VIP (4.17) is equivalent to dynamic user-optimal conditions (4.1). We now define a feasible solution $\{h_p^{rs}(k)\}$ to be the same as the equilibrium flow pattern $\{h_p^{rs*}(k)\}$, except for the two routes $p_1^{rs}$, $p_2^{rs}$. Without loss of generality, we consider two situations that could arise while at equilibrium.

(i) Both route flows are positive, i.e., $h_{p_1}^{rs*}(k)>0$, and $h_{p_2}^{rs*}(k)>0$. We switch a small amount of flow $\Delta_1$ from route $p_1^{rs}$ to $p_2^{rs}$ with $0<\Delta_1\leq h_{p_1}^{rs*}(k)$. That is:

$$h_{p_1}^{rs}(k)=h_{p_1}^{rs*}(k)-\Delta_1 \text{ and} \qquad (4.23)$$

$$h_{p_2}^{rs}(k)=h_{p_2}^{rs*}(k)+\Delta_1 \qquad (4.24)$$

By substituting a new feasible solution $\{h_p^{rs}(k)\}$ into equation (4.17), we have:

$$c_{p_1}^{rs*}(k)\left[h_{p_1}^{rs}(k)-h_{p_1}^{rs*}(k)\right]+c_{p_2}^{rs*}(k)\left[h_{p_2}^{rs}(k)-h_{p_2}^{rs*}(k)\right]\geq 0 \qquad (4.25)$$

By using equations (4.23)~(4.24), one obtains:

$$c_{p_2}^{rs*}(k)\geq c_{p_1}^{rs*}(k) \qquad (4.26)$$

Similarly, by switching a small amount of flow $\Delta_2$ with $0 < \Delta_2 \le h_{p_2}^{rs*}(k)$ from route $p_2^{rs}$ to $p_1^{rs}$, we can obtain:

$$c_{p_1}^{rs*}(k) \ge c_{p_2}^{rs*}(k) \tag{4.27}$$

Equations (4.26) and (4.27) together imply:

$$c_{p_2}^{rs*}(k) = c_{p_1}^{rs*}(k) \tag{4.28}$$

We can repeat this procedure to verify that for each O-D pair, all used routes with positive flow have the same actual route travel time.

(ii) One route flow is positive and the other route flow is nil. We arbitrarily assume, without loss of generality, $h_{p_1}^{rs*}(k) > 0$, and $h_{p_2}^{rs*}(k) = 0$. We switch a small amount of flow $\Delta_1$ from route $p_1^{rs}$ to $p_2^{rs}$ with $0 < \Delta_1 \le h_{p_1}^{rs*}(k)$. By the same argument shown in (i), we have $c_{p_2}^{rs*}(k) \ge c_{p_1}^{rs*}(k)$. We repeat this procedure to verify that for each O-D pair, all unused routes with zero flow have an actual route travel time not lower than the minimal actual route travel time.

Since both (i) and (ii) must hold, it follows that VIP (4.5) implies equilibrium conditions (4.1). This completes the proof.

Note that the equivalence analysis is performed only under the assumption that some equilibrium conditions are obtained; i.e., $\delta_{apk}^{rs}(t) = \delta_{apk}^{rs*}(t)$. Though the proof can not be directly applied to the original problem, we intentionally include the proof for the subproblem here for the sake of conceptual clarification since many proofs associated with other existing dynamic models are not aware of this *tricky* property.

### 4.1.4 Time-Space Network

When the actual link travel time $\tau_a(t)$ is temporarily fixed, the relationships among inflow, exit flow, and number of vehicles on a link are clearly determined through the link flow propagation constraint. If we have a two-link three-node network as shown in Figure 4.1(a), then with the addition of the estimated actual link travel times, its time-space network can be drawn in Figure 4.1(b). The time-space network essentially contains two dimensions: the horizontal axis denotes spatial distance, and the vertical axis represents time interval. At each interval, the static network is reproduced. In addition, dummy links connecting time-dependent destinations $s(t)$ to the time-independent destination $s$ with zero travel time are also created. Note that the static links are not directly used, but instead, a new kind of links called *time-space links* has been created. A time-space link is usually represented by connecting a tail node at a near interval with a head node at a far interval, whereas the actual link travel time for inflows entering the time-space link $a$ during interval $t$ is $\tau_a(t)$. In other words, for any time-space link $a$, the inflows $u_{apk}^{rs}(t)$ during interval $t$ over route $p$ must be equal to the corresponding exit flow $v_{apk}^{rs}(t + \tau_a(t))$ during interval $t + \tau_a(t)$. For example, $v_a(5)$ is the sum of $u_a(2)$ and $u_a(3)$ in Figure 4.1(b).

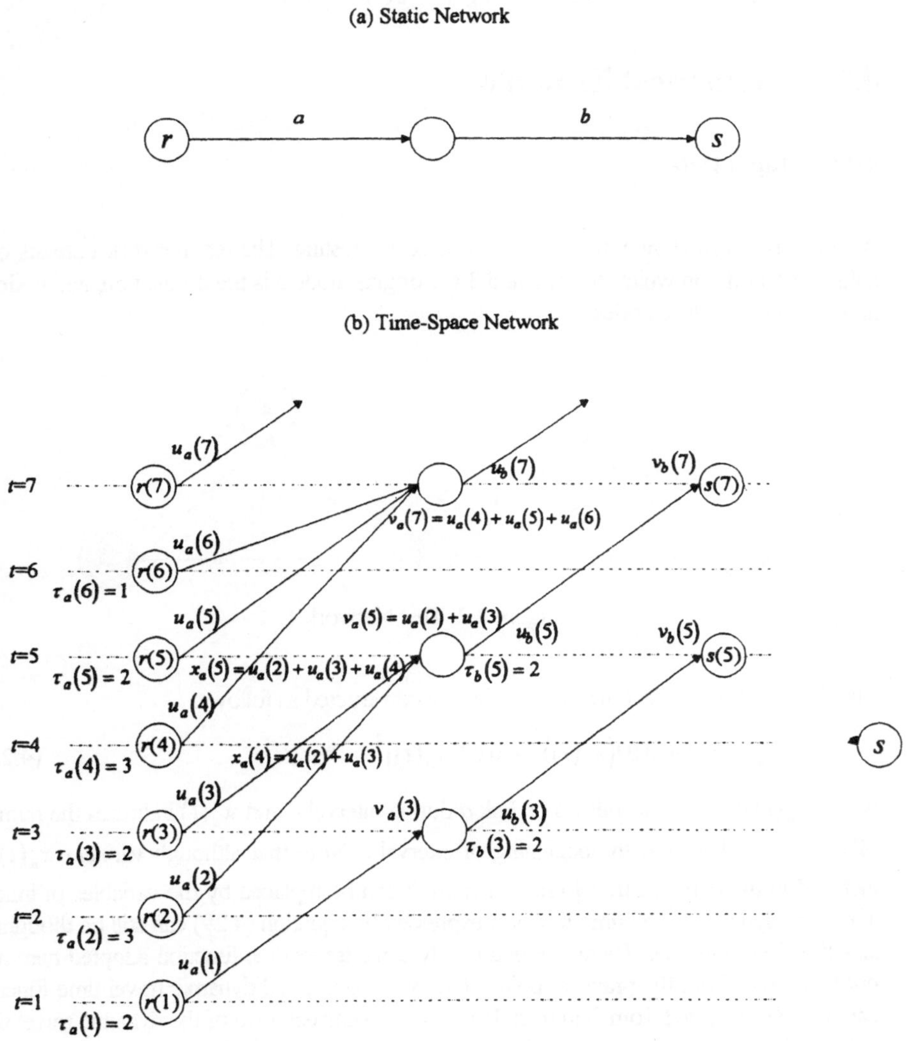


Figure 4.1: Time-Space Network

Furthermore, the slope of a specific time-space link $a$ denotes the inverse of the vehicular traveling speed on that link. The steeper the slope, the lower the vehicular traveling speed, and vice versa. Also when two time-space links interact, the first-in-last-out situation occurs. As to the number of vehicles remaining on link $a$ at the beginning of interval $t$, it can be computed by summing up all of the inflows, $u_a(\bullet)$, passing through link $a$ (except for tail node) during interval $t$. For examples, in Figure 4.1(b), $x_a(5)$ is equal to

$u_a(2)+u_a(3)+u_a(4)$ while $u_a(5)$ is not included, and the number of vehicles on link $a$ at the beginning of interval 7 is equal to $u_a(4)+u_a(5)+u_a(6)$.

## 4.2 Numerical Example

### 4.2.1 Input Data

A simple network shown in Figure 4.2 is used for testing. The test network consists of 6 links and 5 nodes in which nodes 1 and 3 are origins, node 5 is the destination, and nodes 2 and 4 are intermediate nodes.

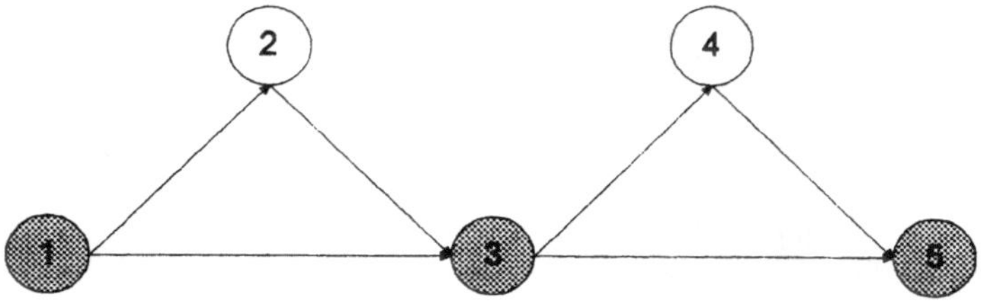


Figure 4.2: Test Network 1

The dynamic travel time function is arbitrarily constructed as follows:

$$c_a(t) = 1 + 0.01(u_a(t))^2 + 0.01(x_a(t))^2 \quad \forall a, t \tag{4.29}$$

where $u_a(t)$ denotes the inflows on link $a$ during interval $t$, and $x_a(t)$ indicates the number of vehicles on link $a$ at the beginning of interval $t$. Note that although variable $x_a(t)$ is included in the dynamic travel time function, it can be replaced by the variables of inflow. Thus, the dynamic travel time function expressed by equation (4.29) can still be thought of as a function of inflows. Be aware that the dynamic travel time function adopted here may not be correct from the practical point of view; in fact, a real dynamic travel time function can only be calibrated from field data. But, since the correct form of the dynamic travel time function is not the main theme of this book and is subject to further study, equation (4.29) should be sufficient for the purpose of providing a demonstration. The assumed origin-destination (O-D) demands are shown in Table 4.1:

Table 4.1: Time-Dependent O-D Demands

| O-D | Time Interval | | | |
|---|---|---|---|---|
| Pair | $k$=1 | $k$=2 | $k$=3 | $k$=4 |
| 1-5 | 15 | 20 | 0 | 0 |
| 3-5 | 0 | 0 | 15 | 20 |

### 4.2.2 Test Results

A computer program coded with Turbo $C^{++}$ solved the DUO route choice model with the given input data. The results are summarized in Table 4.2.

Table 4.2: Results for Test Network 1

| Link | Entering Time Interval | Inflow | Exit Flow | Number of Vehicles | Link Travel Time | Exiting Time Interval |
|---|---|---|---|---|---|---|
| 1→2 | 1 | 3.71 | 0.00 | 0.00 | 1.14 | 2 |
| | 2 | 8.52 | 3.71 | 3.71 | 1.86 | 4 |
| | 3 | 0.00 | 0.00 | 8.52 | 1.73 | - |
| | 4 | 0.00 | 8.52 | 8.52 | 1.73 | - |
| 1→3 | 1 | 11.29 | 0.00 | 0.00 | 2.27 | 3 |
| | 2 | 11.48 | 0.00 | 11.29 | 3.59 | 6 |
| | 3 | 0.00 | 11.29 | 22.76 | 6.18 | - |
| | 4~5 | 0.00 | 0.00 | 11.48 | 2.32 | - |
| | 6 | 0.00 | 11.48 | 11.48 | 2.32 | - |
| 2→3 | 2 | 3.71 | 0.00 | 0.00 | 1.14 | 3 |
| | 3 | 0.00 | 3.71 | 3.71 | 1.14 | - |
| | 4 | 8.52 | 0.00 | 0.00 | 1.73 | 6 |
| | 5 | 0.00 | 0.00 | 8.52 | 1.73 | - |
| | 6 | 0.00 | 8.52 | 8.52 | 1.73 | - |
| 3→4 | 3 | 11.23 | 0.00 | 0.00 | 2.26 | 5 |
| | 4 | 8.28 | 0.00 | 11.23 | 2.95 | 7 |
| | 5 | 0.00 | 11.23 | 19.51 | 4.81 | - |
| | 6 | 19.51 | 0.00 | 8.28 | 5.49 | 11 |
| | 7 | 0.00 | 8.28 | 27.79 | 8.72 | - |
| | 8~10 | 0.00 | 0.00 | 19.51 | 4.80 | - |
| | 11 | 0.00 | 19.51 | 19.51 | 4.80 | - |
| 3→5 | 3 | 18.77 | 0.00 | 0.00 | 4.52 | 8 |
| | 4 | 11.72 | 0.00 | 18.77 | 5.90 | 10 |
| | 5 | 0.00 | 0.00 | 30.49 | 10.29 | - |
| | 6 | 0.49 | 0.00 | 30.49 | 10.30 | 16 |
| | 7 | 0.00 | 0.00 | 30.98 | 10.60 | - |
| | 8 | 0.00 | 18.77 | 30.98 | 10.60 | - |
| | 9 | 0.00 | 0.00 | 12.22 | 2.49 | - |
| | 10 | 0.00 | 11.72 | 12.22 | 2.49 | - |
| | 11~15 | 0.00 | 0.00 | 0.49 | 1.00 | - |
| | 16 | 0.00 | 0.49 | 0.49 | 1.00 | - |

Table 4.2: Results for Test Network 1 (continued)

| Link | Entering Time Interval | Inflow | Exit Flow | Number of Vehicles | Link Travel Time | Exiting Time Interval |
|---|---|---|---|---|---|---|
| 4→5 | 5 | 11.23 | 0.00 | 0.00 | 2.26 | 7 |
| | 6 | 0.00 | 0.00 | 11.23 | 2.26 | - |
| | 7 | 8.28 | 11.23 | 11.23 | 2.95 | 10 |
| | 8~9 | 0.00 | 0.00 | 8.28 | 1.69 | - |
| | 10 | 0.00 | 8.28 | 8.28 | 1.69 | - |
| | 11 | 19.51 | 0.00 | 0.00 | 4.80 | 16 |
| | 12~15 | 0.00 | 0.00 | 19.51 | 4.80 | - |
| | 16 | 0.00 | 19.51 | 19.51 | 4.80 | - |

The sixth column of Table 4.2 denotes the actual link travel time when the inflow is positive during that interval. For example, the actual travel time on link 3→5 during interval 3 is 4.52 time units (rounded off as 5), implying that the inflow of 18.77 units must exit the link during interval 8 (=3+5). This actual link travel time, once determined, cannot be changed in later intervals, even if the traffic conditions are changed dramatically. Note that when no inflow is present during a specific interval, the corresponding actual link travel time shown in the sixth column of Table 4.2 does not have *physical* meaning, and certainly cannot overwrite or replace the actual link travel times determined in previous time intervals. For example, the travel time on link 3→5 during intervals 11~16 is 1.00 time units, which is really a *dummy* link travel time, because no inflow is present during those intervals. Also, refer to equilibrium conditions (4.1) for the definition of a route being used, and the subsequent equation (4.7) for the links in that route.

The rationale of the proposed model and associated solution algorithm can be verified by checking if the resulting actual route travel times satisfy the dynamic user-optimal conditions as defined in Section 4.1.1. Consider route 1→3→5 departing origin 1 during interval 1; the corresponding actual route travel time can be obtained by summing up the actual link travel time on link 1→3 during interval 1 and the actual link travel time on link 3→5 during interval $1+c_{1\to3}(1)$ as follows.

$$\begin{aligned} c_{1\to3\to5}(1) &= c_{1\to3}(1)+c_{3\to5}\left(1+c_{1\to3}(1)\right) \\ &= 2.27+c_{3\to5}(3.27) \\ &\approx 2.27+c_{3\to5}(3) \\ &= 6.79 \end{aligned} \tag{4.30}$$

The remaining used actual route travel times are also computed and summarized in Table 4.3. As can be observed, the travelers departing from the same origin during the same interval also experience approximately the same actual route travel time.

Table 4.3: Actual Route Travel Time for Test Network 1

| Time Interval | Route | | | | | |
|---|---|---|---|---|---|---|
| | 3→5 | 3→4→5 | 1→3→5 | 1→2→3→5 | 1→3→4→5 | 1→2→3→4→5 |
| k=1 | NA | NA | 6.79 | 6.80 | 6.80 | 6.80 |
| k=2 | NA | NA | 13.89 | 13.89 | 13.89 | 13.89 |
| k=3 | 4.52 | 4.52 | NA | NA | NA | NA |
| k=4 | 5.90 | 5.89 | NA | NA | NA | NA |

NA: Not applicable because those routes were not used.

# 4.3 Static Counterpart Approximation

The major difference between the DUO route choice model and its static counterpart is the inclusion of the flow propagation constraint. The flow propagation constraint requires that an inflow can not exit the time-space link unless the actual link travel times elapse. If we mathematically force the actual travel times to equal zeros ($\tau_a = 0$), then our DUO route choice model will reduce to its static counterpart. Setting actual link travel times equal to zero does not mean that flows can traverse the network without delay. This treatment only converts the dynamic user-optimal route choice model, in the mathematical sense, into its static counterpart. To verify this, we take again the hypothesized network data for test network 1, but assume O-D demands depart in a single departure time interval as follows:

Table 4.4: Time-Dependent O-D Demands

| O-D Pair | Time Interval |
|---|---|
| | *k*=1 |
| 1-5 | 35 |
| 3-5 | 35 |

By re-running the computer code with this input data, we obtain the results and actual route travel times as shown in Tables 4.5 and 4.6, respectively.

Table 4.5: Dynamic Results for Test Network 1 (Single Departure Time Interval)

| Link | Entering Time Interval | Inflow | Exit Flow | Number of Vehicles | Link Travel Time | Exiting Time Interval |
|---|---|---|---|---|---|---|
| 1→2 | 1 | 13.5 | 0.0 | 0.0 | 2.82 | 4 |
| | 2~3 | 0.0 | 0.0 | 13.5 | 2.82 | - |
| | 4 | 0.0 | 13.5 | 13.5 | 2.82 | - |
| 1→3 | 1 | 21.5 | 0.0 | 0.0 | 5.63 | 7 |
| | 2~6 | 0.0 | 0.0 | 21.5 | 5.63 | - |
| | 7 | 0.0 | 21.5 | 21.5 | 5.63 | - |
| 2→3 | 4 | 13.5 | 0.0 | 0.0 | 2.82 | 7 |
| | 5~6 | 0.0 | 0.0 | 13.5 | 2.82 | - |
| | 7 | 0.0 | 13.5 | 13.5 | 2.82 | - |
| 3→4 | 1 | 13.5 | 0.0 | 0.0 | 2.82 | 4 |
| | 2~3 | 0.0 | 0.0 | 13.5 | 2.82 | - |
| | 4 | 0.0 | 13.5 | 13.5 | 2.82 | - |
| | 7 | 18.1 | 0.0 | 0.0 | 4.26 | 11 |
| | 8~10 | 0.0 | 0.0 | 18.1 | 4.26 | - |
| | 11 | 0.0 | 18.1 | 18.1 | 4.26 | - |
| 3→5 | 1 | 21.5 | 0.0 | 0.0 | 5.63 | 7 |
| | 2~6 | 0.0 | 0.0 | 21.5 | 5.63 | - |
| | 7 | 16.9 | 21.5 | 21.5 | 8.50 | 16 |
| | 8~15 | 0.0 | 0.0 | 16.9 | 3.87 | - |
| | 16 | 0.0 | 16.9 | 16.9 | 3.87 | - |
| 4→5 | 4 | 13.5 | 0.0 | 0.0 | 2.82 | 7 |
| | 5~6 | 0.0 | 0.0 | 13.5 | 2.82 | - |
| | 7 | 0.0 | 13.5 | 13.5 | 2.82 | - |
| | 11 | 18.1 | 0.0 | 0.0 | 4.26 | 15 |
| | 12~14 | 0.0 | 0.0 | 18.1 | 4.26 | - |
| | 15 | 0.0 | 18.1 | 18.1 | 4.26 | - |

Table 4.6: Actual Route Travel Times for Test Network 1 (Single Departure Time Interval)

| Route | 3→5 | 3→4→5 | 1→3→5 | 1→2→3 →5 | 1→3→4 →5 | 1→2→3 →4→5 |
|---|---|---|---|---|---|---|
| Travel Time ($k$=1) | 5.63 | 5.64 | 14.13 | 14.14 | 14.15 | 14.16 |

By mathematically forcing actual travel times equal to zero during the solution procedure, we can summarize the results and the actual route travel times for the static approximation in Tables 4.7 and 4.8, respectively.

Table 4.7: Static Approximation of the Results for Test Network 1

| Link | Entering Time Interval | Inflow | Exit Flow | Number of Vehicles | Link Travel Time | Exiting Time Interval |
|---|---|---|---|---|---|---|
| 1→2 | 1 | 13.5 | 13.5 | 0.0 | 2.84 | 1 |
| 1→3 | 1 | 21.5 | 21.5 | 0.0 | 5.60 | 1 |
| 2→3 | 1 | 13.5 | 13.5 | 0.0 | 2.84 | 1 |
| 3→4 | 1 | 28.5 | 28.5 | 0.0 | 9.12 | 1 |
| 3→5 | 1 | 41.5 | 41.5 | 0.0 | 18.22 | 1 |
| 4→5 | 1 | 28.5 | 28.5 | 0.0 | 9.12 | 1 |

Table 4.8: Static Approximation of Actual Route Travel Times for Test Network 1

| Route | 3→5 | 3→4→5 | 1→3→5 | 1→2→3 →5 | 1→3→4 →5 | 1→2→3 →4→5 |
|---|---|---|---|---|---|---|
| Travel Time | 18.22 | 18.24 | 23.82 | 23.90 | 23.84 | 23.92 |

By comparing Tables 4.6 and 4.8, it is observed that the actual route travel times for the static counterpart are much higher than that for the DUO route choice model. This is because the DUO route choice model allows inflows entering the same link at different time intervals, yielding lower link travel times. One may be interested in knowing what will happen if the time-dependent O-D demands remain unchanged at each interval; therefore, we assume the time-dependent O-D demands associated with multiple departure time intervals as follows:

Table 4.9: Time-Dependent O-D Demands
(Multiple Departure Time Intervals)

| O-D Pair | Time Interval $k$=1 | $k$=2 |
|---|---|---|
| 1-5 | 35 | 35 |
| 3-5 | 35 | 35 |

By executing the computer program once again, we obtained the actual route travel times indicated in Table 4.10. As compared with Table 4.8, it is observed that with multiple departure time intervals, the actual route travel times for O-D pair (1,5) are higher than for the static counterpart.

Table 4.10: Actual Route Travel Times for Test Network 1
(Multiple Departure Time Intervals)

| Route | 3→5 | 3→4→5 | 1→3→5 | 1→2→3→5 | 1→3→4→5 | 1→2→3→4→5 |
|---|---|---|---|---|---|---|
| Travel Time ($k$=1) | 5.63 | 5.64 | 24.85 | 24.86 | 24.85 | 24.86 |
| Travel Time ($k$=2) | 9.83 | 9.84 | 27.46 | 27.47 | 27.51 | 27.52 |

## 4.4 Multiple Optima

The feasible region of the DUO route choice model delineated by equations (4.4)~(4.10) is nonconvex, implying multiple local solutions could exist. We can demonstrate this by the following numerical example. A simple rectangular network as shown in Figure 4.3 is used for testing. The test network consists of 4 links and 4 nodes. Nodes 1 and 4, respectively, are the origin and destination, whereas nodes 2 and 3 are intermediate.

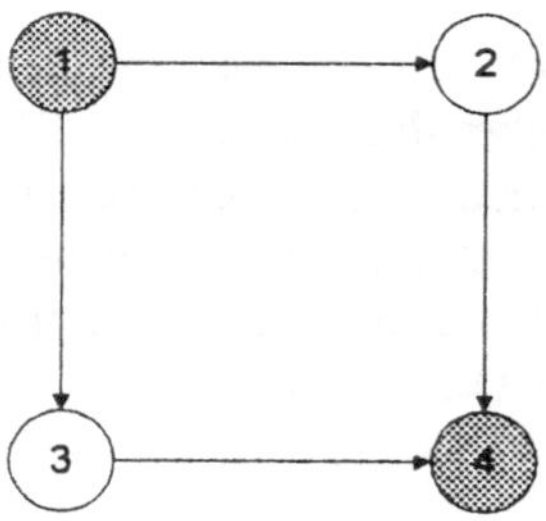


Figure 4.3: Test Network 2

The dynamic travel time function is arbitrarily constructed as follows:

$$c_a(t) = 1 + 0.01(u_a(t))^2 + 0.01(x_a(t))^2 \qquad \forall a, t \tag{4.31}$$

The assumed time-dependent O-D demands are shown in Table 4.11:

Table 4.11: Time-Dependent O-D Demands for Test Network 2

| O-D Pair | Time Interval | | |
|---|---|---|---|
| | $k$=1 | $k$=2 | $k$=3 |
| 1-4 | 40 | 4 | 30 |

Using two different initial feasible solutions, two different local solutions can be obtained.

The first local solution, as shown in Table 4.12, is obtained by the initial feasible solution with all-or-nothing assignment. The second local solution, as shown in Table 4.13, is yielded by the initial feasible solution with incremental assignment (8 partial increments). It can be observed that the link flow pattern is asymmetric for the first local solution, while it is symmetric for the second local solution.

Table 4.12: First Local Solution for Test Network 1

| Link | Entering Time Interval | Inflow | Exit Flow | Number of Vehicles | Link Travel Time | Exiting Time Interval |
|---|---|---|---|---|---|---|
| 1→2 | 1 | 20.0 | 0.0 | 0.0 | 4.99 | 6 |
| | 2 | 1.2 | 0.0 | 20.0 | 5.01 | 7 |
| | 3 | 20.0 | 0.0 | 21.2 | 9.47 | 13 |
| | 4~5 | 0.0 | 0.0 | 41.1 | 17.92 | - |
| | 6 | 0.0 | 20.0 | 41.1 | 17.92 | - |
| | 7 | 0.0 | 1.2 | 21.1 | 5.47 | - |
| | 8~11 | 0.0 | 0.0 | 20.0 | 4.98 | - |
| | 12 | 0.0 | 20.0 | 20.0 | 4.98 | - |
| 1→3 | 1 | 20.0 | 0.0 | 0.0 | 5.01 | 6 |
| | 2 | 2.8 | 0.0 | 20.0 | 5.09 | 7 |
| | 3 | 10.0 | 0.0 | 22.8 | 7.22 | 10 |
| | 4~5 | 0.0 | 0.0 | 32.9 | 11.80 | - |
| | 6 | 0.0 | 20.0 | 32.9 | 11.80 | - |
| | 7 | 0.0 | 2.8 | 12.9 | 2.65 | - |
| | 8~9 | 0.0 | 0.0 | 10.0 | 2.01 | - |
| | 10 | 0.0 | 10.0 | 10.0 | 2.01 | - |
| 2→4 | 6 | 20.0 | 0.0 | 0.0 | 4.99 | 11 |
| | 7 | 1.2 | 0.0 | 20.0 | 5.01 | 12 |
| | 8~10 | 0.0 | 0.0 | 21.2 | 5.49 | - |
| | 11 | 0.0 | 20.0 | 21.2 | 5.49 | - |
| | 12 | 20.0 | 1.2 | 1.2 | 5.00 | 17 |
| | 13~16 | 0.0 | 0.0 | 20.0 | 4.98 | - |
| | 17 | 0.0 | 20.0 | 20.0 | 4.98 | - |
| 3→4 | 6 | 20.0 | 0.0 | 0.0 | 5.01 | 11 |
| | 7 | 2.8 | 0.0 | 20.0 | 5.09 | 12 |
| | 8~9 | 0.0 | 0.0 | 22.8 | 6.21 | - |
| | 10 | 10.0 | 0.0 | 22.8 | 7.22 | 17 |
| | 11 | 0.0 | 20.0 | 32.9 | 11.80 | - |
| | 12 | 0.0 | 2.8 | 12.9 | 2.65 | - |
| | 13~16 | 0.0 | 0.0 | 10.0 | 2.01 | - |
| | 17 | 0.0 | 10.0 | 10.0 | 2.01 | - |

Table 4.13: Second Local Solution for Test Network 2

| Link | Entering Time Interval | Inflow | Exit Flow | Number of Vehicles | Link Travel Time | Exiting Time Interval |
|---|---|---|---|---|---|---|
| 1→2 | 1 | 20.0 | 0.0 | 0.0 | 5.00 | 6 |
| | 2 | 2.0 | 0.0 | 20.0 | 5.04 | 7 |
| | 3 | 14.9 | 0.0 | 22.0 | 8.09 | 11 |
| | 4~5 | 0.0 | 0.0 | 37.0 | 14.68 | - |
| | 6 | 0.0 | 20.0 | 37.0 | 14.68 | - |
| | 7 | 0.0 | 2.0 | 17.0 | 3.88 | - |
| | 8~10 | 0.0 | 0.0 | 14.9 | 3.23 | - |
| | 11 | 0.0 | 14.9 | 14.9 | 3.23 | - |
| 1→3 | 1 | 20.0 | 0.0 | 0.0 | 5.00 | 6 |
| | 2 | 2.0 | 0.0 | 20.0 | 5.04 | 7 |
| | 3 | 15.1 | 0.0 | 22.0 | 8.09 | 11 |
| | 4~5 | 0.0 | 0.0 | 37.0 | 14.70 | - |
| | 6 | 0.0 | 20.0 | 37.0 | 14.70 | - |
| | 7 | 0.0 | 2.0 | 17.0 | 3.90 | - |
| | 8~10 | 0.0 | 0.0 | 15.1 | 3.27 | - |
| | 11 | 0.0 | 15.1 | 15.1 | 3.27 | - |
| 2→4 | 6 | 20.0 | 0.0 | 0.0 | 5.00 | 11 |
| | 7 | 2.0 | 0.0 | 20.0 | 5.04 | 12 |
| | 8~10 | 0.0 | 0.0 | 22.0 | 5.86 | - |
| | 11 | 14.9 | 20.0 | 22.0 | 8.09 | 19 |
| | 12 | 0.0 | 2.0 | 17.0 | 3.88 | - |
| | 13~18 | 0.0 | 0.0 | 14.9 | 3.23 | - |
| | 19 | 0.0 | 14.9 | 14.9 | 3.23 | - |
| 3→4 | 6 | 20.0 | 0.0 | 0.0 | 5.00 | 11 |
| | 7 | 2.0 | 0.0 | 20.0 | 5.04 | 12 |
| | 8~10 | 0.0 | 0.0 | 22.0 | 5.82 | - |
| | 11 | 15.1 | 20.0 | 22.0 | 8.09 | 19 |
| | 12 | 0.0 | 2.0 | 17.0 | 3.90 | - |
| | 13~18 | 0.0 | 0.0 | 15.1 | 3.27 | - |
| | 19 | 0.0 | 15.1 | 15.1 | 3.27 | - |

The actual route travel times for the two local solutions are computed and summarized in Table 4.14.

Table 4.14: Actual Route Travel Times for the Two Local Solutions

| First Local Solution | | | |
|---|---|---|---|
| Route | Time Interval | | |
| | $k$=1 | $k$=2 | $k$=3 |
| 1→2→4 | 9.98 | 10.02 | 14.47 |
| 1→3→4 | 10.02 | 10.18 | 14.44 |
| **Second Local Solution** | | | |
| Route | Time Interval | | |
| | $k$=1 | $k$=2 | $k$=3 |
| 1→2→4 | 10.00 | 10.08 | 16.18 |
| 1→3→4 | 10.00 | 10.08 | 16.18 |

Both local solutions satisfy the dynamic user-optimal conditions in which the actual route travel times experienced by travelers departing during the same interval are equal and minimal. While the actual route travel times in both local solutions are the same for trips departing during intervals 1 and 2, the actual route travel time is clearly different for trips departing during interval 3.

## 4.5 Notes

We used two types of networks for the testing, a double triangular network and a rectangular network. While the latter is commonly seen in research literature, the former, with asymmetric topology property, is specially designed in this book to provide a sufficient testing platform to prevent the generation of ambiguous results from the layout.

A relevant issue associated with the dynamic travel time function concerns the number of intervals $|T|$ needed for solving the discrete model. A basic requirement for determining the number of intervals is to let all O-D trips arrive at their destination within the analysis period $T$. To this end, a trial-and-error technique is commonly employed. In general, the higher the number of intervals adopted (the link travel time functions must be modified accordingly), the more accurate the solution. The least total number of intervals required for the computation may be theoretically estimated as follows:

$$|T| \geq \frac{\max_{r,s,p,k}\left(k + c_p^{rs}(k)\right)}{\Delta t} \tag{4.32}$$

where $k$ denotes the departure time and $c_p^{rs}(k)$ represents the actual route travel time for trips between O-D pair $rs$ over route $p$.

# Chapter 5

# Algorithms for the Dynamic Route Choice Model

In this chapter, algorithms are investigated for the dynamic user-optimal route choice model. All possible link interactions are first discussed in Section 5.1. A nested diagonalization method, embedding the Frank-Wolfe (FW) method, is then presented in Section 5.2. A nested projection method and a modified nested projection method are described in Sections 5.3 and 5.4, respectively. A numerical example is given to validate the correctness of the nested diagonalization method in Section 5.5. Finally, a concluding note is given in Section 5.6.

## 5.1 Link Interactions

The variational inequality approach is a superior approach to perform qualitative analysis and to formulate a general problem with embedded link interactions, especially those with asymmetric link interactions. However, it is an inferior approach for performing solution algorithms. The optimization approach, on the other hand, is weak in representing more complex problems, but strong in handling solution algorithms. To take advantage of the strengths of both of these approaches, it is therefore natural to formulate the problem as a VIP, and then to identify and relax all of link interactions gradually until an optimization subproblem has resulted. The resulting optimization subproblem can then be solved readily through the application of an appropriate algorithm.

For the DUO route choice and other travel choice problems, there are three sources of possible link interactions. The first is from the actual link travel times $\tau_a(t)$, the second is from the temporal interactions among inflows using the same physical link, but appearing at different intervals, and the last is due to topological interference from different physical links which are generally characterized by so-called symmetric and/or asymmetric link travel time functions.

For the DUO route choice model, actual link travel times are not known in advance, and hence, need to be estimated based on a feasible flow pattern. When the actual link travel times are estimated, a new subproblem's feasible region is accordingly determined, or more specifically, a new feasible time-space network is constructed. This new feasible time-space network is then used to determine another feasible flow pattern. Prior to reaching equilibrium, the feasible time-space networks constructed by two sets of estimated actual link travels times in successive iterations are usually different. This procedure is repeated until the estimated actual link travel times converge to the optimal solution.

The second type of link interaction is evident in the real world. This type of link interaction refers to situations where link inflow is affected or even blocked by those previously entered flows. No vehicle heading toward a destination is allowed to *jump* over vehicles ahead of it; however, when the inflows on the same physical link, other than that on the subject time-space link, are temporarily fixed, this type of link interaction can be relaxed.

The last type of link interaction is due to interference between different physical links. For example, consider an undivided two-way street, the affects of flows in one direction on the opposing flows are similar, regardless of their direction, thus yielding a symmetric link interaction. Even more pronounced link interactions can be observed at a signalized intersection characterized by a permitted left-turn movement. It is clear that left-turn movements could be seriously affected by the opposing through traffic, while the opposing through traffic is only slightly affected by the left-turn movements (assuming that drivers obey the priority regulations). Since the degree of interference is not equal on both sides, asymmetric link interactions thus result. As already discussed in many static travel choice models (see Sheffi (1985) for an example) this type of link interactions can be relaxed by temporarily fixing the inflows on the links, other than the subject physical link, at the current level. Note that in order to ensure a unique solution, the Jacobian matrix of the symmetric and/or asymmetric link travel time functions must be positive definite.

Once these three types of link interactions are relaxed as described above, the standard traffic assignment problem results can be expressed. There are several efficient solution algorithms available for this optimization problem; however, in view of the storage requirements for a large urban network, the Frank-Wolfe (FW) method is most commonly used because storage of routes is not required. The FW method typically consists of direction finding, determination of move size, and updating.

Throughout this book, symmetric and/or asymmetric link travel time functions are not really used in the demonstration for the sake of simplicity, even though they can be tackled readily within our proposed solution algorithms; therefore, only the first two link interactions are treated within the proposed solution algorithms. Our proposed solution algorithms are principally motivated by Nagurney (1993), who presented a general iterative scheme to solve the VIP. Special well-known algorithms are deduced from this unified framework; such as the diagonalization method, the projection method and the linearization algorithms. In the following sections, we present in sequence a nested diagonalization method, a nested projection method and a modified nested projection method for solving the DUO route choice model.

## 5.2 Nested Diagonalization Method

A nested diagonalization method is defined as an algorithm that contains the diagonalization method as a subproblem by relaxing some link interactions. The diagonalization method has been successfully adopted in solving static route choice problems with asymmetric link travel time functions. The asymmetric link travel time function is reflected in the link-travel-time Jacobian matrix, which contains different corresponding values for diagonally opposing cells. When the diagonalization method is applied to the asymmetric route choice problem, the inflows on time-space links other than on the subject time-space link are temporarily fixed at the current level. This treatment yields a diagonalized subproblem which can be reformulated as an equivalent optimization problem.

For the DUO route choice problem, two types of link interactions need to be relaxed (the third type of link interaction described in Section 5.1 is not treated in the later numerical examples for the sake of simplicity) before an optimization problem can be obtained and solved, i.e., actual link travel times and the inflows entering the same physical link earlier. To this end, the proposed nested diagonalization method essentially contains three loops; the outermost loop, where the actual link travel times are estimated, the second loop, where the inflows for each physical link other than on the subject time-space link are temporarily fixed, and the innermost loop, where the standard traffic assignment problem is solved. The context of the problem in each loop is described in the following sections.

### 5.2.1 *First Loop* Problem

When the actual link travel times are estimated and temporarily fixed at the current level, the VIP (4.3)~(4.10) is reduced to the following problem:

$$\sum_a \sum_t c_a^*(t)\left[u_a(t) - u_a^*(t)\right] \geq 0 \quad \forall \mathbf{u} \in \overline{\Omega} \tag{5.1}$$

where $\overline{\Omega}$ is a subset of $\Omega$ with $\delta_{apk}^{rs}(t)$ being realized by the estimated actual travel times $\overline{\tau}_a$, i.e., $\delta_{apk}^{rs}(t) = \overline{\delta}_{apk}^{rs}(t), \forall r,s,a,p,k,t$. The symbol $\overline{\Omega}$ denotes the feasible region that is delineated by the following constraints (5.2)~(5.11).

Flow conservation constraint:

$$\sum_p h_p^{rs}(k) = \overline{q}^{rs}(k) \quad \forall r,s,k \tag{5.2}$$

Nonnegativity constraint:

$$h_p^{rs}(k) \geq 0 \quad \forall r,s,p,k \tag{5.3}$$

Definitional Constraints:

$$u_{apk}^{rs}(t) = h_p^{rs}(k)\overline{\delta}_{apk}^{rs}(t) \quad \forall r,s,a,p,k,t \tag{5.4}$$

$$\overline{\delta}_{apk}^{rs}(t) = \{0,1\} \quad \forall r,s,a,p,k,t \tag{5.5}$$

$$u_a(t) = \sum_{rs}\sum_{p}\sum_{k} h_p^{rs}(k)\bar{\delta}_{apk}^{rs}(t) \quad \forall a,t \tag{5.6}$$

$$c_p^{rs}(k) = \sum_{a}\sum_{t} c_a(t)\bar{\delta}_{apt}^{rs}(t) \quad \forall r,s,p,k \tag{5.7}$$

Under the estimated actual travel times, the flow propagation constraints are eliminated and the indicator variables are realized as indices. The remaining type of link interactions is due to link inflows at different intervals.

### 5.2.2 *Second Loop* Problem

The second link interaction can be relaxed by temporarily fixing the previously entered flows in the same physical link $a$. In other words, at iteration $n+1$ during the solution procedure, the link travel times become a function of inflow on the subject time-space link only, as follows:

$$c_a^{n+1}(t) = c_a\left(u_a^n(1), u_a^n(2), \cdots, u_a^n(t-1), u_a^{n+1}(t)\right) \quad \forall a,t \tag{5.8}$$

Since the Jacobian matrix of the above travel time function is symmetric and positive definite, the variational inequality (5.1)~(5.7) can be reformulated as an equivalent optimization problem, as follows:

$$\min \ z(\mathbf{u}) = \sum_{a}\sum_{t} \int_0^{u_a^{n+1}(t)} c_a\left(u_a^n(1), u_a^n(2), \cdots, u_a^n(t-1), \omega\right) d\omega \tag{5.9}$$

Subject to: constraints (5.2)~(5.7)

### 5.2.3 *Third Loop* Problem

Objective function (5.9) and the associated constraints (5.2)~(5.7) essentially construct a convex optimization problem which can be easily solved by the FW method, which was presented in Section 2.3.5. It typically consists of direction finding, line search, and determination of move size, and updating. The search direction is derived based on a linearized subproblem.

$$\min \ z(\mathbf{p}) = \sum_{a}\sum_{t} \bar{c}_a\left(u_a^n(1), u_a^n(2), \cdots, u_a^n(t-1), u_a^n(t)\right) p_a(t) \tag{5.10}$$

Flow conservation constraint:

$$\sum_{p} w_p^{rs}(k) = \bar{q}^{rs}(k) \quad \forall r,s,k \tag{5.11}$$

Nonnegativity constraint:

$$w_p^{rs}(k) \geq 0 \quad \forall r,s,p,k \tag{5.12}$$

Definitional Constraints:

$$p_{apk}^{rs}(t) = w_p^{rs}(k)\bar{\delta}_{apk}^{rs}(t) \quad \forall r,s,a,p,k,t \tag{5.13}$$

$$\bar{\delta}_{apk}^{rs}(t) = \{0,1\} \quad \forall r,s,a,p,k,t \tag{5.14}$$

$$p_a(t) = \sum_{rs}\sum_{p}\sum_{k} w_p^{rs}(k)\bar{\delta}_{apk}^{rs}(t) \quad \forall a,t \tag{5.15}$$

$$\bar{c}_p^{rs}(k) = \sum_{a}\sum_{t} \bar{c}_a(t)\bar{\delta}_{apt}^{rs}(t) \quad \forall r,s,p,k \tag{5.16}$$

The above linear program can be easily solved using an all-or-nothing technique. The obtained subproblem solution accompanied with the main problem solution brings about a descent direction $\mathbf{d}^n = (\mathbf{p}^n - \mathbf{u}^n)$, or in an expanded form, $d_i^n = (p_i^n - u_i^n), \forall i$. Once the descent direction is known, the move size in the direction of $\mathbf{d}^n$ equals the distance to the point along which $\mathbf{d}^n$ minimizes $z(\mathbf{u})$. It follows that:

$$\min_{0 \le \lambda \le 1} z\left[\mathbf{u}^n + \lambda(\mathbf{p}^n - \mathbf{u}^n)\right] \tag{5.17}$$

Once the optimal solution of this line search, $\lambda^n$, is found, the next point can be generated using the following formula:

$$\mathbf{u}^{n+1} = \mathbf{u}^n + \lambda^n(\mathbf{p}^n - \mathbf{u}^n) \tag{5.18}$$

Note that equation (5.18) can be written as:

$$\mathbf{u}^{n+1} = (1-\lambda^n)\mathbf{u}^n + \lambda^n\mathbf{p}^n \tag{5.19}$$

The new solution is thus a convex combination (or a weighted average) of $\mathbf{u}^n$ and $\mathbf{p}^n$. Subsequently, the main problem solution can be updated based on the descent direction and move size. At this stage, a convergence check must be performed. If the convergence criterion is satisfied, the current main problem solution is deemed as the traffic pattern at equilibrium; otherwise, the procedure is repeated.

### 5.2.4 Proposed Solution Algorithm

Based on the above discussion, a nested diagonalization method is formally proposed for solving the discrete DUO route choice problem as follows:

***Nested Diagonalization Method***

**Step 0: Initialization.**

Step 0.1: Let $m$=0. Set $\tau_a^0(t) = NINT\left[c_{a_0}(t)\right], \forall a,t$.

Step 0.2: Let $n$=1. Find an initial feasible solution $\{u_a^1(t)\}$. Compute the associated link travel times $\{c_a^1(t)\}$.

**Step 1: *First Loop* Operation.**

Let $m$=$m$+1. Update the estimated actual link travel times by

$$\tau_a^m(t) = NINT\left[(1-\gamma)\tau_a^{m-1}(t) + \gamma c_a^n(t)\right] \quad \forall a,t \tag{5.20}$$

Construct the corresponding feasible time-space network based on the estimated actual link travel times.

**Step 2: *Second Loop* Operation.**

Step 2.1: Let $n$=1. Compute and reset the initial feasible solution $\{u_a^n(t)\}$, based on the time-space network constructed by the estimated actual link travel times $\{\tau_a^m(t)\}$.

Step 2.2: Fix the inflows for each physical link other than on the subject time-space link at the current level, yielding the optimization problem defined by (5.9) and (5.2)~(5.7).

**Step 3: *Third Loop* Operation.**

Solve for the solution, $\{u_a^{n+1}(t)\}$, in optimization problem (5.9) and (5.2)~(5.7) by the FW method. Compute the resulting link travel times $\{c_a^{n+1}(t)\}$.

**Step 4: Convergence Check for the *Second Loop* Operation.**

If $u_a^{n+1}(t) \approx u_a^n(t), \forall a,t$, go to Step 5; otherwise, set $n$=$n$+1, go to Step 2.2.

**Step 5: Convergence Check for the *First Loop* Operation.**

If $\tau_a^m(t) \approx c_a^{n+1}(t)\ \forall a,t$, stop; the current solution is optimal. Otherwise, set $n$=$n$+1, and go to Step 1.

In Step 0, an initial feasible solution is generated either by the all-or-nothing (AON) assignment method, or by the incremental (INC) assignment method. The AON method is performed based on the link flow free travel times. For each O-D pair, only the route with the shortest travel time will be assigned with all the trips; the other routes will have no assigned flows. The INC method assigns repeatedly a prespecified portion of the O-D trips to the shortest path, and updates the link travel time with the resulting flow pattern.

In Step 1, the actual link travel times are estimated by weighted averages. The adopted formula is general in the sense that many known formulas are constructed for special cases. For example, if $\gamma = \frac{1}{m}$, then $\tau_a^m(t) = NINT\left[\frac{m-1}{m}\tau_a^{m-1}(t) + \frac{1}{m}c_a^n(t)\right]$ and if $\gamma = 1$, then $\tau_a^m(t) = NINT\left[c_a^n(t)\right]$. Note that the weighted averages are prefixed by round-off arithmetic operations, and denoted by the symbol *NINT*. This is because the actual link travel times must be integer-valued as is this case for discrete models. Here, we simply set $NINT[x] = i$, if $i - 0.5 \leq x < i + 0.5$ for the purpose of a demonstration. Other round-off arithmetic operations are also possible. In fact, from an algorithmic point of view, good estimates of actual link travel times are critical to the computational efficiency and quality of the resulted solution, which involves having a good initial solution and an efficient procedure for updating. This issue is left for future research.

In Step 2, the diagonalization method is applied. In Step 2.1, the initial solution associated with the diagonalization method must be generated/refreshed every time, so as to reflect a changed feasible set created by different/updated actual travel times. In other words, the traffic patterns resulting from the previous iteration in Step 3 cannot be directly carried over into the next diagonalization iteration. The reason for this is that different estimated actual link travel times would construct different feasible time-space networks. Unless the estimated actual link travel times are the same in two successive iterations, the traffic pattern resulting from the previous iteration in Step 3 cannot be adopted for later use. This is a major difference between our solution algorithm and those written by others.

In Step 3, the FW algorithm is applied over a feasible time-space network that satisfies the flow propagation constraint.

In Step 5, the convergence criterion is set to be $\tau_a^m(t) \approx c_a^{n+1}(t)\ \forall a,t$, rather than $\tau_a^m(t) \approx \tau_a^{m+1}(t)\ \forall a,t$ as used in Ran and Boyce (1996), to avoid prematured convergence.

In short, our nested diagonalization solution procedure contains three levels, with the actual travel times being estimated in the first (outermost) level, inflows other than those on the subject time-space link being fixed in the second level, and the FW method being applied in the third (innermost) level. Note that the second level treatment is unique to our proposed algorithm, and has not been reported elsewhere.

## 5.3 Nested Projection Method

The previously described steps (2)~(4) are typical of the diagonalization procedure; however, many comparable or even more efficient algorithms, in terms of computational times, are also available. In this section we indicate that by making a minor modification in the nested diagonalization method, a nested projection-type algorithm can be obtained as follows.

**Step 2: *Second Loop* Operation.**

Step 2.1: Let $n$=1. Compute and reset the initial feasible solution $\{u_a^n(t)\}$, based on the time-space network constructed by the estimated actual link travel times $\{\tau_a^m(t)\}$.

Step 2.2: Fix the inflows for each physical link other than on the subject time-space link at the current level, yielding the optimization program defined by (5.21) and (5.2)~(5.7).

$$\min\ Z = \frac{1}{2}\left(\mathbf{u}^{n+1}\right)^T \mathbf{G}\mathbf{u}^{n+1} + \left(\rho\mathbf{c}\left(\mathbf{u}^n\right) - \mathbf{G}\mathbf{u}^n\right)\mathbf{u}^{n+1} \tag{5.21}$$

where matrix **G** is symmetric and positive definite and $\rho$ is a contraction operator.

**Step 3:** ***Third Loop*** **Operation.**

Solve for the solution, $\left\{u_a^{n+1}(t)\right\}$, to optimization problem (5.21) and (5.2)~(5.7) by the FW method. Compute the resulting link travel times $\left\{c_a^{n+1}(t)\right\}$.

Note that a necessary condition for convergence of Step 2 in the nested projection method is that c(**u**) is *strictly* monotone. In case such a condition is not met by the application under consideration, a modified projection method may also be appropriate.

## 5.4 Modified Nested Projection Method

When $c(\mathbf{u})$ is only monotone (versus *strictly* monotone), but with the *Lipschitz* continuity condition holding with constant *L*, the modified nested projection method can be adopted as follows.

**Step 2:** ***Second Loop*** **Operation.**

Step 2.1: Let $n$=1. Compute and reset the initial feasible solution $u_a^n(t), \forall a, t$, based on the time-space network constructed by the estimated actual link travel times $\tau_a^m(t), \forall a, t$.

Step 2.2: Fix the inflows for each physical link other than that on the subject time-space link at the current level.

**Step 3:** ***Third Loop*** **Operation.**

Solve the following VIPs, for $\mathbf{u}^{n+1}$.

Step 3.1: Construction and Computation

Set **G**=**I** and select $\rho$, such that $0 < \rho < \frac{1}{L}$, where *L* is the *Lipschitz* constant for function *c*. Compute $\bar{\mathbf{u}}^n$ by solving the variational inequality subproblem:

$$\left[\bar{\mathbf{u}}^n + \left(\rho\mathbf{c}\left(\mathbf{u}^n\right) - \mathbf{u}^n\right)\right]^T \left(\mathbf{u}' - \bar{\mathbf{u}}^n\right) \geq 0 \quad \forall \mathbf{u}' \in \Omega \tag{5.22}$$

Step 3.2: Adaptation

Compute $\mathbf{u}^{n+1}$ by solving the variational inequality subproblem:

$$\left[\mathbf{u}^{n+1} + \left(\rho\mathbf{c}\left(\bar{\mathbf{u}}^n\right) - \mathbf{u}^n\right)\right]^T \left(\mathbf{u}' - \mathbf{u}^{n+1}\right) \geq 0 \quad \forall \mathbf{u}' \in \Omega \tag{5.23}$$

# 5.5 Numerical Example

## 5.5.1 Input Data

A simple network as shown in Figure 5.1 is used for testing. The test network consists of 6 links and 5 nodes in which nodes 1 and 3 are origins, node 5 is the destination, and nodes 2 and 4 are intermediate nodes.

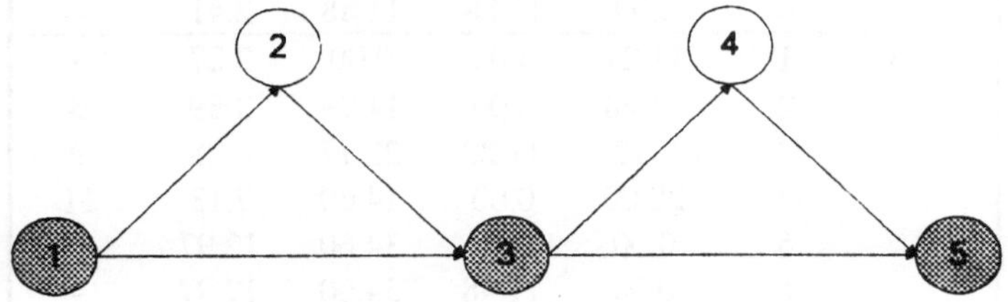


Figure 5.1: Test Network

The adopted dynamic travel time function is arbitrarily constructed as follows.

$$c_a(t) = 1 + 0.01(u_a(t))^2 + 0.01(x_a(t))^2 \quad \forall a,t \tag{5.24}$$

The assumed origin-destination (O-D) demands are shown in Table 5.1.

Table 5.1: Time-Dependent O-D Demands

| O-D | Time Interval | | | | | |
|---|---|---|---|---|---|---|
| Pair | $k$=1 | $k$=2 | $k$=3 | $k$=4 | $k$=8 | $k$=9 |
| 1-5 | 15 | 20 | 15 | 20 | 0 | 0 |
| 3-5 | 15 | 20 | 0 | 0 | 15 | 20 |

## 5.5.2 Test Results

A computer program coded with Turbo $C^{++}$ is used to solve the DUO route choice model with the input data. The obtained results are summarized in Table 5.2. The rationale of the proposed model and associated solution algorithm can be verified by checking if the resulting actual route travel times satisfy the dynamic user-optimal conditions as defined in Section 3.2.1. Based on Table 5.2, the actual route travel times are computed and summarized in Table 5.3. The travelers departing from the same origin during the same interval experience approximately the same actual route travel time.

Table 5.2: Results for the Test Network

| Link | Entering Time Interval | Inflow | Exit Flow | Number of Vehicles | Link Travel Time | Exiting Time Interval |
|---|---|---|---|---|---|---|
| 1→2 | 1 | 3.71 | 0.00 | 0.00 | 1.14 | 2 |
| | 2 | 8.53 | 3.71 | 3.71 | 1.86 | 4 |
| | 3 | 11.88 | 0.00 | 8.53 | 3.14 | 6 |
| | 4 | 0.00 | 8.53 | 20.40 | 5.16 | - |
| | 5 | 0.00 | 0.00 | 11.88 | 2.41 | - |
| | 6 | 0.00 | 11.88 | 11.88 | 2.41 | - |
| 1→3 | 1 | 11.29 | 0.00 | 0.00 | 2.27 | 3 |
| | 2 | 11.48 | 0.00 | 11.29 | 3.59 | 6 |
| | 3 | 3.12 | 11.29 | 22.77 | 6.28 | 9 |
| | 4 | 20.00 | 0.00 | 14.60 | 7.13 | 11 |
| | 5 | 0.00 | 0.00 | 34.60 | 12.97 | - |
| | 6 | 0.00 | 11.48 | 34.60 | 12.97 | - |
| | 7~8 | 0.00 | 0.00 | 23.12 | 6.35 | - |
| | 9 | 0.00 | 3.12 | 23.12 | 6.35 | - |
| | 10 | 0.00 | 0.00 | 20.00 | 5.00 | - |
| | 11 | 0.00 | 20.00 | 20.00 | 5.00 | - |
| 2→3 | 2 | 3.71 | 0.00 | 0.00 | 1.14 | 3 |
| | 3 | 0.00 | 3.71 | 3.71 | 1.14 | - |
| | 4 | 8.53 | 0.00 | 0.00 | 1.73 | 6 |
| | 5 | 0.00 | 0.00 | 8.53 | 1.73 | - |
| | 6 | 11.88 | 8.53 | 8.53 | 3.14 | 9 |
| | 7~8 | 0.00 | 0.00 | 11.88 | 2.41 | - |
| | 9 | 0.00 | 11.88 | 11.88 | 2.41 | - |
| 3→4 | 1 | 3.71 | 0.00 | 0.00 | 1.14 | 2 |
| | 2 | 8.53 | 3.71 | 3.71 | 1.86 | 4 |
| | 3 | 11.88 | 0.00 | 8.53 | 3.14 | 6 |
| | 4 | 0.00 | 8.53 | 20.40 | 5.16 | - |
| | 5 | 0.00 | 0.00 | 11.88 | 2.41 | - |
| | 6 | 5.17 | 11.88 | 11.88 | 2.68 | 9 |
| | 7 | 0.00 | 0.00 | 5.17 | 1.27 | - |
| | 8 | 9.85 | 0.00 | 5.17 | 2.24 | 10 |
| | 9 | 16.55 | 5.17 | 15.02 | 5.99 | 15 |
| | 10 | 0.00 | 9.85 | 26.40 | 7.97 | - |
| | 11 | 20.00 | 0.00 | 16.55 | 7.74 | 19 |
| | 12~14 | 0.00 | 0.00 | 36.55 | 14.36 | - |
| | 15 | 0.00 | 16.55 | 36.55 | 14.36 | - |
| | 16~18 | 0.00 | 0.00 | 20.00 | 5.00 | - |
| | 19 | 0.00 | 20.00 | 20.00 | 5.00 | - |

Table 5.2: Results for the Test Network (continued)

| Link | Entering Time Interval | Inflow | Exit Flow | Number of Vehicles | Link Travel Time | Exiting Time Interval |
|---|---|---|---|---|---|---|
| 3→5 | 1 | 11.29 | 0.00 | 0.00 | 2.27 | 3 |
| | 2 | 11.48 | 0.00 | 11.29 | 3.59 | 6 |
| | 3 | 3.12 | 11.29 | 22.77 | 6.28 | 9 |
| | 4~5 | 0.00 | 0.00 | 14.60 | 3.13 | - |
| | 6 | 14.83 | 11.48 | 14.60 | 5.33 | 11 |
| | 7 | 0.00 | 0.00 | 17.96 | 4.22 | - |
| | 8 | 5.15 | 0.00 | 17.96 | 4.49 | 12 |
| | 9 | 18.45 | 3.12 | 23.10 | 9.74 | 19 |
| | 10 | 0.00 | 0.00 | 38.43 | 15.77 | 26 |
| | 11 | 0.00 | 14.83 | 38.43 | 15.77 | - |
| | 12 | 0.00 | 5.15 | 23.60 | 6.57 | - |
| | 13~18 | 0.00 | 0.00 | 18.45 | 4.40 | - |
| | 19 | 0.00 | 18.45 | 18.45 | 4.40 | - |
| 4→5 | 2 | 3.71 | 0.00 | 0.00 | 1.14 | 3 |
| | 3 | 0.00 | 3.71 | 3.71 | 1.14 | - |
| | 4 | 8.53 | 0.00 | 0.00 | 1.73 | 6 |
| | 5 | 0.00 | 0.00 | 8.53 | 1.73 | - |
| | 6 | 11.88 | 8.53 | 8.53 | 3.14 | 9 |
| | 7~8 | 0.00 | 0.00 | 11.88 | 2.41 | - |
| | 9 | 5.17 | 11.88 | 11.88 | 2.68 | 12 |
| | 10 | 9.85 | 0.00 | 5.17 | 2.24 | 12 |
| | 11 | 0.00 | 0.00 | 15.02 | 3.26 | - |
| | 12 | 0.00 | 15.02 | 15.02 | 3.26 | - |
| | 15 | 16.55 | 0.00 | 0.00 | 3.74 | 19 |
| | 16~18 | 0.00 | 0.00 | 16.55 | 3.74 | - |
| | 19 | 20.00 | 16.55 | 16.55 | 7.74 | 27 |
| | 20~26 | 0.00 | 0.00 | 20.00 | 5.00 | - |
| | 27 | 0.00 | 20.00 | 20.00 | 5.00 | - |

Table 5.3: Actual Route Travel Time for the Test Network

| Route | Time Interval | | | | | |
|---|---|---|---|---|---|---|
| | $k$=1 | $k$=2 | $k$=3 | $k$=4 | $k$=8 | $k$=9 |
| 1→2→3→4→5 | 8.55 | 8.95 | 16.00 | – | – | – |
| 1→2→3→5 | 8.56 | 8.92 | 16.02 | – | – | – |
| 1→3→4→5 | 8.55 | 8.95 | 16.01 | 22.61 | – | – |
| 1→3→5 | 8.55 | 8.92 | 16.02 | – | – | – |
| 3→4→5 | 2.28 | 3.59 | – | – | 4.48 | 9.73 |
| 3→5 | 2.27 | 3.59 | – | – | 4.49 | 9.74 |

## 5.6 Notes

For discrete mathematical programming problems, a discontinuity will always create a non-convergence problem for an iterative algorithm. The algorithms presented in this chapter are iterative procedures with pre-determined step sizes. Although the link travel time functions, and hence the objective function, remain continuous throughout the solution procedure (including the subsequent minimal-travel-time route search), the *actual* link travel times, that are used for exertion of flow propagation over the network, have to be made into discrete integer intervals, i.e., rounded-off to the nearest integer. As a result, the cyclic phenomenon among several non-optimal solutions may occur. To mitigate this non-convergence problem, the application of smaller time intervals and the adoption of a less precise convergence criterion may be appropriate. It is worth noting that if the non-convergence problem cannot be avoided for practical applications and the corresponding cyclic non-optimal solutions did not deviate from the optimal solution by much, then the selection among the cyclic solutions for use may be acceptable.

The discrete nature of the DUO route choice model also affects the AON assignment described in Section 5.2.3. The search for the shortest route is based on the continuous link/route travel times; however, the AON assignment, which was performed in a back-stepping manner, must load flows onto the links during some integer-valued intervals. Our experience shows that in the early iterations of the nested diagonalization solution procedure, the difference between the continuous link travel times and the discrete actual link travel times is usually large, and this undesirable situation (or inconsistency) makes the convergence of the third loop problem very slow, or even impossible. Consequently, a preset number of iterations should used for the convergence check in early iterations of the nested diagonalization method.

A related issue in computer programming is the determination of the actual *route* travel times. Each actual route travel time is comprised of a sequence of actual link travel times in that route. In this regard, two approaches may be possible. One is to round off each link travel time as the actual link travel time over the route and then sum up the links for the actual route travel time. The other is to sum up the link travel times over the route and then round off (or discretize) the actual route travel time. The former, due to round-off errors, may cause a large difference between the *continuous* route travel times and the *discrete* actual travel times. The latter, on the contrary, will restrict the difference between the *continuous* route travel times and the *discrete* actual travel times within one time interval; however, the proofs given in Section 3.7.3.1 may no longer be applicable. Therefore, to preserve the validity of the proofs, the former approach was adopted for all numerical examples shown in this book.

# Chapter 6

# Dynamic User-Optimal Departure Time/Route Choice Model

The purpose of the dynamic user-optimal (DUO) departure time/route choice problem is to determine the best departure time along with the best route choice decision. The DUO departure time/route choice problem is a generalization of the DUO route choice problem, since the departure times for travelers are not prespecified and are allowed to change in response to different levels of network congestion. The relaxation of fixed trip departure times is appropriate; in many situations, travelers tend to choose the departure time that minimizes their *en route* travel times, especially when there is no penalty for either early or late arrival. If we consider a work-to-home trip, the choice of departure time is influenced by the wish of the traveler to avoid unnecessary delay *en route*. In fact, by minimizing an individual's *en route* travel time, more efficient utilization of the total network capacity can be reasonably expected. The day-to-day adjustment for commuters will gradually stabilize the departure time choice, thereby influencing the DUO route choice problem.

Several DUO departure time/route choice models have appeared in research literature. Janson (1992) formulated a DUO route choice model in which trips have variable departure times and scheduled arrival times. Using a heuristic approach based on the Frank-Wolfe (FW) algorithm, Janson (1993) solved a dynamic route choice problem and a combined departure time/route choice problem. Ran et al (1992) formulated a two-level optimal control problem for the DUO departure time/route choice problem in a multiple origin-destination network, proposing a heuristic approach to solve the problem. In the lower level, the O-D departure rates are specified, and therefore exogenous to the upper level problem, while in the upper level, a DUO route choice problem is formulated to find the dynamic trajectories of link states. Friesz et al (1993) presented a joint departure time and route choice model by formulating a link-based VI model for the DUO departure time/route choice problem. And finally, Smith and Ghali (1992) also considered this problem using a microscopic representation of vehicle streams.

In this chapter, the DUO departure time/route choice problem is formulated by the variational inequality approach. A description of the dynamic user-optimal conditions and corresponding VIP along with the equivalence analysis is provided in Section 6.1. A nested diagonalization method is then proposed and elaborated in Section 6.2. A numerical example is given in Section 6.3. The length of the analysis period and the total number of time intervals are analyzed in Section 6.4. Concluding notes are given in Section 6.5.

## 6.1 Equilibrium Conditions and Model Formulation

In this section, the dynamic user-optimal conditions are first defined to characterize the travelers' driving behavior for choosing the best departure time using the minimal travel time route. Due to inherent link interactions, the variational inequality approach is adopted to formulate the DUO departure time/route choice model. The equivalence between the dynamic user-optimal conditions and the variational inequality formulation is then stated as a theorem and verified by a proof.

### 6.1.1 Dynamic User-Optimal Conditions

Given O-D demands that are fixed and time-independent, the dynamic user-optimal conditions state that for each O-D pair, the actual route travel times experienced by travelers, regardless of the departure time, are equal and minimal. At the same time, the actual travel time of any unused route for each O-D pair is greater than or equal to the minimal actual route travel time. In other words, at equilibrium, if the flow departing from origin $r$ during interval $k$ over route $p$ toward destination $s$ is positive, i.e., $h_p^{rs*}(k) \geq 0$, then the corresponding actual route travel time is minimal. On the contrary, if no flow occurs on route $p$ during interval $k$, i.e., $h_p^{rs*}(k) = 0$, then the corresponding actual route travel time is at least as great as the minimal actual route travel time. This equilibrium conditions can be expressed mathematically as follows:

$$c_p^{rs*}(k) \begin{cases} = \pi^{rs} & \text{if } h_p^{rs*}(k) > 0 \\ \geq \pi^{rs} & \text{if } h_p^{rs*}(k) = 0 \end{cases} \quad \forall r,s,p,k \tag{6.1}$$

### 6.1.2 Variational Inequality Formulation

As to which type of mathematical formulation is more appropriate for the DUO departure time/route choice problem, a natural choice is the optimization formulation because many efficient algorithms are already available. Unfortunately, the DUO departure time/route choice problem cannot be formulated as an optimization problem because the dynamic link travel time function does not have a symmetric Jacobian matrix. Consequently, the more general VI approach is adopted to formulate the DUO departure time/route choice problem.

***Theorem 6.2***: The DUO departure time/route choice problem is equivalent to finding a solution $\mathbf{u}^* \in \Omega$ such that the following VIP holds:

$$\mathbf{c}^*[\mathbf{u} - \mathbf{u}^*] \geq 0 \quad \forall \mathbf{u} \in \Omega^* \tag{6.2}$$

Or, alternatively, in expanded form:

$$\sum_a \sum_t c_a^*(t)\left[u_a(t) - u_a^*(t)\right] \geq 0 \quad \forall \mathbf{u} \in \Omega^* \tag{6.3}$$

where $\Omega^*$ is a subset of $\Omega$ with $\delta_{apk}^{rs}(t)$ being realized at equilibrium, i.e., $\delta_{apk}^{rs}(t) = \delta_{apk}^{rs*}(t), \forall r, s, a, p, k, t$. The symbol $\Omega$ denotes the feasible region that is delineated by the following constraints.

Flow conservation constraint:

$$\sum_k \sum_p h_p^{rs}(k) = \bar{q}^{rs} \quad \forall r, s \tag{6.4}$$

Flow propagation constraints:

$$u_{apk}^{rs}(t) = h_p^{rs}(k)\delta_{apk}^{rs}(t) \quad \forall r, s, a, p, k, t \tag{6.5}$$

$$\sum_t \delta_{apk}^{rs}(t) = 1 \quad \forall r, s, p, a \in p, k \tag{6.6}$$

$$\delta_{apk}^{rs}(t) = \{0,1\} \quad \forall r, s, a, p, k, t \tag{6.7}$$

Nonnegativity constraint:

$$h_p^{rs}(k) \geq 0 \quad \forall r, s, p, k \tag{6.8}$$

Definitional Constraints:

$$u_a(t) = \sum_{rs} \sum_p \sum_k h_p^{rs}(k)\delta_{apk}^{rs}(t) \quad \forall a, t \tag{6.9}$$

$$c_p^{rs}(k) = \sum_a \sum_t c_a(t)\delta_{apk}^{rs}(t) \quad \forall r, s, p, k \tag{6.10}$$

Except for equation (6.4), all other constraints are the same as those for the DUO route choice model. Equation (6.4) conserves the time-independent O-D demand in terms of route flows. It states that the sum of the route flows over route $p$ and interval $k$ must equal the total departure flows which originate from $r$ toward destination $s$. Note also that the flow propagation constraints (6.5)~(6.7), as shown in Section 3.7.2, implicitly imply the following relationships:

$$v_a(t) = \sum_i u_a(i)\delta_{a_1}^i(t) \quad \forall a, t \tag{6.11}$$

where

$$\delta_{a_1}^i(t)=\begin{cases}1 & if\ i+\tau_a(i)=t \\ 0 & otherwise\end{cases} \qquad \forall a,i,t \tag{6.12}$$

and

$$x_a(t)=\sum_i u_a(i)\delta_{a_2}^i(t) \qquad \forall a,t \tag{6.13}$$

where

$$\delta_{a_2}^i(t)=\begin{cases}1 & if\ i<t\ and\ i+\tau_a(i)\geq t \\ 0 & otherwise\end{cases} \qquad \forall a,i,t \tag{6.14}$$

### 6.1.3 Equivalence Analysis

The following theorem verifies the equivalence between equilibrium conditions (6.1) and VIP (6.3).

***Theorem 6.2***: Under a certain flow propagation relationship $\left(\delta_{apk}^{rs}(t)=\delta_{apk}^{rs*}(t)\right)$, DUO departure time/route choice conditions (6.1) imply VIP (6.3) and vice versa.

***Proof of necessity***: We need to prove that under a certain propagation relationship $\left(\delta_{apk}^{rs}(t)=\delta_{apk}^{rs*}(t)\right)$, dynamic user-optimal conditions (6.1) can be reformulated as VIP (6.3). We first rearrange equilibrium conditions (6.1) as follows:

$$\left[c_p^{rs*}(k)-\pi^{rs}\right]\left[h_p^{rs}(k)-h_p^{rs*}(k)\right]\geq 0 \qquad \forall r,s,p,k \tag{6.15}$$

Summing the above equation over $r,s,p,k$ yields:

$$\sum_{rs}\sum_k\sum_p c_p^{rs*}(k)\left[h_p^{rs}(k)-h_p^{rs*}(k)\right] - \sum_{rs}\pi^{rs}\sum_k\sum_p\left[h_p^{rs}(k)-h_p^{rs*}(k)\right]\geq 0 \tag{6.16}$$

By making a substitution of $\sum_k\sum_p h_p^{rs}(k)=\sum_k\sum_p h_p^{rs*}(k)=\bar{q}^{rs}$, the second term is eliminated and the remaining term results in the following VIP:

$$\sum_{rs}\sum_k\sum_p c_p^{rs*}(k)\left[h_p^{rs}(k)-h_p^{rs*}(k)\right]\geq 0 \tag{6.17}$$

The remainder of the proof of necessity is same as equations (4.16)~(4.20) in Section 4.1.3.

***Proof of sufficiency***: We next prove that the route-based VIP (6.17) is equivalent to dynamic user-optimal conditions (6.1). We now define a feasible solution $\{h_p^{rs}(k)\}$ to be the same as the equilibrium flow pattern $\{h_p^{rs*}(k)\}$, except for two routes, $p_1^{rs}$ during interval $k_1$, and $p_2^{rs}$ during interval $k_2$. Without loss of generality, we consider two situations that could arise at equilibrium:

(i) Both route flows are positive, i.e., $h_{p_1}^{rs*}(k_1)>0$, and $h_{p_2}^{rs*}(k_2)>0$. We switch a small amount of flow $\Delta_1$ from route $p_1^{rs}$ to $p_2^{rs}$ with $0<\Delta_1\le h_{p_1}^{rs*}(k_1)$. That is:

$$h_{p_1}^{rs}(k_1)=h_{p_1}^{rs*}(k_1)-\Delta_1 \text{ and} \tag{6.18}$$

$$h_{p_2}^{rs}(k_2)=h_{p_2}^{rs*}(k_2)+\Delta_1 \tag{6.19}$$

By substituting a new feasible solution $\{h_p^{rs}(k)\}$ into equation (6.17), we have:

$$c_{p_1}^{rs*}(k_1)\left[h_{p_1}^{rs}(k_1)-h_{p_1}^{rs*}(k_1)\right]+c_{p_2}^{rs*}(k_2)\left[h_{p_2}^{rs}(k_2)-h_{p_2}^{rs*}(k_2)\right]\ge 0 \tag{6.20}$$

By using equations (6.18)~(6.19), one obtains:

$$c_{p_2}^{rs*}(k_2)\ge c_{p_1}^{rs*}(k_1) \tag{6.21}$$

Similarly, by switching a small amount of flow $\Delta_2$ with $0<\Delta_2\le h_{p_2}^{rs*}(k_2)$ from route $p_2^{rs}$ to $p_1^{rs}$, we obtain:

$$c_{p_1}^{rs*}(k_1)\ge c_{p_2}^{rs*}(k_2) \tag{6.22}$$

Equation (6.21) and equation (6.22) together imply:

$$c_{p_2}^{rs*}(k_2)=c_{p_1}^{rs*}(k_1) \tag{6.23}$$

We can repeat this procedure to verify for each time-dependent O-D pair that all used routes with positive flow have the same actual route travel time.

(ii) One route flow is positive and the other route flow is nil. We arbitrarily assume, without loss of generality, $h_{p_1}^{rs*}(k_1)>0$, and $h_{p_2}^{rs*}(k_2)=0$. We switch a small amount of flow $\Delta_1$ from route $p_1^{rs}$ to $p_2^{rs}$ with $0<\Delta_1\le h_{p_1}^{rs*}(k_1)$. By the same argument shown in (i), we have $c_{p_2}^{rs*}(k_2)\ge c_{p_1}^{rs*}(k_1)$. We repeat this procedure to verify for each time-dependent O-D pair that all unused routes with zero flow will have an actual route travel time no lower than the minimal actual route travel time.

Since both (i) and (ii) must hold, it follows that VIP (6.3) implies equilibrium conditions (6.1). This completes the proof.

## 6.2 Nested Diagonalization Method

In this section, we present the nested diagonalization method to solve the DUO departure time/route choice problem. The nested diagonalization method has been proposed and thoroughly discussed in Chapter 5 for the DUO route choice problem. Due to the different form of flow conservation constraints, the adaptation of the nested diagonalization method for the DUO departure time/route choice model is not directly applicable. However, by appropriate network representation, the DUO departure time/route choice model can be treated, in the mathematical sense, as the DUO route choice model.

### 6.2.1 Time-Space Network

By comparing the models' structures, it is observed that the only difference between the DUO departure time/route choice model and the DUO route choice model is in their flow conservation constraints. From the mathematical point of view, this difference can be resolved by an appropriate network representation. We construct the time-space network for the DUO departure time/route choice problem in Figure 6.1. The time-space network starts with a time-independent origin $r$, rather than time-dependent origins $r(k)$ as shown in Figure 4.1. This time-independent origin $r$ is then connected to each of its time-dependent *offspring* origins by a dummy link with travel time equal to zero. The remaining part of the time-space network is exactly the same as that for the DUO route choice problem.

### 6.2.2 Solution Algorithm

A nested diagonalization method is proposed to solve the DUO departure time/route choice model. The nested diagonalization method is defined as an algorithm that consists of the diagonalization method within its solution procedure. The diagonalization method, in turn, has the FW method embedded. This method is extremely useful for problems with two types of link flow interactions because one type of link flow interaction can be relaxed to yield a subproblem that can be solved by the diagonalization method, and the second type of link flow interaction can be further relaxed from the first level subproblem, resulting in a second level diagonalized subproblem that can be solved by the FW method.

For the DUO departure time/route choice problem, without considering the possible asymmetric property of link travel time functions, the source of link interactions is twofold. One is due to the actual link travel time $\tau_a(t)$, which is not known in advance, and hence, needs to be estimated, and the other interactions are accrued from the interference among inflows entering the same physical link, but having different time intervals. Once these two types of link flow interactions are temporarily fixed, a diagonalized subproblem is yielded. This diagonalized subproblem can then be reformulated as a convex optimization problem and solved by the FW method.

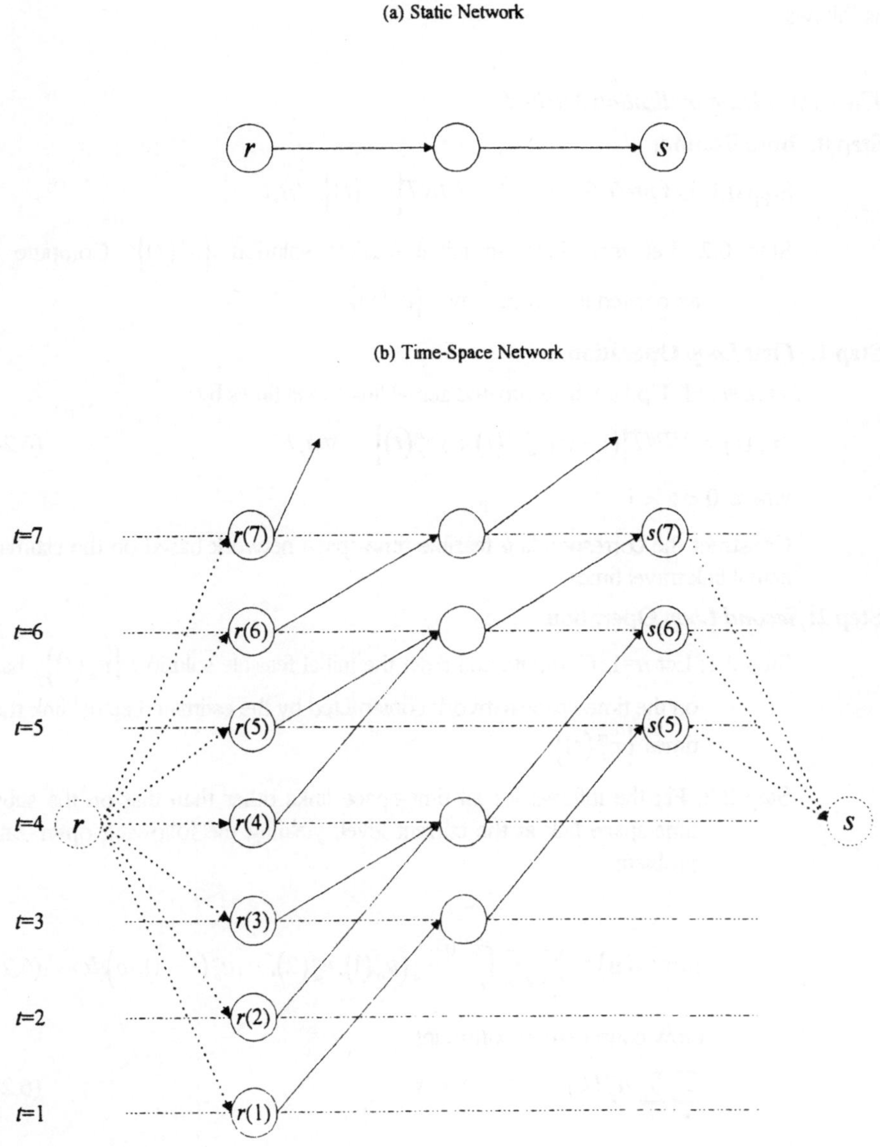


Figure 6.1: Time-Space Network

With the above discussion, the nested diagonalization method is formally described as follows.

***The Nested Diagonalization Method***

**Step 0: Initialization.**

Step 0.1: Let $m$=0. Set $\tau_a^0(t) = NINT\left[c_{a_0}(t)\right], \forall a, t$.

Step 0.2: Let $n$=1. Find an initial feasible solution $\left\{u_a^1(t)\right\}$. Compute the associated link travel times $\left\{c_a^1(t)\right\}$.

**Step 1: *First Loop* Operation.**

Let $m$=$m$+1. Update the estimated actual link travel times by

$$\tau_a^m(t) = NINT\left[(1-\gamma)\tau_a^{m-1}(t) + \gamma c_a^n(t)\right] \quad \forall a, t \tag{6.24}$$

where $0 < \gamma < 1$.

Construct the corresponding feasible time-space network based on the estimated actual link travel times.

**Step 2: *Second Loop* Operation.**

Step 2.1: Let $n$=1. Compute and reset the initial feasible solution $\left\{u_a^n(t)\right\}$, based on the time-space network constructed by the estimated actual link travel times $\left\{\tau_a^m(t)\right\}$.

Step 2.2: Fix the inflows for all time-space links other than that on the subject time-space link at the current level, yielding the following optimization problem.

$$\min\ z(\mathbf{u}) = \sum_a \sum_t \int_0^{u_a^{n+1}(t)} c_a\left(u_a^n(1), u_a^n(2), \cdots, u_a^n(t-1), \omega\right) d\omega \tag{6.25}$$

Flow conservation constraint:

$$\sum_k \sum_p h_p^{rs}(k) = \bar{q}^{rs} \quad \forall r, s \tag{6.26}$$

Nonnegativity constraint:

$$h_p^{rs}(k) \geq 0 \quad \forall r, s, p, k \tag{6.27}$$

Definitional Constraints:

$$u_{apk}^{rs}(t) = h_p^{rs}(k)\bar{\delta}_{apk}^{rs}(t) \quad \forall r, s, a, p, k, t \tag{6.28}$$

$$\bar{\delta}_{apk}^{rs}(t) = \{0,1\} \quad \forall r, s, a, p, k, t \tag{6.29}$$

$$u_a(t) = \sum_{rs}\sum_{p}\sum_{k} h_p^{rs}(k)\bar{\delta}_{apk}^{rs}(t) \quad \forall a, t \tag{6.30}$$

$$c_p^{rs}(k) = \sum_{a}\sum_{t} c_a(t)\bar{\delta}_{apt}^{rs}(t) \quad \forall r, s, p, k \tag{6.31}$$

**Step 3: *Third Loop* Operation.**

Solve for the solution, $\{u_a^{n+1}(t)\}$, in optimization problem (6.25)~(6.31) by the FW method. Compute the resulting link travel times $\{c_a^{n+1}(t)\}$.

**Step 4: Convergence Check for the *Second Loop* Operation.**

If $u_a^{n+1}(t) \approx u_a^n(t), \forall a, t$, go to Step 5; otherwise, set $n=n+1$, go to Step 2.2.

**Step 5: Convergence Check for the *First Loop* Operation.**

If $\tau_a^m(t) \approx c_a^{n+1}(t), \forall a, t$, stop; the current solution is optimal. Otherwise, set $n=n+1$, and go to Step 1.

Note that, in Step 3, the FW method is applied to the time-space network depicted in Figure 6.1. After searching for the shortest route from time-independent origin $r$ toward time-independent destination $s$, the AON assignment is performed in a backward manner. If there is branching from a time-independent origin having a route travel time longer than the preset entire analysis period within the solution procedure, this route must be discarded. Similar to the discussion shown in Section 5.3, a nested projection algorithm can be created by the following modification.

Step 2.2: Fix the inflows for all time-space links other than on the subject time-space link at the current level, yielding the following optimization problem.

$$\min\ z(\mathbf{u}) = \frac{1}{2}\left(\mathbf{u}^{n+1}\right)^T \mathbf{G}\mathbf{u}^{n+1} + \left(\rho \mathbf{c}\left(\mathbf{u}^n\right) - \mathbf{G}\mathbf{u}^n\right)\mathbf{u}^{n+1} \tag{6.32}$$

Subject to: (6.26)~(6.31)

where matrix $\mathbf{G}$ is symmetric and positive definite and $\rho$ is a contraction operator.

## 6.3 Numerical Examples

### 6.3.1 Input Data

A simple network shown in Figure 6.1 is used for testing. The test network consists of 6 links and 5 nodes in which nodes 1 and 3 are origins, node 5 is the destination, and nodes 2 and 4 are intermediate.

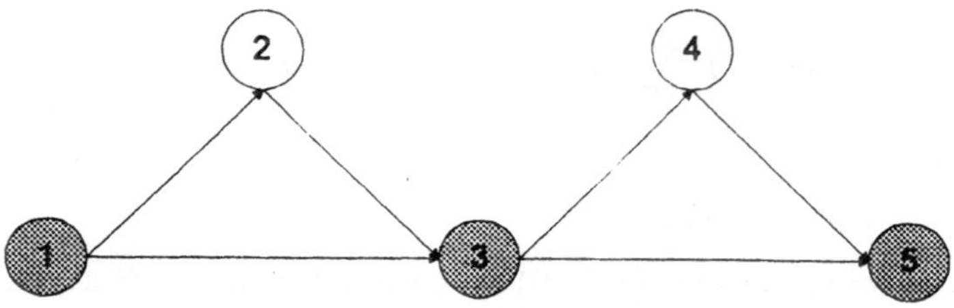


Figure 6.2: Test Network 1

The adopted dynamic travel time function is arbitrarily constructed as follows:

$$c_a(t) = 1 + 0.01(u_a(t))^2 + 0.01(x_a(t))^2 \quad \forall a,\ t \tag{6.33}$$

The assumed time-independent origin-destination (O-D) demands are shown in Table 6.1:

Table 6.1: Time-Independent O-D Demands for Test Network 1

| O-D Pair | Interval |
|---|---|
| | $k$=1~9 |
| 1-5 | 30 |

### 6.3.2 Test Results

The results for the DUO departure time/route choice model with the given data are summarized in Table 6.2.

Table 6.2: Results for Test Network 1

| Link | Entering Time Interval | Inflow | Exit Flow | Number of Vehicles | Link Travel Time | Exiting Time Interval |
|---|---|---|---|---|---|---|
| 1→2 | 1 | 2.20 | 0.00 | 0.00 | 1.05 | 2 |
| | 2 | 0.36 | 2.20 | 2.20 | 1.05 | 3 |
| | 3 | 2.17 | 0.36 | 0.36 | 1.05 | 4 |
| | 4 | 0.57 | 2.17 | 2.17 | 1.05 | 5 |
| | 5 | 2.13 | 0.57 | 0.57 | 1.05 | 6 |
| | 6 | 0.00 | 2.13 | 2.13 | 1.05 | - |

Table 6.2: Results for Test Network 1 (continued)

| Link | Entering Time Interval | Inflow | Exit Flow | Number of Vehicles | Link Travel Time | Exiting Time Interval |
|---|---|---|---|---|---|---|
| 1→3 | 1 | 10.48 | 0.00 | 0.00 | 2.10 | 3 |
| | 2 | 0.58 | 0.00 | 10.48 | 2.10 | 4 |
| | 3 | 0.00 | 10.48 | 11.06 | 2.22 | - |
| | 4 | 10.46 | 0.58 | 0.58 | 2.10 | 6 |
| | 5 | 1.06 | 0.00 | 10.46 | 2.11 | 7 |
| | 6 | 0.00 | 10.46 | 11.51 | 2.33 | - |
| | 7 | 0.00 | 1.06 | 1.06 | 1.01 | - |
| 2→3 | 2 | 2.20 | 0.00 | 0.00 | 1.05 | 3 |
| | 3 | 0.36 | 2.20 | 2.20 | 1.05 | 4 |
| | 4 | 2.17 | 0.36 | 0.36 | 1.05 | 5 |
| | 5 | 0.57 | 2.17 | 2.17 | 1.05 | 6 |
| | 6 | 2.13 | 0.57 | 0.57 | 1.05 | 7 |
| | 7 | 0.00 | 2.13 | 2.13 | 1.05 | - |
| 3→4 | 3 | 2.20 | 0.00 | 0.00 | 1.05 | 4 |
| | 4 | 0.36 | 2.20 | 2.20 | 1.05 | 5 |
| | 5 | 2.17 | 0.36 | 0.36 | 1.05 | 6 |
| | 6 | 0.57 | 2.17 | 2.17 | 1.05 | 7 |
| | 7 | 2.13 | 0.57 | 0.57 | 1.05 | 8 |
| | 8 | 0.00 | 2.13 | 2.13 | 1.05 | - |
| 3→5 | 3 | 10.48 | 0.00 | 0.00 | 2.10 | 5 |
| | 4 | 0.58 | 0.00 | 10.48 | 2.10 | 6 |
| | 5 | 0.00 | 10.48 | 11.06 | 2.22 | - |
| | 6 | 10.46 | 0.58 | 0.58 | 2.10 | 8 |
| | 7 | 1.06 | 0.00 | 10.46 | 2.11 | 9 |
| | 8 | 0.00 | 10.46 | 11.51 | 2.33 | - |
| | 9 | 0.00 | 1.06 | 1.06 | 1.01 | - |
| 4→5 | 4 | 2.20 | 0.00 | 0.00 | 1.05 | 5 |
| | 5 | 0.36 | 2.20 | 2.20 | 1.05 | 6 |
| | 6 | 2.17 | 0.36 | 0.36 | 1.05 | 7 |
| | 7 | 0.57 | 2.17 | 2.17 | 1.05 | 8 |
| | 8 | 2.13 | 0.57 | 0.57 | 1.05 | 9 |
| | 9 | 0.00 | 2.13 | 2.13 | 1.05 | - |

It is observed that all travelers must depart before interval 5; otherwise, they won't be able to arrive at the destination before interval 9 (included). The correctness of the results can be verified by checking if the equilibrated actual route travel times satisfy the dynamic user-optimal conditions defined in Section 6.1.1. If we consider route 1→3→5 departing during interval 1 as an example, we can see that the corresponding actual route travel time can be obtained by summing the actual link travel time for link 1→3 during interval 1 with the actual link travel time for link 3→5 during interval $1 + c_{1\to 3}(1)$ as follows:

$$\begin{aligned} c_{1\to3\to5}(1) &= c_{1\to3}(1) + c_{3\to5}(1 + c_{1\to3}(1)) \\ &= 2.10 + c_{3\to5}(3.10) \approx 2.10 + c_{3\to5}(3) = 4.20 \end{aligned} \tag{6.34}$$

The remaining actual route travel times are also computed and summarized in Table 6.3. For those trips which depart from the same origin, the route travel time is approximately the same. Note that route travel times from $k$=6 to $k$=9 are not applicable because no flow departs the origin during these four time intervals.

Table 6.3: Actual Route Travel Times for Test Network 1

| Route | Time Interval | | | | |
|---|---|---|---|---|---|
| | $k$=1 | $k$=2 | $k$=3 | $k$=4 | $k$=5 |
| 1→2→3→4→5 | 4.20 | 4.20 | 4.20 | 4.20 | 4.20 |
| 1→2→3→5 | 4.20 | 4.20 | – | 4.20 | 4.21 |
| 1→3→4→5 | 4.20 | 4.20 | – | 4.20 | 4.21 |
| 1→3→5 | 4.20 | 4.20 | – | 4.20 | 4.22 |

## 6.4 Analysis Period and Total Number of Time Intervals

The length of the entire analysis period is one of the important formulation issues. Theoretically, the entire analysis period is infinite; however, in practice, we can only analyze a finite time period which must be determined in advance. If the preset analysis period is too short, then some travelers won't be able to reach their destinations within the analysis period (termed infeasible in the mathematical sense). To prevent this situation from taking place, a basic requirement for the length of the analysis period is to let all O-D trips arrive at their destination by the end of the analysis period $T$. Then, the minimal length of analysis period must be determined by the maximum of the sum of the departure time plus the corresponding route travel time for all used routes as follows:

$$T_{\min} \approx \max_{rs,p,k} \left\{ k + c_p^{rs}(k) \right\} \tag{6.35}$$

where $k$ denotes the departure interval, and $c_p^{rs}(k)$ represents the actual route travel time for trips departing from origin $r$ over route $p$ during interval $k$ toward destination $s$.

For the DUO route choice model, the required length of analysis period can be computed by the above formula without any difficulty; however, for the DUO departure time/route choice model, the minimal length of the analysis period is difficult to determine because departure times are flexible. In this case, the analysis period length may be computed by the latest arrival time at destination $s$ minus the earliest departure time at origin $r$, as follows:

$$T_{\min} \approx T_{latest}^{s} - T_{earliest}^{r} \tag{6.36}$$

For the DUO departure time/route choice model, the longer the analysis period, the shorter the corresponding route travel time. Since travelers tend to be flexible in choosing their departure time, the traffic pattern becomes more dispersed. This assertion can be demonstrated by numerical examples with two different analysis periods, ($K$=10) and ($K$=20).

### 6.4.1 Input Data

A simple 2 by 2 grid network shown in Figure 6.3 is used for testing. The test network consists of 4 links and 4 nodes. Nodes 1 and 4 are the origin and destination nodes, respectively, whereas nodes 2 and 3 are intermediate nodes.

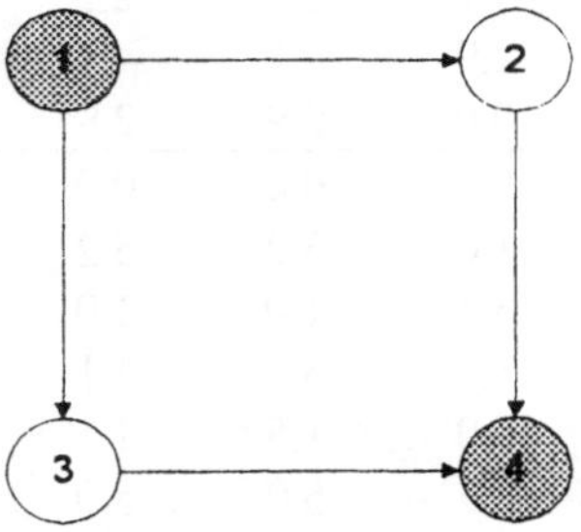


Figure 6.3: Test Network 2

The dynamic travel time function is arbitrarily constructed as follows:

$$c_a(t) = 1 + 0.01(u_a(t))^2 + 0.01(x_a(t))^2 \quad \forall a,\ t \tag{6.37}$$

The time-independent O-D demands are assumed as follows:

Table 6.4: Time-Independent O-D Demands for Test Network 2

| O-D | Number of Time Intervals | |
|---|---|---|
| Pair | $k$=1~10 (\|T\|=10) | $k$=1~20 (\|T\|=20) |
| 1-4 | 55 | 55 |

### 6.4.2 Test Results

The computational results for the DUO departure time/route choice model with the given basic data are summarized in Tables 6.5 and 6.6 for the 10 and 20 time intervals, respectively.

Table 6.5: Results for Test Network 2 (# Time Intervals = 10)

| Link | Entering Time Interval | Inflow | Exit Flow | Number of Vehicles | Link Travel Time | Exiting Time Interval |
|---|---|---|---|---|---|---|
| 1→2 | 1 | 5.2 | 0.0 | 0.0 | 1.27 | 2 |
| | 2 | 0.8 | 5.2 | 5.2 | 1.28 | 3 |
| | 3 | 5.2 | 0.8 | 0.8 | 1.27 | 4 |
| | 4 | 1.5 | 5.2 | 5.2 | 1.29 | 5 |
| | 5 | 5.0 | 1.5 | 1.5 | 1.27 | 6 |
| | 6 | 2.2 | 5.0 | 5.0 | 1.30 | 7 |
| | 7 | 4.7 | 2.2 | 2.2 | 1.27 | 8 |
| | 8 | 2.9 | 4.7 | 4.7 | 1.30 | 9 |
| | 9 | 0.0 | 2.9 | 2.9 | 1.08 | - |
| 1→3 | 1 | 5.2 | 0.0 | 0.0 | 1.27 | 2 |
| | 2 | 1.0 | 5.2 | 5.2 | 1.28 | 3 |
| | 3 | 5.1 | 1.0 | 1.0 | 1.27 | 4 |
| | 4 | 1.5 | 5.1 | 5.1 | 1.28 | 5 |
| | 5 | 5.0 | 1.5 | 1.5 | 1.27 | 6 |
| | 6 | 2.2 | 5.0 | 5.0 | 1.30 | 7 |
| | 7 | 4.7 | 2.2 | 2.2 | 1.26 | 8 |
| | 8 | 2.9 | 4.7 | 4.7 | 1.30 | 9 |
| | 9 | 0.0 | 2.9 | 2.9 | 1.09 | - |
| 2→4 | 2 | 5.2 | 0.0 | 0.0 | 1.27 | 3 |
| | 3 | 0.8 | 5.2 | 5.2 | 1.28 | 4 |
| | 4 | 5.2 | 0.8 | 0.8 | 1.27 | 5 |
| | 5 | 1.5 | 5.2 | 5.2 | 1.29 | 6 |
| | 6 | 5.0 | 1.5 | 1.5 | 1.27 | 7 |
| | 7 | 2.2 | 5.0 | 5.0 | 1.30 | 8 |
| | 8 | 4.7 | 2.2 | 2.2 | 1.27 | 9 |
| | 9 | 2.9 | 4.7 | 4.7 | 1.30 | 10 |
| | 10 | 0.0 | 2.9 | 2.9 | 1.08 | - |
| 3→4 | 2 | 5.2 | 0.0 | 0.0 | 1.27 | 3 |
| | 3 | 1.0 | 5.2 | 5.2 | 1.28 | 4 |
| | 4 | 5.1 | 1.0 | 1.0 | 1.27 | 5 |
| | 5 | 1.5 | 5.1 | 5.1 | 1.28 | 6 |
| | 6 | 5.0 | 1.5 | 1.5 | 1.27 | 7 |
| | 7 | 2.2 | 5.0 | 5.0 | 1.30 | 8 |
| | 8 | 4.7 | 2.2 | 2.2 | 1.26 | 9 |
| | 9 | 2.9 | 4.7 | 4.7 | 1.30 | 10 |
| | 10 | 0.0 | 2.9 | 2.9 | 1.09 | - |

Table 6.6: Results for Test Network 2 (# Time Intervals = 20)

| Link | Entering Time Interval | Inflow | Exit Flow | Number of Vehicles | Link Travel Time | Exiting Time Interval |
|---|---|---|---|---|---|---|
| 1→2 | 1 | 2.1 | 0.0 | 0.0 | 1.05 | 2 |
| | 2 | 0.9 | 2.1 | 2.1 | 1.05 | 3 |
| | 3 | 1.9 | 0.9 | 0.9 | 1.05 | 4 |
| | 4 | 1.1 | 1.9 | 1.9 | 1.05 | 5 |
| | 5 | 1.7 | 1.1 | 1.1 | 1.04 | 6 |
| | 6 | 1.4 | 1.7 | 1.7 | 1.05 | 7 |
| | 7 | 1.5 | 1.4 | 1.4 | 1.04 | 8 |
| | 8 | 1.6 | 1.5 | 1.5 | 1.05 | 9 |
| | 9~10 | 1.6 | 1.6 | 1.6 | 1.05 | 10~11 |
| | 11 | 1.5 | 1.6 | 1.6 | 1.05 | 12 |
| | 12 | 1.5 | 1.5 | 1.5 | 1.05 | 13 |
| | 13~14 | 1.5 | 1.5 | 1.5 | 1.04 | 14~15 |
| | 15 | 1.6 | 1.5 | 1.5 | 1.05 | 16 |
| | 16 | 1.5 | 1.6 | 1.6 | 1.05 | 17 |
| | 17 | 1.5 | 1.5 | 1.5 | 1.04 | 18 |
| | 18 | 1.5 | 1.5 | 1.5 | 1.04 | 19 |
| | 19 | 0.0 | 1.5 | 1.5 | 1.02 | - |
| 1→3 | 1 | 2.1 | 0.0 | 0.0 | 1.05 | 2 |
| | 2 | 0.9 | 2.1 | 2.1 | 1.05 | 3 |
| | 3 | 1.9 | 0.9 | 0.9 | 1.04 | 4 |
| | 4 | 1.2 | 1.9 | 1.9 | 1.05 | 5 |
| | 5 | 1.7 | 1.2 | 1.2 | 1.04 | 6 |
| | 6 | 1.4 | 1.7 | 1.7 | 1.05 | 7 |
| | 7 | 1.5 | 1.4 | 1.4 | 1.04 | 8 |
| | 8 | 1.6 | 1.5 | 1.5 | 1.05 | 9 |
| | 9 | 1.5 | 1.6 | 1.6 | 1.05 | 10 |
| | 10 | 1.6 | 1.5 | 1.5 | 1.05 | 11 |
| | 11 | 1.5 | 1.6 | 1.6 | 1.05 | 12 |
| | 12 | 1.5 | 1.5 | 1.5 | 1.05 | 13 |
| | 13 | 1.6 | 1.5 | 1.5 | 1.05 | 14 |
| | 14 | 1.5 | 1.6 | 1.6 | 1.05 | 15 |
| | 15 | 1.5 | 1.5 | 1.5 | 1.05 | 16 |
| | 16 | 1.4 | 1.5 | 1.5 | 1.04 | 17 |
| | 17 | 1.7 | 1.4 | 1.4 | 1.05 | 18 |
| | 18 | 1.4 | 1.7 | 1.7 | 1.05 | 19 |
| | 19 | 0.0 | 1.4 | 1.4 | 1.02 | - |

Table 6.6: Results for Test Network 2 (# Time Intervals = 20) (continued)

| Link | Entering Time Interval | Inflow | Exit Flow | Number of Vehicles | Link Travel Time | Exiting Time Interval |
|---|---|---|---|---|---|---|
| 2→4 | 2 | 2.1 | 0.0 | 0.0 | 1.05 | 3 |
| | 3 | 0.9 | 2.1 | 2.1 | 1.05 | 4 |
| | 4 | 1.9 | 0.9 | 0.9 | 1.05 | 5 |
| | 5 | 1.1 | 1.9 | 1.9 | 1.05 | 6 |
| | 6 | 1.7 | 1.1 | 1.1 | 1.04 | 7 |
| | 7 | 1.4 | 1.7 | 1.7 | 1.05 | 8 |
| | 8 | 1.5 | 1.4 | 1.4 | 1.04 | 9 |
| | 9 | 1.6 | 1.5 | 1.5 | 1.05 | 10 |
| | 10~11 | 1.6 | 1.6 | 1.6 | 1.05 | 11~12 |
| | 12 | 1.5 | 1.6 | 1.6 | 1.05 | 13 |
| | 13 | 1.5 | 1.5 | 1.5 | 1.05 | 14 |
| | 14~15 | 1.5 | 1.5 | 1.5 | 1.04 | 15~16 |
| | 16 | 1.6 | 1.5 | 1.5 | 1.05 | 17 |
| | 17 | 1.5 | 1.6 | 1.6 | 1.05 | 18 |
| | 18~19 | 1.5 | 1.5 | 1.5 | 1.04 | 19~20 |
| | 20 | 0.0 | 1.5 | 1.5 | 1.02 | - |
| 3→4 | 2 | 2.1 | 0.0 | 0.0 | 1.05 | 3 |
| | 3 | 0.9 | 2.1 | 2.1 | 1.05 | 4 |
| | 4 | 1.9 | 0.9 | 0.9 | 1.04 | 5 |
| | 5 | 1.2 | 1.9 | 1.9 | 1.05 | 6 |
| | 6 | 1.7 | 1.2 | 1.2 | 1.04 | 7 |
| | 7 | 1.4 | 1.7 | 1.7 | 1.05 | 8 |
| | 8 | 1.5 | 1.4 | 1.4 | 1.04 | 9 |
| | 9 | 1.6 | 1.5 | 1.5 | 1.05 | 10 |
| | 10 | 1.5 | 1.6 | 1.6 | 1.05 | 11 |
| | 11 | 1.6 | 1.5 | 1.5 | 1.05 | 12 |
| | 12 | 1.5 | 1.6 | 1.6 | 1.05 | 13 |
| | 13 | 1.5 | 1.5 | 1.5 | 1.05 | 14 |
| | 14 | 1.6 | 1.5 | 1.5 | 1.05 | 15 |
| | 15 | 1.5 | 1.6 | 1.6 | 1.05 | 16 |
| | 16 | 1.5 | 1.5 | 1.5 | 1.05 | 17 |
| | 17 | 1.4 | 1.5 | 1.5 | 1.04 | 18 |
| | 18 | 1.7 | 1.4 | 1.4 | 1.05 | 19 |
| | 19 | 1.4 | 1.7 | 1.7 | 1.05 | 20 |
| | 20 | 0.0 | 1.4 | 1.4 | 1.02 | - |

The validity of the results can be verified by checking if the equilibrated actual route travel times satisfy the dynamic user-optimal conditions defined in Section 6.1.1. If we consider route 1→3→4 departing during interval 1 as an example, we can see that the corresponding actual route travel time can be obtained by summing up the actual link travel time for link 1→3 during interval 1 and the actual link travel time for link 3→4 during interval $1+c_{1\to3}(1)$ as follows:

$$\begin{aligned} c_{1\to3\to4}(1) &= c_{1\to3}(1)+c_{3\to4}\left(1+c_{1\to3}(1)\right) \\ &= 1.27+c_{3\to4}(2.27) \approx 1.27+c_{3\to4}(2) = 2.54 \end{aligned} \tag{6.38}$$

Following the same procedure shown above, the remaining actual route travel times are also computed and summarized in Table 6.7 for the total number of time intervals equal to 10, and in Table 6.8, for the total number of time intervals equal to 20. For those trips which depart the same origin regardless of the departure times, the route travel time is approximately the same. Note that in Table 6.7, the route travel times for *k*=9 and *k*=10 are not applicable because no flow departs the origin during these two time intervals.

Table 6.7: Actual Route Travel Times for Test Network 2 (# Time Intervals = 10)

| Route | Time Interval | | | | | | | |
|---|---|---|---|---|---|---|---|---|
| | *k*=1 | *k*=2 | *k*=3 | *k*=4 | *k*=5 | *k*=6 | *k*=7 | *k*=8 |
| 1→2→4 | 2.54 | 2.56 | 2.54 | 2.58 | 2.54 | 2.60 | 2.54 | 2.60 |
| 1→3→4 | 2.54 | 2.56 | 2.54 | 2.56 | 2.54 | 2.60 | 2.52 | 2.60 |

Table 6.8: Actual Route Travel Times for Test Network 2 (# Time Intervals = 20)

| Route | Time Interval | | | | | | | | |
|---|---|---|---|---|---|---|---|---|---|
| | *k*=1 | *k*=2 | *k*=3 | *k*=4 | *k*=5 | *k*=6 | *k*=7 | *k*=8 | *k*=9 |
| 1→2→4 | 2.1 | 2.1 | 2.1 | 2.1 | 2.08 | 2.1 | 2.08 | 2.1 | 2.1 |
| 1→3→4 | 2.1 | 2.1 | 2.08 | 2.1 | 2.08 | 2.1 | 2.08 | 2.1 | 2.1 |
| Route | *k*=10 | *k*=11 | *k*=12 | *k*=13 | *k*=14 | *k*=15 | *k*=16 | *k*=17 | *k*=18 |
| 1→2→4 | 2.1 | 2.1 | 2.1 | 2.08 | 2.08 | 2.1 | 2.1 | 2.08 | 2.08 |
| 1→3→4 | 2.1 | 2.1 | 2.1 | 2.1 | 2.1 | 2.1 | 2.08 | 2.1 | 2.1 |

By comparing Table 6.7 with Table 6.8, the assertion that a longer analysis period results in a shorter actual route travel time is confirmed.

## 6.5 Notes

A relevant issue associated with discrete-time dynamic models is the determination of time interval length $\Delta t$. The time interval length $\Delta t$ is normally set as the minimum of free flow travel times on all of the links. The reason for using this method to determine the time interval length is to avoid a possible *link jump*, which is critical, especially for reactive dynamic models, because travelers are not prohibited from changing their subroute at any intermediate node based on *en route* information. In general, the shorter the time interval length, the closer the approximation of a discrete-time model is to its continuous-time counterpart. Also, the appropriate form of dynamic travel time functions would be greatly affected by the time interval length chosen. With the given analysis period and time interval length, the total number of time intervals can be computed by the following formula:

$$|T| \approx \frac{T_{\min}}{\Delta t} \tag{6.39}$$

However, a trial-and-error technique is commonly employed to determine the total number of time intervals.

The DUO departure time/route choice model is extremely useful for cases where time windows are imposed. One example would be the case of a trucking company which is required to deliver the commodities within a certain time period upon the clients' request. Referring back to our time-space network, these problems can be easily handled by connecting only those *required* time-dependent destinations to the time-independent destination. This DUO departure time/route choice problem with time window constraints (involving both departure and arrival time windows) has been studied to some extent by the authors.

There is a wide range of applications which could be adopted in the area of Logistics, or effective traffic control measures such as ramp controls on freeway entrances, or priority control settings for designated transit/emergency vehicles, where the benefit DUO departure time/route choice model is even more pronounced. Such a concept could lead to the development of combined dynamic traffic control and travel choice models.

# Chapter 7

# Dynamic User-Optimal Models With Variable Demand

In contrast to the DUO models with fixed demand, the DUO models with variable demand consider both elastic traffic demand and route choice within a mathematical program framework. The essence of this framework is the existence of feedback on network conditions, where the amount of O-D demand is determined by the corresponding minimal route travel times, which are in turn affected by the O-D demand through the traffic assignment procedure. Since the interactions between the variable demand and route choice decisions are treated within a unified framework, the *internal* inconsistencies that are apt to characterize conventional transportation planning procedures can be avoided. In consideration of the departure time choices, two types of models can be further identified, i.e., the DUO variable demand/route choice model, and the DUO variable demand/departure time/route choice model.

In this chapter, the DUO variable demand/route choice model, including equilibrium conditions and model formulation, the nested diagonalization method and a numerical example, are first provided in Section 7.1. A similar discussion for the DUO variable demand/departure time/route choice model is provided in Section 7.2. Finally, concluding notes are given in Section 7.3.

## 7.1 Dynamic User-Optimal Variable Demand/Route Choice Model

In this section, we discuss the equilibrium conditions and model formulation, the nested diagonalization method and a numerical example for the dynamic user-optimal variable demand/route choice model.

### 7.1.1 Equilibrium Conditions and Model Formulation

The dynamic user-optimal conditions are first defined to characterize the travelers' driving behavior for choosing the best departure time using the minimal travel time route. Due to inherent link interactions, the variational inequality approach is adopted to formulate the DUO variable demand/route choice model. The equivalence between the dynamic user-optimal conditions and the variational inequality formulation is then stated by a theorem and verified by a proof.

#### 7.1.1.1 Dynamic user-optimal conditions

When the O-D demands are variable and time-dependent, the corresponding dynamic user-optimal conditions may be characterized by equilibrium conditions both on route choice behavior and on trip demand functions. For route choice behavior, they state for each O-D pair that the actual route travel times experienced by travelers departing during the same interval are equal and minimal, or no traveler would be better off by unilaterally changing his/her route. In other words, the actual route travel time of any unused route for each O-D pair is greater than or equal to the minimal actual route travel time. Therefore, for each O-D pair *rs*, if the flow over route *p* during interval *k* is positive, i.e., $h_p^{rs*}(k) > 0$, then the corresponding actual route travel time is minimal. However, if no flow occurs on route *p*, i.e., $h_p^{rs*}(k) = 0$, then the corresponding actual route travel time is at least as great as the minimal actual route travel time. This equilibrium conditions can be mathematically expressed as follows.

$$c_p^{rs*}(k)\begin{cases} = \pi^{rs}(k) & \text{if } h_p^{rs*}(k) > 0 \\ \geq \pi^{rs}(k) & \text{if } h_p^{rs*}(k) = 0 \end{cases} \quad \forall r, s, p, k \tag{7.1}$$

For the trip demand functions, it states for each O-D pair that if the departure flow rate during interval *k* is positive, i.e., $q^{rs*}(k) > 0$, then the corresponding minimal route travel time is equal to the inverse demand function $\left(D^{rs*}(k)\right)^{-1}$. However, if no traffic demand is induced during interval *k*, i.e., $q^{rs*}(k) = 0$, then the corresponding minimal route travel time is at least as great as the inverse demand function $\left(D^{rs*}(k)\right)^{-1}$. This equilibrium conditions can be mathematically expressed as follows:

$$\pi^{rs}(k)\begin{cases} = \left(D^{rs*}(k)\right)^{-1} & \text{if } q^{rs*}(k) > 0 \\ \geq \left(D^{rs*}(k)\right)^{-1} & \text{if } q^{rs*}(k) = 0 \end{cases} \quad \forall r, s, k \tag{7.2}$$

### 7.1.1.2 Variational inequality formulation

The DUO variable demand/route choice problem can be formulated using the variational inequality approach.

***Theorem 7.1***: The DUO variable demand/route choice problem is equivalent to finding a solution $(\mathbf{u}^*, \mathbf{q}^*) \in \Omega$ such that the following VIP holds:

$$\mathbf{c}^*(\mathbf{u}-\mathbf{u}^*) - \mathbf{D}^{-1^*}(\mathbf{q}-\mathbf{q}^*) \geq 0 \quad \forall(\mathbf{u},\mathbf{q}) \in \Omega^* \tag{7.3}$$

or, alternatively, in expanded form:

$$\sum_a \sum_t c_a^*(t)\left[u_a(t) - u_a^*(t)\right] \\ -\sum_{rs}\sum_k \left(D^{rs*}(k)\right)^{-1}\left[q^{rs}(k) - q^{rs*}(k)\right] \geq 0 \quad \forall(\mathbf{u},\mathbf{q}) \in \Omega^* \tag{7.4}$$

where $\Omega^*$ is a subset of $\Omega$ with $\delta_{apk}^{rs}(t)$ being realized at equilibrium, i.e., $\delta_{apk}^{rs}(t) = \delta_{apk}^{rs*}(t), \forall r,s,a,p,k,t$. The symbol $\Omega$ denotes the feasible region that is delineated by the following constraints:

Flow conservation constraint:

$$\sum_p h_p^{rs}(k) = q^{rs}(k) \quad \forall r,s,k \tag{7.5}$$

Flow propagation constraints:

$$u_{apk}^{rs}(t) = h_p^{rs}(k)\delta_{apk}^{rs}(t) \quad \forall r,s,a,p,k,t \tag{7.6}$$

$$\sum_t \delta_{apk}^{rs}(t) = 1 \quad \forall r,s,p,a \in p,k \tag{7.7}$$

$$\delta_{apk}^{rs}(t) = \{0,1\} \quad \forall r,s,a,p,k,t \tag{7.8}$$

Nonnegativity constraint:

$$h_p^{rs}(k) \geq 0 \quad \forall r,s,p,k \tag{7.9}$$

Definitional constraints:

$$u_a(t) = \sum_{rs}\sum_p\sum_k h_p^{rs}(k)\delta_{apk}^{rs}(t) \quad \forall a,t \tag{7.10}$$

$$c_p^{rs}(k) = \sum_a\sum_t c_a(t)\delta_{apk}^{rs}(t) \quad \forall r,s,p,k \tag{7.11}$$

The constraints for the DUO variable demand/route choice model are almost the same as that for the DUO route choice model, except for equation (7.5). Equation (7.5) expresses the time-dependent O-D demand in terms of route flows. Note that the time-dependent O-D demand shown in equation (7.5) is defined as a variable, rather than a fixed number.

#### 7.1.1.3 Equivalence analysis

***Theorem** 7.2*: Under a certain flow propagation relationship $\left(\delta_{apk}^{rs}(t)=\delta_{apk}^{rs*}(t)\right)$, DUO variable demand/route choice conditions (7.1)~(7.2) imply VIP (7.4) and vice versa.

***Proof of necessity***: We need to prove that under a certain flow propagation relationship $\left(\delta_{apk}^{rs}(t)=\delta_{apk}^{rs*}(t)\right)$, DUO variable demand/route choice conditions (7.1)~(7.2) imply VIP (7.4). We first rearrange equilibrium conditions (7.1)~(7.2) as follows:

$$\left[c_p^{rs*}(k)-\pi^{rs}(k)\right]\left[h_p^{rs}(k)-h_p^{rs*}(k)\right]\geq 0 \quad \forall r,s,p,k \tag{7.12}$$

$$\left[\pi^{rs}(k)-\left(D^{rs*}(k)\right)^{-1}\right]\left[q^{rs}(k)-q^{rs*}(k)\right]\geq 0 \quad \forall r,s,k \tag{7.13}$$

By summing equation (7.12) over *r,s,p,k* and equation (7.13) over *r,s,k*, and then making a substitution of $\sum_p h_p^{rs}(k)=q^{rs}(k)$, we obtain:

$$\sum_{rs}\sum_k\sum_p c_p^{rs*}(k)\left[h_p^{rs}(k)-h_p^{rs*}(k)\right] -\sum_{rs}\sum_k\left(D^{rs*}(k)\right)^{-1}\left[q^{rs}(k)-q^{rs*}(k)\right]\geq 0 \tag{7.14}$$

By applying equation (7.11), we obtain:

$$\sum_{rs}\sum_k\sum_p\left[\sum_a\sum_t c_a^*(t)\delta_{apk}^{rs}(t)\right]\left[h_p^{rs}(k)-h_p^{rs*}(k)\right] -\sum_{rs}\sum_k\left(D^{rs*}(k)\right)^{-1}\left[q^{rs}(k)-q^{rs*}(k)\right]\geq 0 \tag{7.15}$$

By changing the order of the summation on the first term, using equation (7.10), we have:

$$\sum_a\sum_t c_a^*(t)\left[u_a(t)-u_a^*(t)\right] -\sum_{rs}\sum_k\left(D^{rs*}(k)\right)^{-1}\left[q^{rs}(k)-q^{rs*}(k)\right]\geq 0 \tag{7.16}$$

Equation (7.16) is identical to equation (7.4).

***Proof of sufficiency***: We show next that VIP (7.4) implies DUO variable demand/route choice conditions (7.1)~(7.2). Note that equations (7.14)~(7.16) are essentially reversible.

The remaining steps of the proof are to show that the expression (7.14) is equivalent to the DUO variable demand/route choice conditions. First we demonstrate that VIP (7.14) implies dynamic user-optimal conditions (7.1). We then define a feasible solution $\{h_p^{rs}(k),\ q^{rs*}(k)\}$ to be the same as $\{h_p^{rs*}(k),\ q^{rs*}(k)\}$ except for two routes, $p_1^{rs}$ and $p_2^{rs}$. We consider two situations that could arise at equilibrium:

(i) Both route flows are positive, i.e., $h_{p_1}^{rs*}(k) > 0$ and $h_{p_2}^{rs*}(k) > 0$. We switch a small amount of flow $\Delta_1$ from route $p_1^{rs}$ to $p_2^{rs}$ with $0 < \Delta_1 \le h_{p_1}^{rs*}(k)$. That is:

$$h_{p_1}^{rs}(k) = h_{p_1}^{rs*}(k) - \Delta_1 \tag{7.17}$$

and

$$h_{p_2}^{rs}(k) = h_{p_2}^{rs*}(k) + \Delta_1 \tag{7.18}$$

Substituting the feasible solution $\{h_p^{rs}(k),\ q^{rs*}(k)\}$ into equation (7.14) yields:

$$c_{p_1}^{rs*}(k)\left[h_{p_1}^{rs}(k) - h_{p_1}^{rs*}(k)\right] + c_{p_2}^{rs*}(k)\left[h_{p_2}^{rs}(k) - h_{p_2}^{rs*}(k)\right] \ge 0 \tag{7.19}$$

Applying equation (7.17) and equation (7.18) and division by $\Delta_1$, we have:

$$c_{p_2}^{rs*}(k) \ge c_{p_1}^{rs*}(k) \tag{7.20}$$

Similarly, by switching a small amount of flow $\Delta_2$ with $0 < \Delta_2 \le h_{p_2}^{rs*}(k)$ from route $p_2^{rs}$ to $p_1^{rs}$, we can obtain:

$$c_{p_1}^{rs*}(k) \ge c_{p_2}^{rs*}(k) \tag{7.21}$$

Equations (7.20) and (7.21) together imply:

$$c_{p_2}^{rs*}(k) = c_{p_1}^{rs*}(k) \tag{7.22}$$

We repeat this procedure to verify for each O-D pair that all used routes with a positive flow have the same actual route travel time.

(ii) One route flow is positive and the other route flow is nil. We arbitrarily assume, without loss of generality, $h_{p_1}^{rs*}(k) > 0$ and $h_{p_2}^{rs*}(k) = 0$. We switch a small amount of flow $\Delta_1$ from route $p_1^{rs}$ to $p_2^{rs}$ with $0 < \Delta_1 \le h_{p_1}^{rs*}(k)$. By the same argument shown in (i), we have $c_{p_2}^{rs*}(k) \ge c_{p_1}^{rs*}(k)$. We repeat this procedure to verify for each O-D pair that all unused routes with zero flow have higher or equal actual route travel times. By cases (i) and (ii), it follows that VIP (7.4) implies expression (7.1).

We conclude the proof by showing that VIP (7.14) implies expression (7.2). We now define a feasible solution $\{h_p^{rs}(k),\ q^{rs}(k)\}$ to be the same as $\{h_p^{rs*}(k),\ q^{rs*}(k)\}$ except for one O-D demand $q^{\hat{r}\hat{s}}(k)$ and one route flow $h_p^{\hat{r}\hat{s}}(k)$, for which $q^{\hat{r}\hat{s}}(k) = q^{\hat{r}\hat{s}*}(k) - \Delta$, $h_p^{\hat{r}\hat{s}}(k) = h_p^{\hat{r}\hat{s}*}(k) - \Delta$, where $0 < |\Delta| \le h_p^{\hat{r}\hat{s}*}(k)$. Substituting this feasible solution into VIP (7.14), one obtains:

$$\left[c_p^{\bar{r}\bar{s}*}(k) - \left(D^{\bar{r}\bar{s}*}(k)\right)^{-1}\right]\Delta \le 0 \tag{7.23}$$

For $q^{\bar{r}\bar{s}*}(k) > 0$, we can find a route $p$ such that $h_p^{\bar{r}\bar{s}*}(k) > 0$. Since the value of $\Delta$ could be either positive or negative, the equality must therefore hold. Applying expression (7.1) and dividing by $\Delta$ implies:

$$\pi^{\bar{r}\bar{s}}(k) = \left(D^{\bar{r}\bar{s}*}(k)\right)^{-1} \tag{7.24}$$

For $q^{\bar{r}\bar{s}*}(k) = 0$ and $h_p^{\bar{r}\bar{s}*}(k) = 0$, the value of $\Delta$ can only be negative, which implies:

$$\left[c_p^{\bar{r}\bar{s}*}(k) - \left(D^{\bar{r}\bar{s}*}(k)\right)^{-1}\right] \ge 0 \tag{7.25}$$

We repeat this procedure for all possible routes between O-D pair *rs* during interval *k*. By using $\pi^{\bar{r}\bar{s}}(k) = \min_p\left\{c_p^{\bar{r}\bar{s}*}(k)\right\}$, we obtain:

$$\pi^{\bar{r}\bar{s}}(k) \ge \left(D^{\bar{r}\bar{s}*}(k)\right)^{-1} \tag{7.26}$$

Equation (7.24) and inequality (7.26) together imply expression (7.2). This completes the proof.

### 7.1.2 Nested Diagonalization Method

In this section, we first show that by an appropriate network representation, the DUO variable demand/route choice problem can be treated, in the mathematical sense, as the DUO route choice problem. Then, the nested diagonalization method is formally stated.

#### 7.1.2.1 Time-space network

In the DUO variable demand/route choice model, an O-D demand is defined by a function of minimal actual route travel time, rather than a fixed number, as was in the DUO route choice model. However, after introducing a time-dependent maximal number of departure flows for each O-D pair by allowing the *excess* O-D demand to flow from the time-dependent origin directly to the time-independent destination, the DUO variable demand/route choice model can then be deemed as the DUO route choice model. In other words, while all of the other constraints remain unchanged, variational inequality (7.4) and flow conservation constraint (7.5) can be replaced, respectively, as:

$$\sum_a \sum_t c_a^*(t)\left[u_a(t) - u_a^*(t)\right] + \sum_{rs} \sum_k \left(D^{rs*}(k)\right)^{-1}\left[e^{rs}(k) - e^{rs*}(k)\right] \ge 0 \tag{7.27}$$

$$\forall (\mathbf{u}, \mathbf{q}) \in \Omega^*$$

$$\sum_p h_p^{rs}(k) + e^{rs}(k) = \bar{q}_{\max}^{rs}(k) \quad \forall r, s, k \tag{7.28}$$

This concept can be illustrated by a time-space network representation. In Figure 7.1(a), we first assume a two-link three-node static network with an additional dummy link transporting the *excess* O-D demand from origin $r$ to destination $s$. In Figure 7.1(b), the corresponding time-space network is constructed with two additional types of artificial time-space links. The first type of artificial time-space links connects each time-dependent destination $s(k)$ to the time-independent destination $s$. The corresponding travel time function is defined as $c^{ss}(k)=0$. The second type of time-space links directly connects each time-dependent origin $r(k)$ to the time-independent destination $s$. The associated flow rate is realized as the *excess* O-D demand, i.e., $e^{rs}(k)=\bar{q}^{rs}_{\max}(k)-q^{rs}(k)$, and the corresponding travel time function is defined as $c^{rs}(k)=\left(D^{rs}(k)\right)^{-1}$. The following theorem states that the DUO variable demand/route choice problem is equivalent to, in the mathematical sense, the DUO route choice problem.

***Theorem 7.3***: The DUO variable demand/route choice problem with the time-dependent maximal departure rate by $\bar{q}^{rs}_{\max}(k)$ is equivalent to the DUO route choice problem with the time-space network shown in Figure 7.1(b).

***Proof***: For the DUO route choice problem with the time-space network shown in Figure 7.1(b), the corresponding VIP is as follows:

$$\sum_a\sum_t c_a^*(t)\left[u_a(t)-u_a^*(t)\right] + \sum_{rs}\sum_k\left(D^{rs*}(k)\right)^{-1}\left[e^{rs}(k)-e^{rs*}(k)\right]\geq 0 \quad \forall(\mathbf{u},\mathbf{e})\in\Omega^* \tag{7.29}$$

Since $e^{rs}(k)=\bar{q}^{rs}_{\max}(k)-q^{rs}(k)$, it follows:

$$\sum_a\sum_t c_a^*(t)\left[u_a(t)-u_a^*(t)\right]+\sum_{rs}\sum_k\left(D^{rs*}(k)\right)^{-1} \left[\left(\bar{q}^{rs}_{\max}(k)-q^{rs}(k)\right)-\left(\bar{q}^{rs}_{\max}(k)-q^{rs*}(k)\right)\right]\geq 0 \quad \forall(\mathbf{u},\mathbf{q})\in\Omega^* \tag{7.30}$$

Eliminating $\bar{q}^{rs}_{\max}(k)$ yields:

$$\sum_a\sum_t c_a^*(t)\left[u_a(t)-u_a^*(t)\right] - \sum_{rs}\sum_k\left(D^{rs*}(k)\right)^{-1}\left[q^{rs}(k)-q^{rs*}(k)\right]\geq 0 \quad \forall(\mathbf{u},\mathbf{q})\in\Omega^* \tag{7.31}$$

Inequality (7.31) is the exact VIP for the DUO variable demand/route choice problem as shown in expression (7.4). The steps shown above are essentially reversible. This completes the proof.

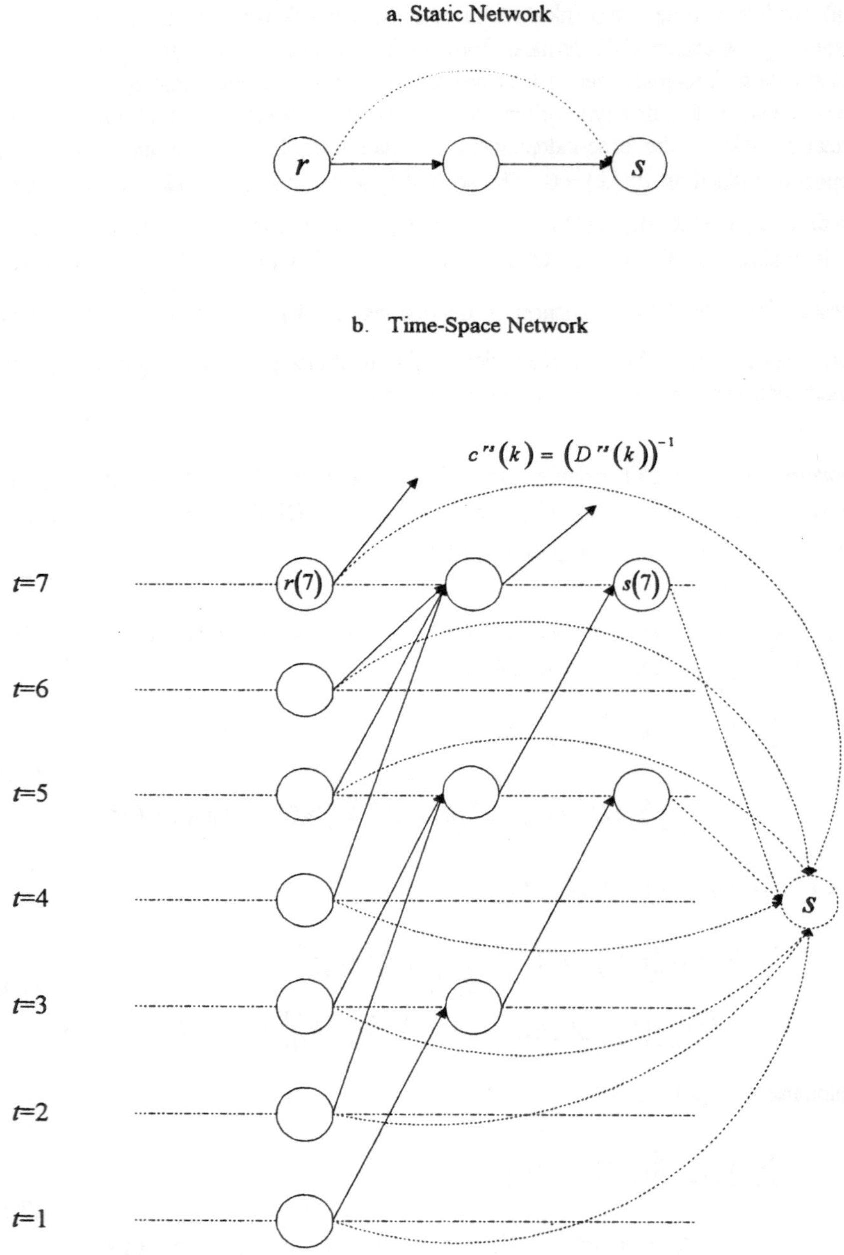


Figure 7.1: Time-Space Network for the Dynamic User-Optimal Variable Demand/Route Choice Model

#### 7.1.2.2 Solution algorithm

We adopt the nested diagonalization method to solve for the DUO variable demand/route choice model.

***Nested Diagonalization Method***

**Step 0: Initialization.**

Step 0.1: Let $m$=0. Set $\tau_a^0(t) = NINT\left[c_{a_0}(t)\right], \forall a, t$.

Step 0.2: Let $n$=1. Find an initial feasible solution $\left\{u_a^1(t), q^{rs}(k)^1\right\}$. Compute the associated link travel times $\left\{c_a^1(t)\right\}$.

**Step 1: *First Loop* Operation.**

Let $m$=$m$+1. Update the estimated actual link travel times by

$$\tau_a^m(t) = NINT\left[(1-\gamma)\tau_a^{m-1}(t) + \gamma c_a^n(t)\right] \quad \forall a, t \tag{7.32}$$

where $0 < \gamma \le 1$.

Construct the corresponding feasible time-space network based on the estimated actual link travel times.

**Step 2: *Second Loop* Operation.**

Step 2.1: Let $n$=1. Compute and reset the initial feasible solution $\left\{u_a^n(t), q^{rs}(k)^n\right\}$, based on the time-space network constructed by the estimated actual link travel times $\left\{\tau_a^m(t)\right\}$.

Step 2.2: Fix the inflows for all time-space links other than on the subject time-space link at the current level, yielding the following optimization problem.

$$\begin{aligned} \min\ z(\mathbf{u}, \mathbf{q}) = &\sum_a \sum_t \int_0^{u_a^{n+1}(t)} c_a\left(u_a^n(1), u_a^n(2), \cdots, u_a^n(t-1), \omega\right) d\omega \\ &- \sum_{rs} \sum_k \int_0^{q^{rs}(k)^{n+1}} \left(D^{rs}(\omega)\right)^{-1} d\omega \end{aligned} \tag{7.33}$$

Flow conservation constraint:

$$\sum_p h_p^{rs}(k) + e^{rs}(k) = \bar{q}_{\max}^{rs}(k) \quad \forall r, s, k \tag{7.34}$$

Nonnegativity constraints:

$$h_p^{rs}(k) \ge 0 \quad \forall r, s, p, k \tag{7.35}$$

$$e^{rs}(k) \ge 0 \quad \forall r, s, k \tag{7.36}$$

Definitional constraints:

$$u_{apk}^{rs}(t) = h_p^{rs}(k)\bar{\delta}_{apk}^{rs}(t) \quad \forall r,s,a,p,k,t \tag{7.37}$$

$$\bar{\delta}_{apk}^{rs}(t) = \{0,1\} \quad \forall r,s,a,p,k,t \tag{7.38}$$

$$\sum_p h_p^{rs}(k) = q^{rs}(k) \quad \forall r,s,k \tag{7.39}$$

$$u_a(t) = \sum_{rs}\sum_p\sum_k h_p^{rs}(k)\bar{\delta}_{apk}^{rs}(t) \quad \forall a,t \tag{7.40}$$

$$c_p^{rs}(k) = \sum_a\sum_t c_a(t)\bar{\delta}_{apt}^{rs}(t) \quad \forall r,s,p,k \tag{7.41}$$

**Step 3: *Third Loop* Operation.**

Solve for the solution, $\{u_a^{n+1}(t), q^{rs}(k)^{n+1}\}$, in optimization problem (7.33)~(7.41) by the FW method. Compute the resulting link travel times $\{c_a^{n+1}(t)\}$.

**Step 4: Convergence Check for the *Second Loop* Operation.**

If $u_a^{n+1}(t) \approx u_a^n(t), \forall a,t$ and $q^{rs}(k)^{n+1} \approx q^{rs}(k)^n, \forall r,s,k$, go to Step 5; otherwise, set $n=n+1$, go to Step 2.2.

**Step 5: Convergence Check for the *First Loop* Operation.**

If $\tau_a^m(t) \approx c_a^{n+1}(t), \forall a,t$, stop; the current solution is optimal. Otherwise, set $n=n+1$, and go to Step 1.

Analogous to the discussion in Section 5.3, the nested projection algorithm results from the following modification:

Step 2.2: Fix the inflows for all time-space links other than on the subject time-space link at the current level, yielding the following optimization problem:

$$\begin{aligned}\min\ z(\mathbf{u},\mathbf{q}) = {} & \frac{1}{2}\left(\mathbf{u}^{n+1}\right)^T \mathbf{G}_1\mathbf{u}^{n+1} + \left(\rho_1\mathbf{c}\left(\mathbf{u}^n\right) - \mathbf{G}_1\mathbf{u}^n\right)\mathbf{u}^{n+1} \\ & -\frac{1}{2}\left(\mathbf{q}^{n+1}\right)^T \mathbf{G}_2\mathbf{q}^{n+1} - \left(\rho_2\left(\mathbf{D}\left(\mathbf{q}^n\right)\right)^{-1} - \mathbf{G}_2\mathbf{q}^n\right)\mathbf{q}^{n+1}\end{aligned} \tag{7.42}$$

Subject to: Equations (7.34)~(7.41)

where matrices $\mathbf{G}_1$ and $\mathbf{G}_2$ are symmetric and positive definite, and $\rho_1$ and $\rho_2$ are contraction operators.

### 7.1.3 Numerical Examples

#### 7.1.3.1 Input data

A simple network shown in Figure 7.2 is used for testing. The test network consists of 8 links (including artificial links 1→5 and 2→5) and 5 nodes, in which nodes 1 and 2 are the origins, node 5 is the destination, and nodes 3 and 4 are intermediate nodes.

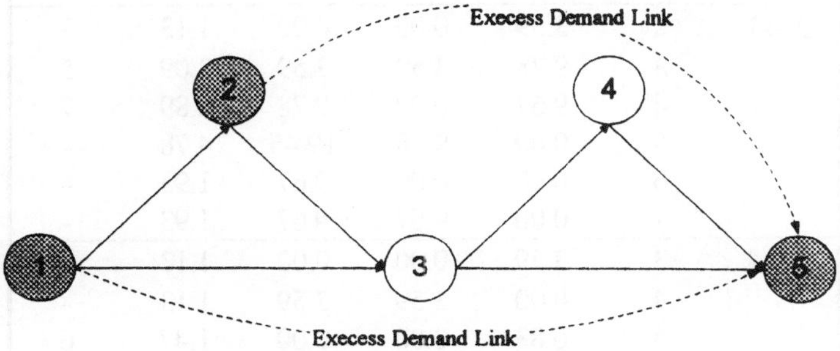


Figure 7.2: Test Network 1

The adopted dynamic travel time functions are arbitrarily constructed as follows:

$$c_a(t) = 1 + 0.01(u_a(t))^2 + 0.01(x_a(t))^2 \quad \forall a,\ t \tag{7.43}$$

The demand functions are assumed as follows:

$$q^{rs}(k) = 10(6 - \pi^{rs}(k)) \quad \forall r, s, k \tag{7.44}$$

The time-dependent maximal origin-destination (O-D) demands are assumed in Table 7.1:

Table 7.1: Time-Dependent Maximal O-D Demands for Test Network 1

| O-D | Time Interval | | | |
|---|---|---|---|---|
| Pair | *k*=1 | *k*=2 | *k*=3 | *k*=4 |
| 1-5 | 25 | 30 | - | - |
| 2-5 | - | - | 25 | 30 |

#### 7.1.3.2 Test results

The resulting flow pattern and time-dependent O-D demands for the DUO variable demand/route choice model with the given data are summarized in Tables 7.2 and 7.3, respectively.

Table 7.2: Results for Test Network 1

| Link | Entering Time Interval | Inflow | Exit Flow | Number of Vehicles | Link Travel Time | Exiting Time Interval |
|---|---|---|---|---|---|---|
| 1→2 | 1 | 3.59 | 0.00 | 0.00 | 1.13 | 2 |
| | 2 | 0.00 | 3.59 | 3.59 | 1.13 | - |
| 1→3 | 1 | 11.22 | 0.00 | 0.00 | 2.26 | 3 |
| | 2 | 5.28 | 0.00 | 11.22 | 2.54 | 5 |
| | 3 | 0.00 | 11.22 | 16.51 | 3.72 | - |
| | 4 | 0.00 | 0.00 | 5.28 | 1.28 | - |
| | 5 | 0.00 | 5.28 | 5.28 | 1.28 | - |
| 2→3 | 2 | 3.59 | 0.00 | 0.00 | 1.13 | 3 |
| | 3 | 9.78 | 3.59 | 3.59 | 2.09 | 5 |
| | 4 | 9.67 | 0.00 | 9.78 | 2.89 | 7 |
| | 5 | 0.00 | 9.78 | 19.45 | 4.78 | - |
| | 6 | 0.00 | 0.00 | 9.67 | 1.93 | - |
| | 7 | 0.00 | 9.67 | 9.67 | 1.93 | - |
| 3→4 | 3 | 3.59 | 0.00 | 0.00 | 1.13 | 4 |
| | 4 | 0.00 | 3.59 | 3.59 | 1.13 | - |
| | 5 | 6.83 | 0.00 | 0.00 | 1.47 | 6 |
| | 6 | 0.00 | 6.83 | 6.83 | 1.47 | - |
| | 7 | 2.84 | 0.00 | 0.00 | 1.08 | 8 |
| | 8 | 0.00 | 2.84 | 2.84 | 1.08 | - |
| 3→5 | 3 | 11.22 | 0.00 | 0.00 | 2.26 | 5 |
| | 4 | 0.00 | 0.00 | 11.22 | 2.26 | - |
| | 5 | 8.23 | 11.22 | 11.22 | 2.94 | 8 |
| | 6 | 0.00 | 0.00 | 8.23 | 1.68 | - |
| | 7 | 6.82 | 0.00 | 8.23 | 2.14 | 9 |
| | 8 | 0.00 | 8.23 | 15.05 | 3.27 | - |
| | 9 | 0.00 | 6.82 | 6.82 | 1.47 | - |
| 4→5 | 4 | 3.59 | 0.00 | 0.00 | 1.13 | 5 |
| | 5 | 0.00 | 3.59 | 3.59 | 1.13 | - |
| | 6 | 6.83 | 0.00 | 0.00 | 1.47 | 7 |
| | 7 | 0.00 | 6.83 | 6.83 | 1.47 | - |
| | 8 | 2.84 | 0.00 | 0.00 | 1.08 | 9 |
| | 9 | 0.00 | 2.84 | 2.84 | 1.08 | - |

Table 7.3: Resulting Time-Dependent O-D Demands

| O-D Pair | Time Interval | | | |
|---|---|---|---|---|
| | $k$=1 | $k$=2 | $k$=3 | $k$=4 |
| 1-5 | 14.82 (10.18)* | 5.28 (24.72) | - | - |
| 2-5 | - | - | 9.78 (15.22) | 9.67 (20.33) |

* denotes excess flow

The corresponding actual route travel times are computed and summarized in Table 7.4.

Table 7 4: Actual Route Travel Times for Test Network 1

| Route | Time Interval | | | |
|---|---|---|---|---|
| | $k$=1 | $k$=2 | $k$=3 | $k$=4 |
| 1→2→3→4→5 | 4.52 (x)* | – | – | – |
| 1→2→3→5 | 4.52 (3.59-x)* | – | – | – |
| 1→3→4→5 | 4.52 (3.59-x)* | 5.48 (y) | – | – |
| 1→3→5 | 4.52 (7.61+x) | 5.48 (5.28-y) | – | – |
| 1→5 (Excess Demand) | 4.52 (10.08) | 5.47 (24.72) | – | – |
| 2→3→4→5 | NA | NA | 5.03 (6.83-y) | 5.05 (2.84) |
| 2→3→5 | NA | NA | 5.03 (2.95+y) | 5.03 (6.82) |
| 2→5 (Excess Demand) | NA | NA | 5.02 (15.22) | 5.03 (20.36) |

* route flow, where x≤3.59 and y≤ 5.28

# 7.2 Dynamic User-Optimal Variable Demand/Departure Time/Route Choice Model

In the following section, we discuss the equilibrium conditions and model formulation, the nested diagonalization method, and a numerical example for the dynamic user-optimal variable demand/departure time/route choice model.

## 7.2.1 Equilibrium Conditions and Model Formulation

The dynamic user-optimal conditions are first defined to characterize the travelers' driving behavior for choosing the best departure time using the minimal travel time route. Once again, due to inherent link interactions, the variational inequality approach is adopted to formulate the DUO variable demand/departure time/route choice model. The equivalence between the dynamic user-optimal conditions and the variational inequality formulation is then stated by a theorem and verified by a proof.

#### 7.2.1.1 Dynamic user-optimal conditions

When the O-D demands are variable and time-independent, the corresponding dynamic user-optimal conditions may be characterized by equilibrium conditions both on route choice behavior and on trip demand functions. For the route choice behavior, they state for each O-D pair that the actual route travel times experienced by travelers regardless of departure time are equal and minimal; no traveler would be better off by unilaterally changing his/her route. In contrast, the actual route travel time of any unused route for each O-D pair is greater than or equal to the minimal actual route travel time. Therefore, for each O-D pair *rs*, if the flow over route *p* during interval *k* is positive, i.e., $h_p^{rs*}(k) > 0$, then the corresponding actual route travel time is minimal. However, if no flow occurs on route *p*, i.e., $h_p^{rs*}(k) = 0$, then the corresponding actual route travel time is at least as great as the minimal actual route travel time. This equilibrium conditions can be mathematically expressed as follows:

$$c_p^{rs*}(k)\begin{cases} = \pi^{rs} & \text{if } h_p^{rs*}(k) > 0 \\ \geq \pi^{rs} & \text{if } h_p^{rs*}(k) = 0 \end{cases} \quad \forall r,s,p,k \tag{7.45}$$

For the trip demand functions, the conditions state for each O-D pair that if the departure flow rate during the entire analysis period is positive, i.e., $q^{rs*} > 0$, then the corresponding minimal route travel time is equal to the inverse demand function $\left(D^{rs*}\right)^{-1}$. On the other hand, if no traffic demand is induced during the entire analysis period, i.e., $q^{rs*} = 0$, then the corresponding minimal route travel time is at least as great as the inverse demand function $\left(D^{rs*}\right)^{-1}$. This equilibrium conditions can be mathematically expressed as follows:

$$\pi^{rs}\begin{cases} = \left(D^{rs*}\right)^{-1} & \text{if } q^{rs*} > 0 \\ \geq \left(D^{rs*}\right)^{-1} & \text{if } q^{rs*} = 0 \end{cases} \quad \forall r,s \tag{7.46}$$

#### 7.2.1.2 Variational inequality formulation

The DUO variable demand/departure time/route choice problem can be formulated using the variational inequality approach.

***Theorem** 7.4*: The DUO variable demand/departure time/route choice problem is equivalent to finding a solution $(\mathbf{u}^*, \mathbf{q}^*) \in \Omega$ such that the following VIP holds:

$$\mathbf{c}^*(\mathbf{u}-\mathbf{u}^*) - \mathbf{D}^{-1^*}(\mathbf{q}-\mathbf{q}^*) \geq 0 \quad \forall(\mathbf{u},\mathbf{q}) \in \Omega^* \tag{7.47}$$

Or, alternatively, in expanded form:

$$\sum_a \sum_t c_a^*(t)\left[u_a(t) - u_a^*(t)\right] - \sum_{rs} \left(D^{rs*}\right)^{-1}\left[q^{rs} - q^{rs*}\right] \geq 0 \tag{7.48}$$
$$\forall(\mathbf{u},\mathbf{q}) \in \Omega^*$$

where $\Omega^*$ is a subset of $\Omega$ with $\delta_{apk}^{rs}(t)$ being realized at equilibrium, i.e., $\delta_{apk}^{rs}(t) = \delta_{apk}^{rs*}(t), \forall r,s,a,p,k,t$. The symbol $\Omega$ denotes the feasible region that is delineated by the following constraints:

Flow conservation constraint:

$$\sum_k \sum_p h_p^{rs}(k) = q^{rs} \quad \forall r,s \tag{7.49}$$

Flow propagation constraints:

$$u_{apk}^{rs}(t) = h_p^{rs}(k)\delta_{apk}^{rs}(t) \quad \forall r,s,a,p,k,t \tag{7.50}$$

$$\sum_t \delta_{apk}^{rs}(t) = 1 \quad \forall r,s,p,a \in p,k \tag{7.51}$$

$$\delta_{apk}^{rs}(t) = \{0,1\} \quad \forall r,s,a,p,k,t \tag{7.52}$$

Nonnegativity constraint:

$$h_p^{rs}(k) \geq 0 \quad \forall r,s,p,k \tag{7.53}$$

Definitional constraints:

$$u_a(t) = \sum_{rs} \sum_p \sum_k h_p^{rs}(k)\delta_{apk}^{rs}(t) \quad \forall a,t \tag{7.54}$$

$$c_p^{rs}(k) = \sum_a \sum_t c_a(t)\delta_{apk}^{rs}(t) \quad \forall r,s,p,k \tag{7.55}$$

The constraint set for the DUO variable demand/departure time/route choice model is almost the same as that for the DUO departure time/route choice model except for equation (7.49). Equation (7.49) expresses the time-independent O-D demand in terms of route flows. Note that the time-independent O-D demand shown in equation (7.49) is defined as a variable, rather than a fixed number.

#### 7.2.1.3 Equivalence analysis

***Theorem 7.5***: Under a certain flow propagation relationship $\left(\delta_{apk}^{rs}(t) = \delta_{apk}^{rs*}(t)\right)$, DUO variable demand/departure time/route choice conditions (7.45)~(7.46) imply VIP (7.48) and vice versa.

***Proof of necessity***: We need to prove that under a certain flow propagation relationship $\left(\delta_{apk}^{rs}(t)=\delta_{apk}^{rs*}(t)\right)$, DUO variable demand/departure time/route choice conditions (7.45)~(7.46) imply VIP (7.48). We first rearrange equilibrium conditions (7.45)~(7.46) as follows:

$$\left[c_p^{rs*}(k)-\pi^{rs}\right]\left[h_p^{rs}(k)-h_p^{rs*}(k)\right]\geq 0 \quad \forall r,s,p,k \tag{7.56}$$

$$\left[\pi^{rs}-\left(D^{rs*}\right)^{-1}\right]\left[q^{rs}-q^{rs*}\right]\geq 0 \quad \forall r,s \tag{7.57}$$

By summing equation (7.56) over $r,s,p,k$ and equation (7.57) over $r,s$, and then making a substitution of $\sum_p\sum_k h_p^{rs}(k)=q^{rs}$, we obtain:

$$\sum_{rs}\sum_k\sum_p c_p^{rs*}(k)\left[h_p^{rs}(k)-h_p^{rs*}(k)\right]-\sum_{rs}\left(D^{rs*}\right)^{-1}\left[q^{rs}-q^{rs*}\right]\geq 0 \tag{7.58}$$

By applying equation (7.55), we have:

$$\sum_{rs}\sum_k\sum_p\left[\sum_a\sum_t c_a^*(t)\delta_{apk}^{rs}(t)\right]\left[h_p^{rs}(k)-h_p^{rs*}(k)\right]$$
$$-\sum_{rs}\left(D^{rs*}\right)^{-1}\left[q^{rs}-q^{rs*}\right]\geq 0 \tag{7.59}$$

By changing the order of the summation on the first term, using equation (7.54), we obtain:

$$\sum_a\sum_t c_a^*(t)\left[u_a(t)-u_a^*(t)\right]-\sum_{rs}\left(D^{rs*}\right)^{-1}\left[q^{rs}-q^{rs*}\right]\geq 0 \tag{7.60}$$

Inequality (7.60) is identical to expression (7.48).

***Proof of sufficiency***: We show next VIP (7.48) implies DUO variable demand/route choice conditions (7.45)~(7.46). Note that equations (7.58)~(7.60) are essentially reversible. The remaining steps of the proof are to show that expression (7.58) is equivalent to the DUO variable demand/route choice conditions. First we demonstrate that VIP (7.58) implies dynamic user-optimal conditions (7.45). We then define a feasible solution $\left\{h_p^{rs}(k),\ q^{rs*}\right\}$ to be the same as $\left\{h_p^{rs*}(k),\ q^{rs*}\right\}$ except for two routes $p_1^{rs}$ during interval $k_1$ and $p_2^{rs}$ during interval $k_2$. We consider two situations that could arise at equilibrium:

(i) Both route flows are positive, i.e., $h_{p_1}^{rs*}(k_1)>0$ and $h_{p_2}^{rs*}(k_2)>0$. We switch a small amount of flow $\Delta_1$ from route $p_1^{rs}$ to $p_2^{rs}$ with $0<\Delta_1\leq h_{p_1}^{rs*}(k_1)$. That is:

$$h_{p_1}^{rs}(k_1)=h_{p_1}^{rs*}(k_1)-\Delta_1 \quad \text{and} \tag{7.61}$$

$$h_{p_2}^{rs}(k_2)=h_{p_2}^{rs*}(k_2)+\Delta_1 \tag{7.62}$$

Substituting the feasible solution $\{h_p^{rs}(k),\ q^{rs*}\}$ into equation (7.58) yields:

$$c_{p_1}^{rs*}(k_1)[h_{p_1}^{rs}(k_1)-h_{p_1}^{rs*}(k_1)]+c_{p_2}^{rs*}(k_2)[h_{p_2}^{rs}(k_2)-h_{p_2}^{rs*}(k_2)]\geq 0 \qquad (7.63)$$

By applying equation (7.61) and equation (7.62), we have:

$$c_{p_2}^{rs*}(k_2)\geq c_{p_1}^{rs*}(k_1) \qquad (7.64)$$

Similarly, by switching a small amount of flow $\Delta_2$ with $0<\Delta_2\leq h_{p_2}^{rs*}(k_2)$ from route $p_2^{rs}$ to $p_1^{rs}$, one obtains:

$$c_{p_1}^{rs*}(k_1)\geq c_{p_2}^{rs*}(k_2) \qquad (7.65)$$

Equations (7.64) and (7.65) together implies:

$$c_{p_2}^{rs*}(k_2)=c_{p_1}^{rs*}(k_1) \qquad (7.66)$$

We repeat this procedure to verify for each O-D pair that all used routes with a positive flow have the same actual route travel time.

(ii) One route flow is positive and the other route flow is nil. We arbitrarily assume, without loss of generality, $h_{p_1}^{rs*}(k_1)>0$ and $h_{p_2}^{rs*}(k_2)=0$. We switch a small amount of flow $\Delta_1$ from route $p_1^{rs}$ to $p_2^{rs}$ with $0<\Delta_1\leq h_{p_1}^{rs*}(k_1)$. By the same argument shown in (i), we have $c_{p_2}^{rs*}(k_2)\geq c_{p_1}^{rs*}(k_1)$. We repeat this procedure to verify for each O-D pair that all unused routes with zero flow have higher or equal actual route travel times. By cases (i) and (ii), using expression (7.1), it follows that VIP (7.58) implies expression (7.45).

We conclude the proof by showing that VIP (7.58) implies expression (7.46). We define now a feasible solution $\{h_p^{rs}(k),\ q^{rs}\}$ be the same as $\{h_p^{rs*}(k),\ q^{rs*}\}$ except for one O-D demand $q^{\bar{r}\bar{s}}$ and one route flow $h_p^{\bar{r}\bar{s}}(k)$, for which $q^{\bar{r}\bar{s}}=q^{\bar{r}\bar{s}*}-\Delta$, $h_p^{\bar{r}\bar{s}}(k)=h_p^{\bar{r}\bar{s}*}(k)-\Delta$, where $0<|\Delta|\leq h_p^{\bar{r}\bar{s}*}(k)$. Substituting this feasible solution into VIP (7.58), one obtains:

$$\left[c_p^{\bar{r}\bar{s}*}(k)-\left(D^{\bar{r}\bar{s}*}\right)^{-1}\right]\Delta\leq 0 \qquad (7.67)$$

For $q^{\bar{r}\bar{s}*}>0$, we can find a route $p$ such that $h_p^{\bar{r}\bar{s}*}(k)>0$. Since the value of $\Delta$ could be either positive or negative, the equality must therefore hold. Applying expression (7.45) and dividing by $\Delta$ implies:

$$\pi^{\bar{r}\bar{s}}=\left(D^{\bar{r}\bar{s}*}\right)^{-1} \qquad (7.68)$$

For $q^{\bar{r}\bar{s}*}=0$ and $h_p^{\bar{r}\bar{s}*}(k)=0$, the value of $\Delta$ can only be negative, which implies:

$$\left[c_p^{\bar{r}\bar{s}*}(k)-\left(D^{\bar{r}\bar{s}*}\right)^{-1}\right]\geq 0 \qquad (7.69)$$

We repeat this procedure for all possible routes between O-D pair *rs* during interval *k*. By using $\pi^{\bar{rs}} = \min_{k,p}\left\{c_p^{\bar{rs}*}(k)\right\}$, we obtain:

$$\pi^{\bar{rs}} \geq \left(D^{\bar{rs}*}\right)^{-1} \tag{7.70}$$

Equation (7.68) and inequality (7.70) together imply expression (7.46). This completes the proof.

### 7.2.2 Nested Diagonalization Method

In this section, we first show that by an appropriate network representation, the DUO variable demand/departure time/route choice problem can be treated, in the mathematical sense, as the DUO route choice problem. Afterwards, the nested diagonalization method is formally stated.

#### 7.2.2.1 Time-space network

In the DUO variable demand/departure time/route choice model, an O-D demand is defined by a function of minimal actual route travel time, rather than a fixed number, as was in the DUO departure time/route choice model. However, after introducing a time-independent maximal number of departure flow rates for each O-D pair by allowing the *excess* O-D demand to flow from the time-independent origin directly to the time-independent destination, the DUO variable demand/departure time/route choice model can then be deemed as the DUO route choice model. In other words, while all of the other constraints remain unchanged, variational inequality (7.48) and flow conservation constraint (7.49) can be replaced, respectively, as:

$$\sum_a \sum_t c_a^*(t)\left[u_a(t) - u_a^*(t)\right] + \sum_{rs}\left(D^{rs*}\right)^{-1}\left(e^{rs} - e^{rs*}\right) \geq 0 \quad \forall(\mathbf{u}, \mathbf{e}) \in \Omega^* \tag{7.71}$$

$$\sum_k \sum_p h_p^{rs}(k) + e^{rs} = \bar{q}_{\max}^{rs} \quad \forall r, s \tag{7.72}$$

This concept can be illustrated by a time-space network representation. In Figure 7.3(a), we first assume a two-link three-node static network with an additional dummy link transporting *excess* O-D demand from origin $r$ to destination $s$. In Figure 7.3(b), the corresponding time-space network is constructed with three additional types of artificial time-space links. The first type of artificial time-space links directly connects each time-independent origin $r$ to the time-independent destination $s$. The associated flow rate is realized as the *excess* O-D demand, i.e., $e^{rs} = \bar{q}_{\max}^{rs} - q^{rs}$, and the corresponding travel time function is defined as $c^{rs} = \left(D^{rs}\right)^{-1}$. The second type of artificial time-space links joins the time-independent origin $r$ to the time-dependent origin $r(k)$. The corresponding travel time function is defined as $c^{rr(k)} = 0$. The third type of artificial time-space links connects each time-dependent destination $s(k)$ to the time-independent destination $s$. The corresponding travel time function is defined as $c^{ss}(k) = 0$.

The following theorem states that the DUO variable demand/departure time/route choice problem can be treated as the DUO route choice problem.

***Theorem 7.6***: The DUO variable demand/departure time/route choice problem with the time-independent maximal departure rate by $\bar{q}^{rs}_{\max}$ is equivalent to the DUO route choice problem with the time-space network representation in Figure 7.3(b).

***Proof***: For the DUO route choice problem with the time-space network in Figure 7.3(b), the corresponding VIP is as follows:

$$\sum_a \sum_t c_a^*(t)\left[u_a(t) - u_a^*(t)\right] + \sum_{rs} \left(D^{rs*}\right)^{-1}\left[e^{rs} - e^{rs*}\right] \geq 0 \quad \forall(\mathbf{u}, \mathbf{e}) \in \Omega^* \tag{7.73}$$

Since $e^{rs} = \bar{q}^{rs}_{\max} - q^{rs}$, it follows:

$$\begin{aligned}&\sum_a \sum_t c_a^*(t)\left[u_a(t) - u_a^*(t)\right] \\ &\quad + \sum_{rs} \left(D^{rs*}\right)^{-1}\left[\left(\bar{q}^{rs}_{\max} - q^{rs}\right) - \left(\bar{q}^{rs}_{\max} - q^{rs*}\right)\right] \geq 0 \quad \forall(\mathbf{u}, \mathbf{q}) \in \Omega^*\end{aligned} \tag{7.74}$$

Eliminating $\bar{q}^{rs}_{\max}$ yields

$$\begin{aligned}&\sum_a \sum_t c_a^*(t)\left[u_a(t) - u_a^*(t)\right] \\ &\quad - \sum_{rs} \left(D^{rs*}\right)^{-1}\left[q^{rs} - q^{rs*}\right] \geq 0 \quad \forall(\mathbf{u}, \mathbf{q}) \in \Omega^*\end{aligned} \tag{7.75}$$

Inequality (7.75) is the exact VI formulation for the DUO variable demand/departure time/route choice problem as shown in expression (7.48). The steps from expression (7.73) to (7.75) are essentially reversible. This completes the proof.

#### 7.2.2.2 Solution algorithm

The nested diagonalization method is adopted to solve the DUO variable demand/departure time/route choice model as follows.

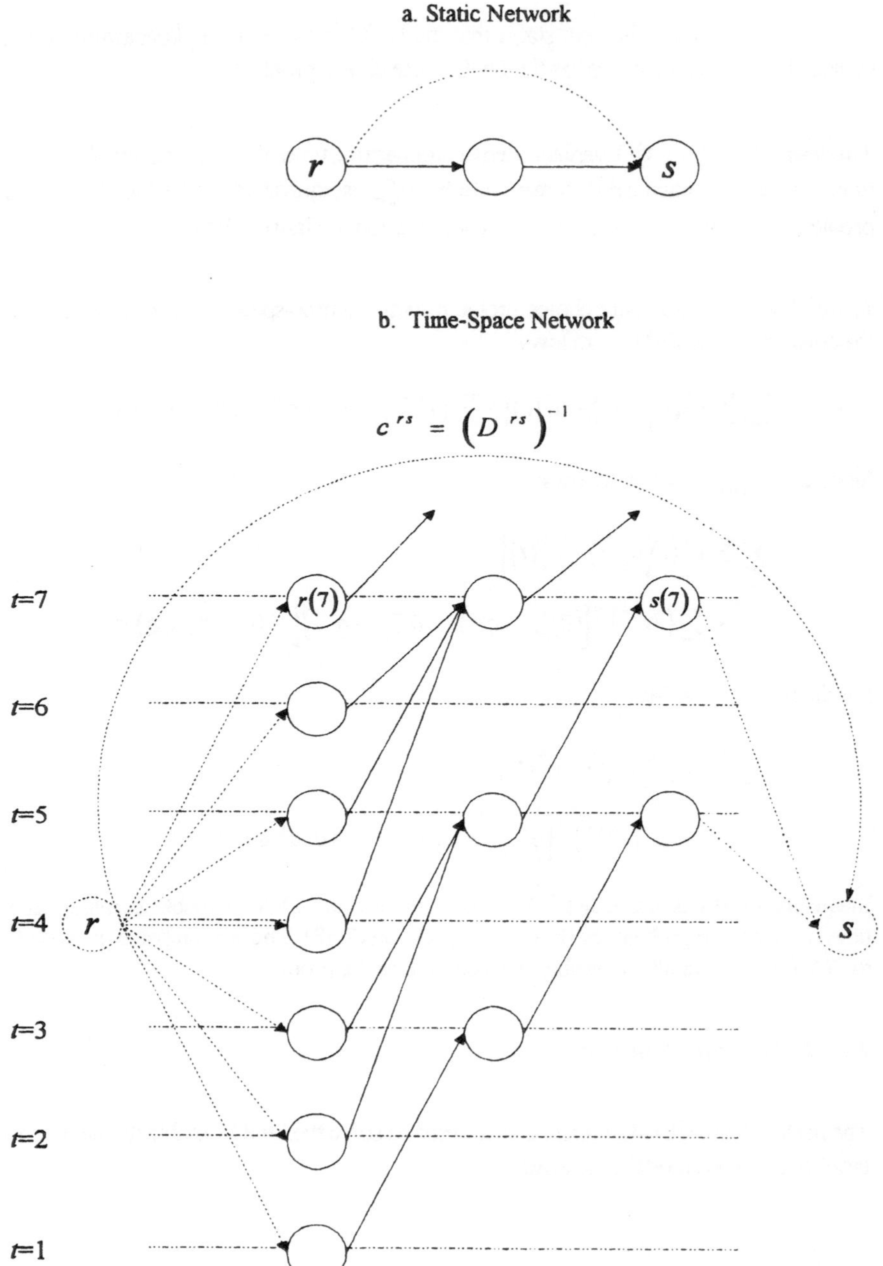


Figure 7.3: Time-Space Network for the Dynamic User-Optimal Variable Demand /Departure Time/Route Choice Model

***Nested Diagonalization Method***

**Step 0: Initialization.**

Step 0.1: Let $m$=0. Set $\tau_a^0(t) = NINT\left[c_{a_0}(t)\right], \forall a, t$.

Step 0.2: Let $n$=1. Find an initial feasible solution $\left\{u_a^1(t), \left(q^{rs}\right)^1\right\}$. Compute the associated link travel times $\left\{c_a^1(t)\right\}$.

**Step 1: *First Loop* Operation.**

Let $m$=$m$+1. Update the estimated actual link travel times by

$$\tau_a^m(t) = NINT\left[(1-\gamma)\tau_a^{m-1}(t) + \gamma c_a^n(t)\right] \quad \forall a, t \tag{7.76}$$

where $0 < \gamma \leq 1$.

Construct the corresponding feasible time-space network based on the estimated actual link travel times.

**Step 2: *Second Loop* Operation.**

Step 2.1: Let $n$=1. Compute and reset the initial feasible solution $\left(u_a^n(t), \left(q^{rs}\right)^n\right)$, based on the time-space network constructed by the estimated actual link travel times $\left\{\tau_a^m(t)\right\}$.

Step 2.2: Fix the inflows for all time-space links other than on the subject time-space link at the current level, yielding the following optimization problem.

$$\begin{aligned} \min\ z(\mathbf{u}, \mathbf{q}) = & \sum_a \sum_t \int_0^{u_a^{n+1}(t)} c_a\left(u_a^n(1), u_a^n(2), \cdots, u_a^n(t-1), \omega\right) d\omega \\ & - \sum_r \sum_s \int_0^{\left(q^{rs}\right)^{n+1}} \left(D^{rs}(\omega)\right)^{-1} d\omega \end{aligned} \tag{7.77}$$

Flow conservation constraint:

$$\sum_k \sum_p h_p^{rs}(k) + e^{rs} = \bar{q}_{\max}^{rs} \quad \forall r, s \tag{7.78}$$

Nonnegativity constraints:

$$h_p^{rs}(k) \geq 0 \quad \forall r, s, p, k \tag{7.79}$$

$$e^{rs} \geq 0 \quad \forall r, s \tag{7.80}$$

Definitional constraints:

$$u_{apk}^{rs}(t) = h_p^{rs}(k)\bar{\delta}_{apk}^{rs}(t) \quad \forall r, s, a, p, k, t \tag{7.81}$$

$$\bar{\delta}_{apk}^{rs}(t) = \{0,1\} \quad \forall r,s,a,p,k,t \tag{7.82}$$

$$\sum_k \sum_p h_p^{rs}(k) = q^{rs} \quad \forall r,s \tag{7.83}$$

$$u_a(t) = \sum_{rs} \sum_p \sum_k h_p^{rs}(k) \bar{\delta}_{apk}^{rs}(t) \quad \forall a,t \tag{7.84}$$

$$c_p^{rs}(k) = \sum_a \sum_t c_a(t) \bar{\delta}_{apt}^{rs}(t) \quad \forall r,s,p,k \tag{7.85}$$

**Step 3: *Third Loop* Operation.**

Solve for the solution, $\left\{u_a^{n+1}(t), \left(q^{rs}\right)^{n+1}\right\}$, in optimization problem (7.77)~(7.85) by the FW method. Compute the resulting link travel times $\left\{c_a^{n+1}(t)\right\}$.

**Step 4: Convergence Check for the *Second Loop* Operation.**

If $u_a^{n+1}(t) \approx u_a^n(t), \forall a,t$ and $\left(q^{rs}\right)^{n+1} \approx \left(q^{rs}\right)^n, \forall r,s,k$, go to Step 5; otherwise, set $n=n+1$, go to Step 2.2.

**Step 5: Convergence Check for the *First Loop* Operation.**

If $\tau_a^m(t) \approx c_a^{n+1}(t), \forall a,t$, stop; the current solution is optimal. Otherwise, set $n=n+1$, and go to Step 1.

Analogous to the discussion shown in Section 5.3, the nested projection algorithm results from the following modification.

Step 2.2: Fix the inflows for all time-space links other than that on the subject time-space link at the current level, yielding the following optimization problem:

$$\begin{aligned} \min\ z(\mathbf{u},\mathbf{q}) = &\frac{1}{2}\left(\mathbf{u}^{n+1}\right)^T \mathbf{G}_1 \mathbf{u}^{n+1} + \left(\rho_1 \mathbf{c}\left(\mathbf{u}^n\right) - \mathbf{G}_1 \mathbf{u}^n\right)\mathbf{u}^{n+1} \\ &- \frac{1}{2}\left(\mathbf{q}^{n+1}\right)^T \mathbf{G}_2 \mathbf{q}^{n+1} - \left(\rho_2 \left(\mathbf{D}(\mathbf{q}^n)\right)^{-1} - \mathbf{G}_2 \mathbf{q}^n\right)\mathbf{q}^{n+1} \end{aligned} \tag{7.86}$$

Subject to: Equations (7.78)~(7.85)

where matrices $\mathbf{G}_1$ and $\mathbf{G}_2$ are symmetric and positive definite, and $\rho_1$ and $\rho_2$ are contraction operators.

### 7.2.3 Numerical Example

#### 7.2.3.1 Input data

A simple network shown in Figure 7.4 is used for testing. The test network consists of 8 links (including artificial links 1→5 and 2→5) and 5 nodes, in which nodes 1 and 2 are the origins, node 5 is the destination, and nodes 3 and 4 are intermediate nodes.

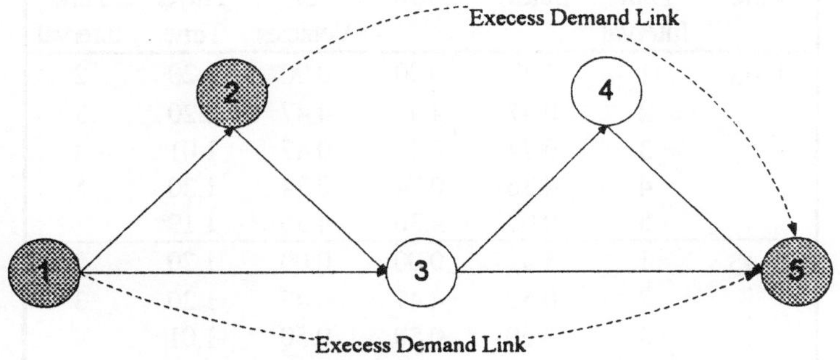


Figure 7.4: Test Network 2

The adopted dynamic travel time function is arbitrarily constructed as follows:

$$c_a(t) = 1 + 0.01(u_a(t))^2 + 0.01(x_a(t))^2 \quad \forall a,\ t \tag{7.87}$$

The demand functions are assumed as follows:

$$q^{rs} = 10(4 - \pi^{rs}) \quad \forall r, s \tag{7.88}$$

The assumed time-independent origin-destination (O-D) demands are given in Table 7.5:

Table 7.5: Time-Independent Maximal O-D Demands for Test Network 2

| O-D | Time Interval |
|---|---|
| Pair | $k$=1~3 |
| 1-5 | 30 |
| 2-5 | 20 |

#### 7.2.3.2 Test results

The resulting flow pattern and time-dependent O-D demands for the DUO variable demand/departure time/route choice model with the given data are summarized in Tables 7.6 and 7.7, respectively.

Table 7.6: Results for Test Network 2

| Link | Entering Time Interval | Inflow | Exit Flow | Number of Vehicles | Link Travel Time | Exiting Time Interval |
|---|---|---|---|---|---|---|
| 1→3 | 1 | 4.47 | 0.00 | 0.00 | 1.20 | 2 |
| | 2 | 0.47 | 4.47 | 4.47 | 1.20 | 3 |
| | 3 | 0.74 | 0.47 | 0.47 | 1.01 | 4 |
| | 4 | 4.36 | 0.74 | 0.74 | 1.20 | 5 |
| | 5 | 0.00 | 4.36 | 4.36 | 1.19 | - |
| 2→3 | 1 | 4.47 | 0.00 | 0.00 | 1.20 | 2 |
| | 2 | 0.50 | 4.47 | 4.47 | 1.20 | 3 |
| | 3 | 0.69 | 0.50 | 0.50 | 1.01 | 4 |
| | 4 | 4.37 | 0.69 | 0.69 | 1.20 | 5 |
| | 5 | 0.00 | 4.37 | 4.37 | 1.19 | - |
| 3→4 | 4 | 0.37 | 0.00 | 0.00 | 1.00 | 5 |
| | 5 | 0.00 | 0.37 | 0.37 | 1.00 | - |
| 3→5 | 2 | 8.93 | 0.00 | 0.00 | 1.80 | 4 |
| | 3 | 0.96 | 0.00 | 8.93 | 1.81 | 5 |
| | 4 | 1.05 | 8.93 | 9.90 | 1.99 | 6 |
| | 5 | 8.73 | 0.96 | 2.02 | 1.80 | 7 |
| | 6 | 0.00 | 1.05 | 9.79 | 1.96 | - |
| | 7 | 0.00 | 8.73 | 8.74 | 1.76 | - |
| 4→5 | 5 | 0.37 | 0.00 | 0.00 | 1.00 | 6 |
| | 6 | 0.00 | 0.37 | 0.37 | 1.00 | - |

Table 7.7: Resulting Time-Dependent O-D Demands

| O-D Pair | Time Interval $k$=1~7 |
|---|---|
| 1-5 | 10.03 (19.97)* |
| 2-5 | 10.03 (9.97) |

* denotes excess flow

The corresponding actual route travel times are computed and summarized in Table 7.8.

Table 7.8: Actual Route Travel Times for Test Network 2

| Route | Time Interval | | | |
|---|---|---|---|---|
| | $k$=1 | $k$=2 | $k$=3 | $k$=4 |
| 1→3→4→5 | NA | NA | 3.01 (0.37) | NA |
| 1→3→5 | 3.00 (4.47) | 3.01 (0.47) | 3.00 (0.37) | 3.00 (4.36) |
| 1→5 (Excess Demand) | 3.00 (19.97) | NA | NA | NA |
| 2→3→5 | 3.00 (4.47) | 3.01 (0.5) | 3.00 (0.69) | 3.00 (4.37) |
| 2→3→4→5 | NA | NA | 3.01 | NA |
| 2→5 (Excess Demand) | 3.00 (9.97) | NA | NA | NA |

* route flow

# 7.3 Notes

In this chapter, we presented two types of DUO route choice models with variable demand, i.e., the DUO variable demand/route choice model and the DUO variable demand/departure time/route choice model. These models consider the traffic demand as a function of minimal actual route travel times. One may argue that traffic demand is influenced by travel habits, and don't change over short periods of time; thus the DUO models with variable demand are not applicable. However, while this comment may seem to be true, we argue that in a dynamic world, for certain trip purposes, such as non-home-based or home-based entertainment trips, where habitual behavior are not prominent, DUO models with variable demand may have valid applications.

We also adopt the nested diagonalization method to solve the DUO models with variable demand by a network representation. There are several formulations available for this purpose, including two approaches presented by Gartner (1980), i.e., the zero-cost overflow formulation and the excess-demand formulation. We adopt the dynamic counterpart of the latter formulation in this chapter as it is more efficient in terms of computation time. For the former approach, the reader is referred to Gartner (1980).

# Chapter 8

# Dynamic User-Optimal Mode Choice Models

Dynamic user-optimal mode choice models assume that the total number of trips are fixed, but travelers are allowed to choose their transportation mode according to personal preference. This situation may be exemplified by work trips, where the total number of work trips is usually prespecified; however, the mode of transportation is decided by the behavior and characteristics of the travelers; in particular, certain travelers will be captive transit riders. The choice of mode of transportation may be determined in such a way that the necessary *en route* travel time (cost) is also minimal. In consideration of departure time choices, two types of models can be identified, i.e., the DUO mode choice/route choice model and the DUO mode choice/departure time/route choice model.

In the real world, travelers can choose from a number of transportation modes. Examples are automobile, public transportation (bus, transit, subway, train), para-transit (taxi, dial-a-bus) among others. However, for simplicity, this chapter only considers binary mode choice problems.

In this chapter, the DUO mode choice/route choice model, including equilibrium conditions and model formulation, the nested diagonalization method and a numerical example are first explored in Section 8.1. A similar discussion for the DUO mode choice/departure time/route choice model is provided in Section 8.2. Finally, concluding notes are given in Section 8.3.

## 8.1 Dynamic User-Optimal Mode Choice/Route Choice Model

In this section, we discuss the equilibrium conditions and model formulation, the nested diagonalization method and a numerical example for the dynamic user-optimal mode choice/route choice model.

### 8.1.1 Equilibrium Conditions and Model Formulation

The dynamic user-optimal conditions are first defined to characterize the travelers' travel behavior for choosing the best transportation mode and minimal travel time route. Due to inherent link interactions, the variational inequality approach is adopted to formulate the DUO mode choice/route choice model. The equivalence between the dynamic user-optimal conditions and the variational inequality formulation is then stated by a theorem and verified by a proof.

#### 8.1.1.1 Dynamic user-optimal conditions

Suppose we have a binary choice of transportation modes, where the auto mode is denoted by $m_1$ and the transit mode is denoted by $m_2$. When mode and route choices are considered, the corresponding dynamic user-optimal conditions may be characterized by equilibrium conditions on both route choice behavior and mode choice functions.

For route choice behavior, the conditions for each O-D pair and transportation mode state that the actual route travel times experienced by travelers departing from the same origin during the same interval are equal and minimal, or no traveler would be better off by unilaterally changing his/her route. In contrast, the actual route travel time of any unused route for each O-D pair and mode is greater than or equal to the minimal actual route travel time among all modes. Therefore, if the flow of mode $m$ departing from origin $r$ during interval $k$ over route $p$ toward destination $s$ is positive, i.e., $h_{mp}^{rs*}(k) > 0$, then the corresponding actual route travel time is minimal among all modes. However, if no flow of mode $m$ occurs on route $p$, i.e., $h_{mp}^{rs*}(k) = 0$, then the corresponding actual route travel time is at least as great as the minimal actual route travel time among all modes. These equilibrium conditions can be mathematically expressed as follows:

$$c_{mp}^{rs*}(k) \begin{cases} = \pi_m^{rs}(k) & \text{if } h_{mp}^{rs*}(k) > 0 \\ \geq \pi_m^{rs}(k) & \text{if } h_{mp}^{rs*}(k) = 0 \end{cases} \quad \forall r, s, m, p, k \tag{8.1}$$

For the mode choice functions, the auto choice function is assumed to be a strictly decreasing function of the difference between the auto and transit minimal route travel times. Thus, the inverse auto choice function has a strictly monotone mapping. The general form can be expressed mathematically as follows:

$$\pi_{m_1}^{rs}(k) - \pi_{m_2}^{rs}(k) = \left(D_{m_1}^{rs*}(k)\right)^{-1} \quad \forall r, s, k \tag{8.2}$$

The well known inverse auto choice function is as follows:

$$\left(D_{m_1}^{rs}(k)\right)^{-1} = \frac{1}{\theta} \ln \frac{\bar{q}^{rs}(k) - q_{m_1}^{rs}(k)}{q_{m_1}^{rs}(k)} + \varphi^{rs}(k) \quad \forall r, s, k \tag{8.3}$$

where $\varphi^{rs}(k)$ represents the preference for auto trips that depart from origin $r$ during interval $k$ toward destination $s$, and $\theta$ indicates a dispersion/variance parameter. Note that when equation (8.3) is inserted into equation (8.2), the following mode choice functions result.

$$q_{m_1}^{rs}(k) = \bar{q}^{rs}(k)\frac{1}{1+e^{\theta\left(\pi_{m_1}^{rs}(k)-\pi_{m_2}^{rs}(k)-\varphi^{rs}(k)\right)}} \quad \forall r,s,k \tag{8.4}$$

$$q_{m_2}^{rs}(k) = \bar{q}^{rs}(k) - q_{m_1}^{rs}(k) \quad \forall r,s,k \tag{8.5}$$

#### 8.1.1.2 Variational inequality formulation

The DUO mode choice/route choice problem can be formulated using the variational inequality approach.

***Theorem 8.1***: The DUO mode choice/route choice problem is equivalent to finding a solution $(\mathbf{u}^*, \mathbf{q}^*) \in \Omega$ such that the following VI formulation holds:

$$\mathbf{c}^*(\mathbf{u}-\mathbf{u}^*) - (\mathbf{D}^*)^{-1}(\mathbf{q}-\mathbf{q}^*) \geq 0 \quad \forall(\mathbf{u},\mathbf{q}) \in \Omega^* \tag{8.6}$$

or, alternatively, in expanded form:

$$\begin{aligned}&\sum_m\sum_a\sum_t c_{ma}^*(t)\left[u_{ma}(t) - u_{ma}^*(t)\right]\\ &\quad -\sum_{rs}\sum_k \left(D_{m_1}^{rs*}(k)\right)^{-1}\left[q_{m_1}^{rs}(k) - q_{m_1}^{rs*}(k)\right] \quad \forall(\mathbf{u},\mathbf{q}) \in \Omega^*\end{aligned} \tag{8.7}$$

where $c_{m_1a}$, $c_{m_2a}$ are the link travel time functions for auto and transit respectively, and $\left(D_{m_1}^{rs*}(k)\right)^{-1}$ is the inverse auto mode choice function. The link travel time functions may depend on the flows of both modes, $u_{m_1a}$, $u_{m_2a}$, but are separable by links. The symbol $\Omega^*$ is a subset of $\Omega$ with $\delta_{mapk}^{rs}(t)$ being realized at equilibrium, i.e., $\delta_{mapk}^{rs}(t) = \delta_{mapk}^{rs*}(t), \forall r,s,m \in (m_1,m_2),a,p,k,t$. The symbol $\Omega$ denotes the feasible region that is delineated by the following constraints:

Flow conservation constraint:

$$\sum_m\sum_p h_{mp}^{rs}(k) = \bar{q}^{rs}(k) \quad \forall r,s,k \tag{8.8}$$

Flow propagation constraints:

$$u_{mapk}^{rs}(t) = h_{mp}^{rs}(k)\delta_{mapk}^{rs}(t) \quad \forall r,s,m,a,p,k,t \tag{8.9}$$

$$\sum_t \delta^{rs}_{mapk}(t) = 1 \quad \forall r,s,p,m,a\in p,k \tag{8.10}$$

$$\delta^{rs}_{mapk}(t) = \{0,1\} \quad \forall r,s,m,a,p,k,t \tag{8.11}$$

Nonnegativity constraint:

$$h^{rs}_{mp}(k) \geq 0 \quad \forall r,s,p,m,k \tag{8.12}$$

Definitional constraints:

$$\sum_p h^{rs}_{mp}(k) = q^{rs}_m(k) \quad \forall r,s,m,k \tag{8.13}$$

$$u_{ma}(t) = \sum_{rs}\sum_p\sum_k h^{rs}_{mp}(k)\delta^{rs}_{mapk}(t) \quad \forall m,a,t \tag{8.14}$$

$$c^{rs}_{mp}(k) = \sum_a\sum_t c_{ma}(t)\delta^{rs}_{mapk}(t) \quad \forall r,s,m,p,k \tag{8.15}$$

Equation (8.8) expresses the time-dependent O-D demand in terms of route flows among all modes, and conserves the fixed total number of trips during interval $k$. Equation (8.13) is definitional, and expresses time-dependent O-D demand in terms of the time-dependent route flows among all modes. The other constraints are essentially the same as those for the DUO variable demand/route choice model, except that mode choice is also considered.

#### 8.1.1.3 Equivalence analysis

***Theorem 8.2***: Under a certain flow propagation relationship $\delta^{rs}_{mapk}(t) = \delta^{rs*}_{mapk}(t)$, $\forall r,s,m,a,p,k,t$, DUO mode choice/route choice conditions (8.1)~(8.2) imply VIP (8.7) and vice versa.

***Proof of necessity***: We need to prove that under a certain flow propagation relationship $\left(\delta^{rs}_{mapk}(t) = \delta^{rs*}_{mapk}(t),\ \forall r,s,m,a,p,k,t\right)$, the DUO mode choice/route choice conditions (8.1)~(8.2) imply VIP (8.7). We first rearrange the equilibrium conditions (8.1) as follows:

$$\left[c^{rs*}_{mp}(k) - \pi^{rs}_m(k)\right]\left[h^{rs}_{mp}(k) - h^{rs*}_{mp}(k)\right] \geq 0 \quad \forall r,s,m,p,k \tag{8.16}$$

By summing over $r,s,m,p,k$, and then making a substitution of $\sum_p h^{rs}_{mp}(k) = q^{rs}_m(k), \forall m$, we obtain:

$$\begin{aligned}&\sum_{rs}\sum_m\sum_p\sum_k c^{rs*}_{mp}(k)\left[h^{rs}_{mp}(k) - h^{rs*}_{mp}(k)\right]\\ &\quad -\sum_{rs}\sum_m\sum_k \pi^{rs}_m(k)\left[q^{rs}_m(k) - q^{rs*}_m(k)\right] \geq 0\end{aligned} \tag{8.17}$$

By applying equation (8.14), we obtain:

$$\sum_{rs}\sum_{m}\sum_{k}\sum_{p}\left[\sum_{a}\sum_{t}c^{*}_{ma}(t)\delta^{rs*}_{mapk}(t)\right]\left[h^{rs}_{mp}(k)-h^{rs*}_{mp}(k)\right] \\ -\sum_{rs}\sum_{m}\sum_{k}\pi^{rs}_{m}(k)\left[q^{rs}_{m}(k)-q^{rs*}_{m}(k)\right]\geq 0 \tag{8.18}$$

By changing the order of the summation on the first term, using equation (8.14), it follows:

$$\sum_{m}\sum_{a}\sum_{t}c^{*}_{ma}(t)\left[u_{ma}(t)-u^{*}_{ma}(t)\right] \\ -\sum_{rs}\sum_{m}\sum_{k}\pi^{rs}_{m}(k)\left[q^{rs}_{m}(k)-q^{rs*}_{m}(k)\right]\geq 0 \tag{8.19}$$

Since $q^{rs}_{m_2}(k)=\bar{q}^{rs}(k)-q^{rs}_{m_1}(k)$, using equation (8.2), we have:

$$\sum_{m}\sum_{a}\sum_{t}c^{*}_{ma}(t)\left[u_{ma}(t)-u^{*}_{ma}(t)\right] \\ -\sum_{rs}\sum_{k}\left(D^{rs*}_{m_1}(k)\right)^{-1}\left[q^{rs}_{m_1}(k)-q^{rs*}_{m_1}(k)\right]\geq 0 \tag{8.20}$$

The inequality (8.20) is identical to VIP (8.7).

***Proof of sufficiency:*** We next prove that the VIP (8.7) implies mode choice/route choice equilibrium conditions (8.1)~(8.2). First we demonstrate that VIP (8.7) implies expression (8.1). Without loss of generality, we now define a feasible solution $\left\{h^{rs}_{mp}(k),q^{rs*}_{m}(k)\right\}$, $\forall r,s,m,p,k$ to be the same as $\left\{h^{rs*}_{mp}(k),q^{rs*}_{m}(k)\right\}$, $\forall r,s,m,p,k$, except for two routes, $p^{rs}_1$ and $p^{rs}_2$, for mode $m_1$. We consider two situations that could arise at equilibrium:

(i) Both route flows are positive, i.e., $h^{rs}_{m_1p_1}(k)>0$ and $h^{rs}_{m_1p_2}(k)>0$. We switch a small amount of flow $\Delta_1$ from route $p^{rs}_1$ to $p^{rs}_2$ with $0<\Delta_1\leq h^{rs*}_{m_1p_1}(k)$. That is:

$$h^{rs}_{m_1p_1}(k)=h^{rs*}_{m_1p_1}(k)-\Delta_1 \quad \text{and} \tag{8.21}$$

$$h^{rs}_{m_1p_2}(k)=h^{rs*}_{m_1p_2}(k)+\Delta_1 \tag{8.22}$$

Substituting the feasible solution $\left\{h^{rs}_{mp}(k),q^{rs*}_{m}(k)\right\}$, $\forall r,s,m,p,k$ into equation (8.7) yields:

$$c^{rs*}_{m_1p_1}(k)\left[h^{rs}_{m_1p_1}(k)-h^{rs*}_{m_1p_1}(k)\right]+c^{rs*}_{m_1p_2}(k)\left[h^{rs}_{m_1p_2}(k)-h^{rs*}_{m_1p_2}(k)\right]\geq 0 \tag{8.23}$$

By applying equations (8.21) and (8.22), we have:

$$c^{rs*}_{m_1p_2}(k)\geq c^{rs*}_{m_1p_1}(k) \tag{8.24}$$

Similarly, by switching a small amount of flow $\Delta_2$ with $0<\Delta_2\leq h^{rs*}_{m_1p_2}(k)$ from route $p^{rs}_2$ to $p^{rs}_1$, we obtain:

$$c_{m_1 p_1}^{rs*}(k) \geq c_{m_1 p_2}^{rs*}(k) \tag{8.25}$$

Equations (8.24) and (8.25) together imply:

$$c_{m_1 p_2}^{rs*}(k) = c_{m_1 p_1}^{rs*}(k) \tag{8.26}$$

We repeat this procedure to verify for each O-D pair and mode that all used routes with positive flow have the same actual route travel time.

(ii) One route flow is positive and the other route flow is nil. We arbitrarily assume, without loss of generality, $h_{m_1 p_1}^{rs*}(k) > 0$ and $h_{m_1 p_2}^{rs*}(k) = 0$. We switch a small amount of flow $\Delta_1$ from route $p_1^{rs}$ to $p_2^{rs}$ with $0 < \Delta_1 \leq h_{m_1 p_1}^{rs*}(k)$. By the same argument shown in (i), we have $c_{m_1 p_2}^{rs*}(k) \geq c_{m_1 p_1}^{rs*}(k)$. We repeat this procedure to verify for each O-D pair and mode that all unused routes with zero flow have higher or equal actual route travel times. By cases (i) and (ii), it follows that VIP (8.7) implies expression (8.1).

We now conclude the proof by showing that VIP (8.7) implies equation (8.2). As before, we construct a feasible solution $\{h_{mp}^{rs}(k),\ q_m^{rs}(k)\}, \forall r, s, m, p, k$ that differs from $\{h_{mp}^{rs*}(k),\ q_m^{rs*}(k)\}$, $\forall r, s, m, p, k$ only for the flows of the two used routes, $p_1$ of mode $m_1$, and $p_2$ of mode $m_2$, for which $h_{m_1 p_1}^{rs}(k) = h_{m_1 p_1}^{rs*}(k) - \Delta$, $h_{m_2 p_2}^{rs}(k) = h_{m_2 p_2}^{rs*}(k) + \Delta$, where $0 < |\Delta| \leq \min\left(h_{m_1 p_1}^{rs*}(k),\ h_{m_2 p_2}^{rs*}(k)\right)$. Since $q_{m_1}^{rs}(k) = q_{m_1}^{rs*}(k) - \Delta$ and $q_{m_2}^{rs}(k) = q_{m_2}^{rs*}(k) + \Delta$, therefore applying VIP (8.7) yields $c_{m_1 p_1}^{rs}(k)(-\Delta) + c_{m_2 p_2}^{rs}(k)\Delta + \left(D_{m_1}^{rs*}(k)\right)^{-1}\Delta \geq 0$, which after dividing by $\Delta$ and using equation (8.1), results in $-\pi_{m_1}^{rs}(k) + \pi_{m_2}^{rs}(k) + \left(D_{m_1}^{rs*}(k)\right)^{-1} \geq 0$, for $\Delta > 0$, and $-\pi_{m_1}^{rs}(k) + \pi_{m_2}^{rs}(k) + \left(D_{m_1}^{rs*}(k)\right)^{-1} \leq 0$, for $\Delta < 0$; hence, equation (8.2) is satisfied. This completes the proof.

### 8.1.2 Nested Diagonalization Method

In this section, we first show the network representation for the DUO mode choice/route choice problem. Then, the nested diagonalization method is formally stated.

#### 8.1.2.1 Time-space network

The network representation for the DUO mode choice/route choice model is shown in Figure 8.1, of which auto and transit modes share a basic network, but are separated by links.

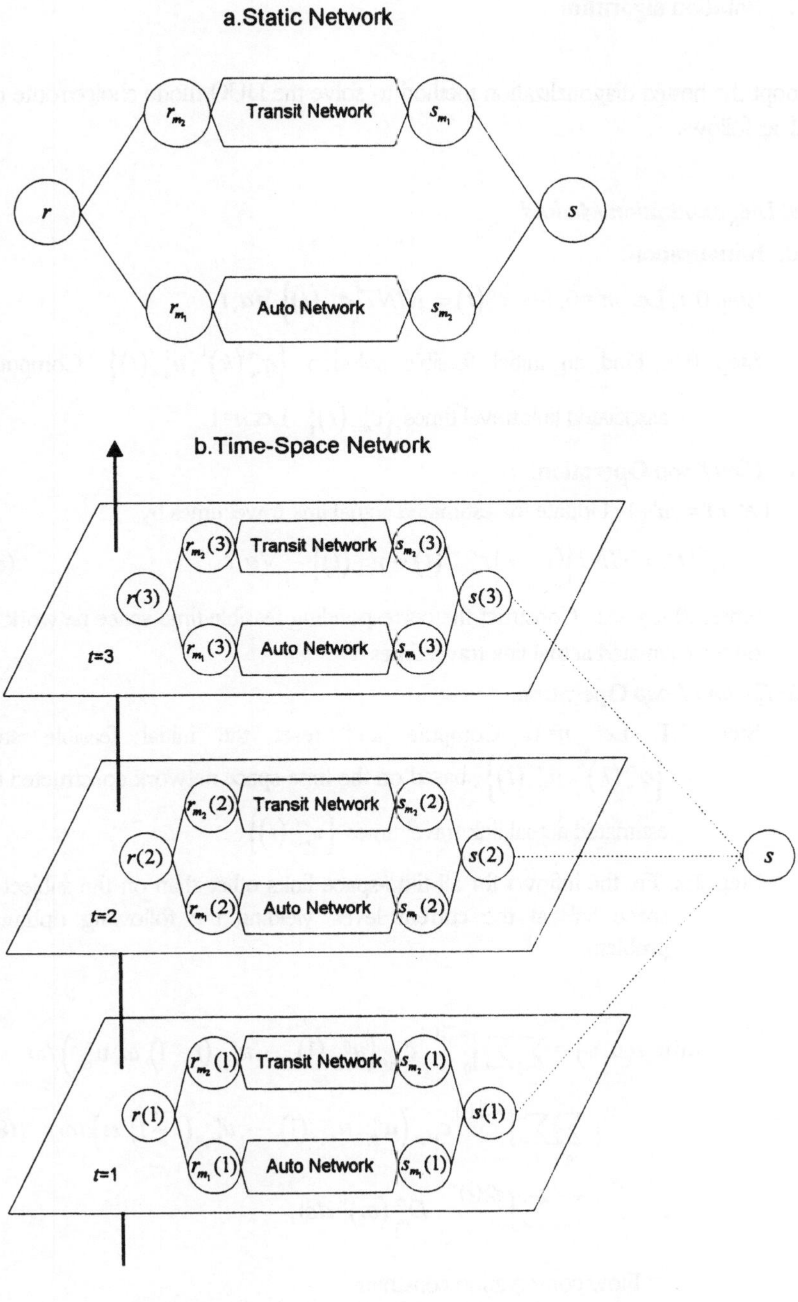


Figure 8.1: Time-Space Network for the Dynamic User-Optimal Mode Choice/Route Choice Model

#### 8.1.2.2 Solution algorithm

We adopt the nested diagonalization method to solve the DUO mode choice/route choice model, as follows:

*Nested Diagonalization Method*

**Step 0: Initialization.**

Step 0.1: Let $m'$=0. Set $\tau_a^0(t) = NINT[c_{a_0}(t)], \forall a, t$.

Step 0.2: Find an initial feasible solution $\{q_m^{rs}(k)^1, u_{ma}^1(t)\}$. Compute the associated link travel times $\{c_{ma}^1(t)\}$. Let $n$=1.

**Step 1: *First Loop* Operation.**

Let $m' = m'+1$. Update the estimated actual link travel times by

$$\tau_a^{m'}(t) = NINT[(1-\gamma)\tau_a^{m'-1}(t) + \gamma c_a^n(t)] \quad \forall a, t \tag{8.27}$$

where $0 < \gamma < 1$. Construct the corresponding feasible time-space network based on the estimated actual link travel times.

**Step 2: *Second Loop* Operation.**

Step 2.1: Let $n$=1. Compute and reset the initial feasible solution $\{q_m^{rs}(k)^n, u_{ma}^n(t)\}$, based on the time-space network constructed by the estimated actual link travel times $\{\tau_{ma}^{m'}(t)\}$.

Step 2.2: Fix the inflows for all time-space links other than on the subject time-space link at the current level, yielding the following optimization problem.

$$\begin{aligned}\min\ z(\mathbf{u},\mathbf{q}) = &\sum_a\sum_t\int_0^{u_{m_1a}^{n+1}(t)} c_{m_1a}\left(u_{m_1a}^n(1),\cdots,u_{m_1a}^n(t-1),\omega,\mathbf{u}_{m_2}^n\right)d\omega\\ &+\sum_a\sum_t\int_0^{u_{m_2a}^{n+1}(t)} c_{m_2a}\left(\mathbf{u}_{m_1}^n,u_{m_2a}^n(1),\cdots,u_{m_2a}^n(t-1),\omega\right)d\omega\\ &-\sum_{rs}\sum_k\int_0^{q_{m_1}^{rs}(k)^{n+1}} D_{m_1}^{rs}(\omega)^{-1}d\omega\end{aligned} \tag{8.28}$$

Flow conservation constraint:

$$\sum_m\sum_p h_{mp}^{rs}(k) = \bar{q}^{rs}(k) \quad \forall r, s, k \tag{8.29}$$

Nonnegativity constraint:

$$h_{mp}^{rs}(k) \geq 0 \quad \forall r,s,m,p,k \tag{8.30}$$

Definitional constraints:

$$u_{mapk}^{rs}(t) = h_{mp}^{rs}(k)\bar{\delta}_{mapk}^{rs}(t) \quad \forall r,s,m,a,p,k,t \tag{8.31}$$

$$\bar{\delta}_{mapk}^{rs}(t) = \{0,1\} \quad \forall r,s,m,a,p,k,t \tag{8.32}$$

$$\sum_p h_{mp}^{rs}(k) = q_m^{rs}(k) \quad \forall r,s,m,k \tag{8.33}$$

$$u_{ma}(t) = \sum_{rs}\sum_p\sum_k h_{mp}^{rs}(k)\bar{\delta}_{mapk}^{rs}(t) \quad \forall m,a,t \tag{8.34}$$

$$c_{mp}^{rs}(k) = \sum_a\sum_t c_{ma}(t)\bar{\delta}_{mapt}^{rs}(t) \quad \forall r,s,m,p,k \tag{8.35}$$

**Step 3: *Third Loop* Operation.**

Solve for the solution, $\{q_m^{rs}(k)^{n+1},\ u_{ma}^{n+1}(t)\}$, in optimization problem (8.28)~(8.35) by the partial linearization technique, also known as the Evans' method. Compute the resulting link travel times $\{c_{ma}^{n+1}(t)\}$.

**Step 4: Convergence Check for the *Second Loop* Operation.**

If $u_a^{n+1}(t) \approx u_a^n(t), \forall a,t$ and $q_m^{rs}(k)^{n+1} \approx q_m^{rs}(k)^n, \forall r,s,m,k$, go to Step 5; otherwise, set $n=n+1$, go to Step 2.2.

**Step 5: Convergence Check for the *First Loop* Operation.**

If $\tau_{ma}^{mt}(t) \approx c_{ma}^{n+1}(t), \forall m,a,t$, stop; the current solution is optimal. Otherwise, set $n=n+1$, and go to Step 1.

Analogous to the discussion in Section 5.3, the nested projection algorithm can be obtained by the following modification.

Step 2.2: Fix the inflows for all time-space links other than on the subject time-space link at the current level, yielding the following optimization problem.

$$\begin{aligned} \min\ z(\mathbf{u},\mathbf{q}) = &\frac{1}{2}(\mathbf{u}^{n+1})^T\mathbf{G}_1\mathbf{u}^{n+1} + \left(\rho_1\mathbf{c}(\mathbf{u}^n) - \mathbf{G}_1\mathbf{u}^n\right)\mathbf{u}^{n+1} \\ &-\frac{1}{2}(\mathbf{q}^{n+1})^T\mathbf{G}_2\mathbf{q}^{n+1} - \left(\rho_2(\mathbf{D}(\mathbf{q}^n))^{-1} - \mathbf{G}_2\mathbf{q}^n\right)\mathbf{q}^{n+1} \end{aligned} \tag{8.36}$$

Subject to: Equations (8.29)~(8.35)

where matrices $\mathbf{G}_1$ and $\mathbf{G}_2$ are symmetric and positive definite, and $\rho_1$

and $\rho_2$ are contraction operators.

#### 8.1.2.3 Single transit link with constant travel time

In many circumstances, a single transit line, such as a subway, is operated on a closed system with constant travel time $\bar{\pi}^{rs}_{m_2}$. By an appropriate network representation, this specific DUO mode choice/route choice problem can be treated as the DUO route choice problem. In Figure 8.2(a), we first assume that the network consists of a basic auto network and a single transit link with constant travel time. In Figure 8.2(b), the corresponding time-space network contains two types of artificial time-space links. The first type of artificial time-space links connects with the time-dependent origins *r*(*k*) to the time-independent destination $s$. The associated travel time functions are set as $c^{rs}(k) = \bar{\pi}^{rs}_{m_2} + \left(D^{rs}_{m_1}(k)\right)^{-1}$. The second type of artificial time-space links directly joins each auto-specific time-dependent destination $s_{m_1}(k)$ to the time-independent destination *s*, where the corresponding travel time function is defined as $c^{s_m s}(k) = 0$. The following theorem states that this specific DUO mode choice/route choice problem is equivalent to, in the mathematical sense, the DUO route choice problem.

***Theorem 8.3***: The DUO mode choice/route choice problem is equivalent to the DUO route choice problem with the time-space network shown in Figure 8.2(b).

***Proof***: For the DUO route choice problem with time-space network shown in Figure 8.2(b), the corresponding VIP is as follows:

$$\sum_a \sum_t c^*_{m_1 a}(t)\left[u_{m_1 a}(t) - u^*_{m_1 a}(t)\right] + \sum_{rs} \sum_k \left[\bar{\pi}^{rs}_{m_2} + \left(D^{rs*}_{m_1}(k)\right)^{-1}\right]\left[q^{rs}_{m_2}(k) - q^{rs*}_{m_2}(k)\right] \geq 0 \quad \forall (\mathbf{u}, \mathbf{q}) \in \Omega^* \tag{8.37}$$

By rearranging the second term, using $q^{rs}_{m_2}(k) = \bar{q}^{rs}(k) - q^{rs}_{m_1}(k)$, one obtains:

$$\sum_a \sum_t c^*_{m_1 a}(t)\left[u_{m_1 a}(t) - u^*_{m_1 a}(t)\right] + \sum_{rs} \bar{\pi}^{rs}_{m_2} \sum_k \left[q^{rs}_{m_2}(k) - q^{rs*}_{m_2}(k)\right] - \sum_{rs} \sum_k \left(D^{rs*}_{m_1}(k)\right)^{-1}\left[q^{rs}_{m_1}(k) - q^{rs*}_{m_1}(k)\right] \geq 0 \quad \forall (\mathbf{u}, \mathbf{q}) \in \Omega^* \tag{8.38}$$

Since the second term is equivalent to $\sum_a \sum_t c^*_{m_2 a}(t)\left[u_{m_2 a}(t) - u^*_{m_2 a}(t)\right]$, we therefore have:

$$\sum_m \sum_a \sum_t c^*_{ma}(t)\left[u_{ma}(t) - u^*_{ma}(t)\right] - \sum_{rs} \sum_k \left(D^{rs*}_{m_1}(k)\right)^{-1}\left[q^{rs}_{m_1}(k) - q^{rs*}_{m_1}(k)\right] \geq 0 \quad \forall(\mathbf{u},\mathbf{q}) \in \Omega^* \tag{8.39}$$

Inequality (8.39) is the exact VIP for the DUO mode choice/route choice problem shown in expression (8.7). The steps shown above are essentially reversible. This completes the proof.

When the mode choice decisions are appropriately described by the following logit formula:

$$q^{rs}_{m_1}(k) = \bar{q}^{rs}(k)\frac{1}{1+e^{\theta\left(\pi^{rs}_{m_1}(k)-\bar{\pi}^{rs}_{m_2}-\varphi^{rs}(k)\right)}} \quad \forall r,s,k \tag{8.40}$$

and we further assume that the dispersion parameter is set as $\theta = 1$, and the auto preference parameters $\varphi^{rs}(k) = 0, \forall r,s,k$, then the nested diagonalization method can still be adopted for solutions with the following modifications.

Step 2.2: Fix the inflows for all time-space links other than on the subject time-space link at the current level, yielding the following optimization problem.

$$\min\ z(\mathbf{u},\mathbf{q}) = \sum_a \sum_t \int_0^{u^{n+1}_{m_1a}(t)} c_{m_1a}\left(u^n_{m_1a}(1),\cdots,u^n_{m_1a}(t-1),\omega,\mathbf{u}^n_{m_2}\right)d\omega + \sum_{rs}\sum_k \left(q^{rs}_{m_2}(k)\right)^{n+1}\bar{\pi}^{rs\,n+1}_{m_2} - \sum_{rs}\sum_k \int_0^{q^{rs}_{m_1}(k)^{n+1}} \ln\frac{\bar{q}^{rs}(k)-\omega}{\omega}d\omega \tag{8.41}$$

Subject to: Equations (8.29)~(8.35)

**Step 4: Convergence Check for the *Second Loop* Operation.**

If the following convergence criteria are met, go to Step 5; otherwise, set $n=n+1$, and go to Step 2.2.

$$u^{n+1}_a(t) \approx u^n_a(t) \quad \forall a,t \tag{8.42}$$

$$\frac{q^{rs}_{m_1}(k)^{n+1}}{\bar{q}^{rs}(k)} \approx \frac{1}{1+e^{\left(c^{rs}_{m_1}(k)^{n+1}-\bar{\pi}^{rs\,n+1}_{m_2}\right)}} \quad \forall r,s,k \tag{8.43}$$

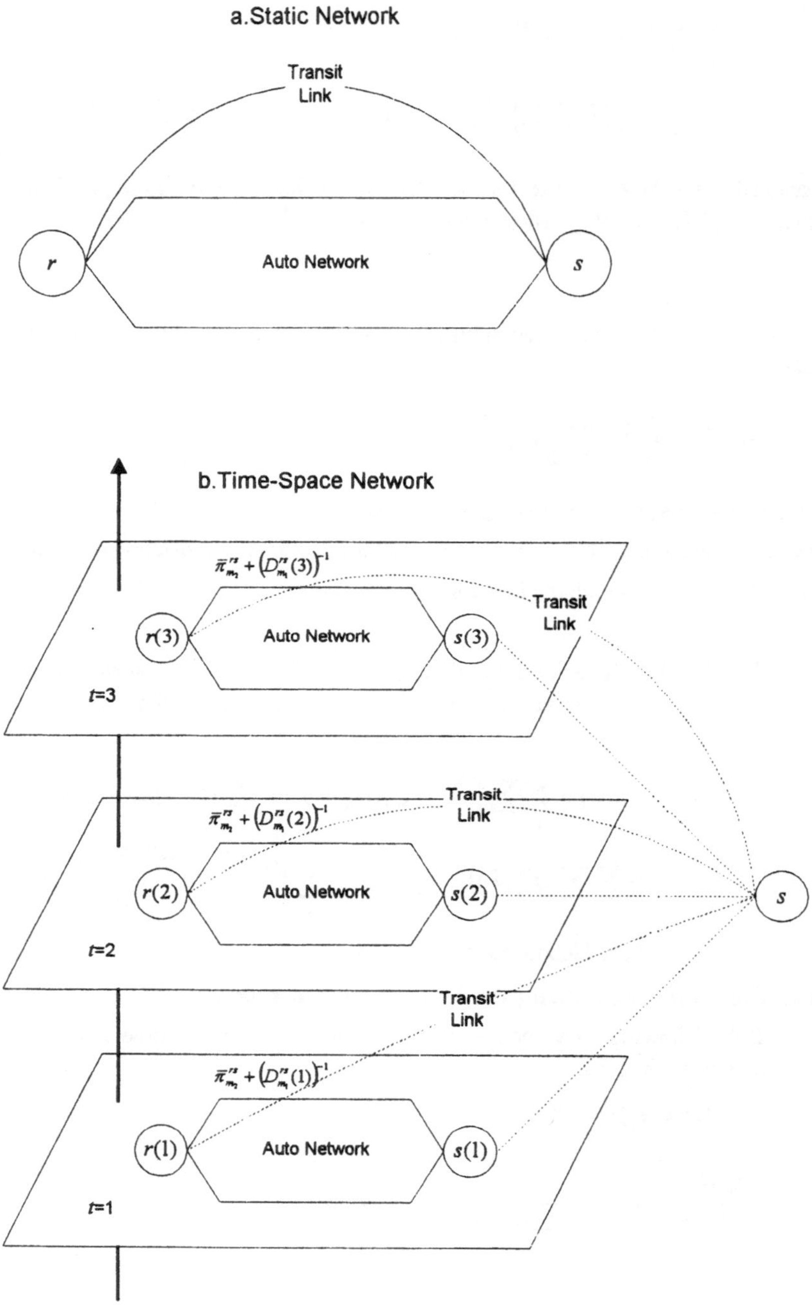


Figure 8.2: Time-Space Network for the Dynamic User-Optimal Mode Choice/Route Choice Model (Single Transit Link with Constant Travel Time)

### 8.1.3 Numerical Example

#### 8.1.3.1 Input data

A simple network shown in Figure 8.3 is used for testing. The test network consists of 8 links (of which links 1→5 and 2→5 denote the transit links), and 5 nodes, in which nodes 1 and 2 are origin nodes, node 5 is the destination, and nodes 3 and 4 are intermediate nodes.

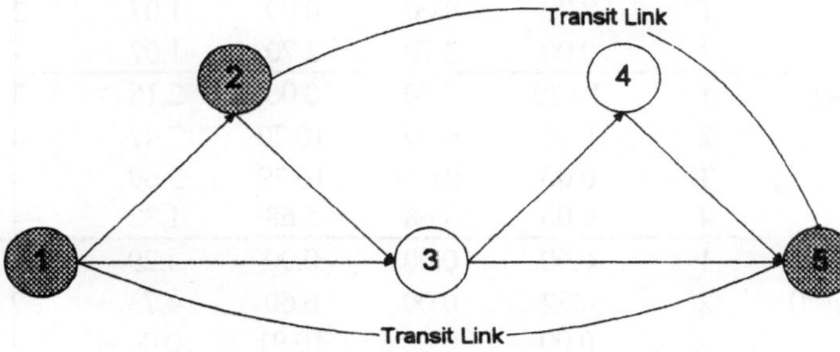


Figure 8.3: Test Network 1

The adopted dynamic travel time functions are arbitrarily constructed as follows:

$$c_{m_1a}(t) = 1 + 0.01\left(u_{m_1a}(t)\right)^2 + 0.01\left(x_{m_1a}(t)\right)^2 \quad \forall a,t \tag{8.44}$$

$$c_{m_2a}(t) = 5 + \left(D_{m_1}^{rs}(t)\right)^{-1} \quad \forall a,t \tag{8.45}$$

The time-dependent origin-destination (O-D) demands are assumed in Table 8.1:

Table 8.1: Time-Dependent O-D Demands for Test Network 1

| O-D | Time Interval | | | |
|---|---|---|---|---|
| Pair | $k$=1 | $k$=2 | $k$=3 | $k$=4 |
| 1-5 | 20 | 10 | - | - |
| 2-5 | - | - | 20 | 10 |

The mode choice functions are assumed to be the following logit formulas:

$$q_{m_1}^{rs}(k) = \bar{q}^{rs}(k)\frac{1}{1+e^{\left(\bar{\pi}_{m_1}^{rs} - \bar{\pi}_{m_2}^{rs}(k)\right)}} \quad \forall r,s,k \tag{8.46}$$

$$q_{m_2}^{rs}(k) = \bar{q}^{rs}(k)\frac{1}{1+e^{\left(\bar{\pi}_{m_2}^{rs} - \pi_{m_1}^{rs}(k)\right)}} \quad \forall r,s,k \tag{8.47}$$

#### 8.1.3.2 Test results

The resulting flow pattern and time-dependent O-D demands among all modes for the DUO mode choice/route choice model are summarized in Tables 8.2 and 8.3, respectively.

Table 8.2: Resulting Flow Pattern for Test Network 1

| Link | Entering Time Interval | Inflow | Exit Flow | Number of Vehicles | Link Travel Time | Exiting Time Interval |
|---|---|---|---|---|---|---|
| 1→2 | 1 | 2.70 | 0.00 | 0.00 | 1.07 | 2 |
| | 2 | 0.00 | 2.70 | 2.70 | 1.07 | - |
| 1→3 | 1 | 10.70 | 0.00 | 0.00 | 2.15 | 3 |
| | 2 | 5.68 | 0.00 | 10.70 | 2.47 | 4 |
| | 3 | 0.00 | 10.70 | 16.39 | 3.69 | - |
| | 4 | 0.00 | 5.68 | 5.68 | 1.32 | - |
| 1→5 | 1 | 6.60 | 0.00 | 0.00 | 4.29 | 5 |
| (Transit) | 2 | 4.32 | 0.00 | 6.60 | 4.73 | 7 |
| | 3 | 0.00 | 0.00 | 10.91 | 0.00 | - |
| | 4 | 0.00 | 0.00 | 10.91 | 0.00 | - |
| | 5 | 0.00 | 6.60 | 10.91 | 0.00 | - |
| | 6 | 0.00 | 0.00 | 4.32 | 0.00 | - |
| | 7 | 0.00 | 4.32 | 4.32 | 0.00 | - |
| 2→3 | 2 | 2.70 | 0.00 | 0.00 | 1.07 | 3 |
| | 3 | 9.71 | 2.70 | 2.70 | 2.02 | 5 |
| | 4 | 6.82 | 0.00 | 9.71 | 2.41 | 6 |
| | 5 | 0.00 | 9.71 | 16.53 | 3.73 | - |
| | 6 | 0.00 | 6.82 | 6.82 | 1.47 | - |
| 2→5 | 3 | 10.29 | 0.00 | 0.00 | 5.06 | 8 |
| (Transit) | 4 | 3.18 | 0.00 | 10.29 | 4.24 | 8 |
| | 5 | 0.00 | 0.00 | 13.47 | 0.00 | - |
| | 6 | 0.00 | 0.00 | 13.47 | 0.00 | - |
| | 7 | 0.00 | 0.00 | 13.47 | 0.00 | - |
| | 8 | 0.00 | 13.47 | 13.47 | 0.00 | - |
| 3→4 | 3 | 2.70 | 0.00 | 0.00 | 1.07 | 4 |
| | 4 | 2.35 | 2.70 | 2.70 | 1.13 | 5 |
| | 5 | 7.02 | 2.35 | 2.35 | 1.55 | 7 |
| | 6 | 0.00 | 0.00 | 7.02 | 1.49 | - |
| | 7 | 0.00 | 7.02 | 7.02 | 1.49 | - |
| 3→5 | 3 | 10.70 | 0.00 | 0.00 | 2.15 | 5 |
| | 4 | 3.33 | 0.00 | 10.70 | 2.26 | 6 |
| | 5 | 2.69 | 10.70 | 14.03 | 3.04 | 8 |
| | 6 | 6.82 | 3.33 | 6.02 | 1.83 | 8 |
| | 7 | 0.00 | 0.00 | 9.51 | 1.90 | - |
| | 8 | 0.00 | 9.51 | 9.51 | 1.90 | - |

Table 8.2: Resulting Flow Pattern for Test Network 1 (continued)

| Link | Entering Time Interval | Inflow | Exit Flow | Number of Vehicles | Link Travel Time | Exiting Time Interval |
|---|---|---|---|---|---|---|
| 4→5 | 4 | 2.70 | 0.00 | 0.00 | 1.07 | 5 |
| | 5 | 2.35 | 2.70 | 2.70 | 1.13 | 6 |
| | 6 | 0.00 | 2.35 | 2.35 | 1.06 | - |
| | 7 | 7.02 | 0.00 | 0.00 | 1.49 | 8 |
| | 8 | 0.00 | 7.02 | 7.02 | 1.49 | - |

Table 8.3: Resulting Time-Dependent Mode-Specific O-D Demands for Test Network 1

| O-D Pair | Time Interval *k*=1 | *k*=2 | *k*=3 | *k*=4 |
|---|---|---|---|---|
| 1-5 | 13.40 (6.60)* | 5.68 (4.32) | - | - |
| 2-5 | - | - | 9.71 (10.29) | 6.82 (3.18) |

* "Numbers" in brackets denote transit O-D demands

The corresponding actual route travel times are computed and summarized in Table 8.4.

Table 8.4: Actual Route Travel Times for Test Network 1

| Route | Time Interval *k*=1 | *k*=2 | *k*=3 | *k*=4 |
|---|---|---|---|---|
| 1→2→3→4→5 | 4.29 (2.70)* | – | – | – |
| 1→3→4→5 | NA | 4.73 (2.35) | – | – |
| 1→3→5 | 4.29 (10.70) | 4.73 (3.33) | – | – |
| 1→5 (Transit Demand) | 4.29 (6.60) | 4.73 (4.32) | – | – |
| 2→3→4→5 | – | – | 5.06 (7.02) | – |
| 2→3→5 | – | – | 5.06 (2.69) | 4.24 (3.18) |
| 2→5 (Transit Demand) | – | – | 5.06 (10.29) | 4.24 (6.82) |

* "Numbers" in brackets refer to "route flow"

# 8.2 Dynamic User-Optimal Mode Choice/Departure Time/Route Choice Model

In this section, we discuss the equilibrium conditions and model formulation, the nested diagonalization method, and a numerical example for the dynamic user-optimal mode choice/departure time/route choice model.

## 8.2.1 Equilibrium Conditions and Model Formulation

The dynamic user-optimal equilibrium conditions are first defined to characterize the travelers' travel behavior for choosing the best transportation mode using the minimal travel time route. Due to inherent link interactions, the variational inequality approach is adopted to formulate the DUO mode choice/departure time/route choice model. The equivalence between the dynamic user-optimal conditions and the variational inequality formulation is then stated by a theorem and verified by a proof.

### 8.2.1.1 Dynamic user-optimal conditions

Suppose we have binary choice modes, where the auto mode denoted by $m_1$ and the transit mode is denoted by $m_2$. When mode and the route choices are considered, the corresponding dynamic user-optimal conditions may be characterized by equilibrium conditions on both route choice behavior and mode choice functions.

For route choice behavior, these conditions state for each O-D pair and transportation mode that the actual route travel times experienced by travelers departing from the same origin, regardless of the departure time are equal and minimal, or no traveler would be better off by unilaterally changing his/her route. In contrast, the actual route travel time of any unused route for each O-D pair and mode is greater than or equal to the minimal actual route travel time among all modes. Therefore, if the flow of mode $m \in \{m_1, m_2\}$ departing from origin $r$ during interval $k$ over route $p$ toward destination $s$ is positive, i.e., $h_{mp}^{rs*}(k) > 0$, then the corresponding actual route travel time is minimal among all modes. However, if no flow of mode $m \in \{m_1, m_2\}$ occurs on route $p$, i.e., $h_{mp}^{rs*}(k) = 0$, then the corresponding actual route travel time is at least as great as the minimal actual route travel time among all modes. These equilibrium conditions can be mathematically expressed as follows.

$$c_{mp}^{rs*}(k) \begin{cases} = \pi_m^{rs} & \text{if } h_{mp}^{rs*}(k) > 0 \\ \geq \pi_m^{rs} & \text{if } h_{mp}^{rs*}(k) = 0 \end{cases} \quad \forall r, s, m, p, k \tag{8.48}$$

For the mode choice functions, the auto choice function is assumed to be a strictly decreasing function of the difference between the auto and transit minimal route travel times. Thus, the inverse auto choice function has a strictly monotone mapping. The general form can be expressed mathematically as follows:

$$\pi_{m_1}^{rs} - \pi_{m_2}^{rs} = \left(D_{m_1}^{rs*}\right)^{-1} \quad \forall r,s \tag{8.49}$$

The widely used inverse auto choice function is as follows:

$$\left(D_{m_1}^{rs*}\right)^{-1} = \frac{1}{\theta}\ln\frac{\bar{q}^{rs} - q_{m_1}^{rs}}{q_{m_1}^{rs}} + \varphi^{rs} \quad \forall r,s \tag{8.50}$$

where $\varphi^{rs}$ represents the preference for auto trips that depart from origin $r$ toward destination $s$, and $\theta$ indicates a dispersion/variance parameter. Note that when equation (8.50) is inserted into equation (8.49), the following mode choice functions result.

$$q_{m_1}^{rs*} = \bar{q}^{rs}\frac{1}{1+e^{\theta\left(\pi_{m_1}^{rs} - \pi_{m_2}^{rs} - \varphi^{rs}\right)}} \quad \forall r,s \tag{8.51}$$

$$q_{m_2}^{rs*} = \bar{q}^{rs} - q_{m_1}^{rs*} \quad \forall r,s \tag{8.52}$$

#### 8.2.1.2 Variational inequality formulation

The DUO mode choice/departure time/route choice problem can be formulated using the variational inequality approach.

***Theorem 8.4***: The DUO mode choice/departure time/route choice problem is equivalent to finding a solution $(\mathbf{u}^*, \mathbf{q}^*) \in \Omega$ such that the following VI formulation holds:

$$\mathbf{c}^*(\mathbf{u} - \mathbf{u}^*) - (\mathbf{D}^*)^{-1}(\mathbf{q} - \mathbf{q}^*) \geq 0 \quad \forall(\mathbf{u},\mathbf{q}) \in \Omega^* \tag{8.53}$$

or, alternatively, in expanded form:

$$\begin{aligned}&\sum_m\sum_a\sum_t c_{ma}^*(t)\left[u_{ma}(t) - u_{ma}^*(t)\right] \\ &\quad - \sum_{rs}\left(D_{m_1}^{rs*}\right)^{-1}\left(q_{m_1}^{rs} - q_{m_1}^{rs*}\right) \quad \forall(\mathbf{u},\mathbf{q}) \in \Omega^*\end{aligned} \tag{8.54}$$

where $c_{m_1a}$, $c_{m_2a}$ are the link travel time functions for auto and transit respectively, and $\left(D_{m_1}^{rs*}\right)^{-1}$ is the inverse auto choice function. The link travel time functions may depend on the flows of both modes, $u_{m_1a}$, $u_{m_2a}$, but are separable by links. The symbol $\Omega^*$ is a subset of $\Omega$ with $\delta_{mapk}^{rs}(t)$ being realized at equilibrium, i.e., $\delta_{mapk}^{rs}(t) = \delta_{mapk}^{rs*}(t)$, $\forall r,s,m \in (m_1,m_2),a,p,k,t$. The symbol $\Omega$ denotes the feasible region that is delineated by the following constraints:

Flow conservation constraint:

$$\sum_m \sum_k \sum_p h_{mp}^{rs}(k) = \bar{q}^{rs} \quad \forall r,s \tag{8.55}$$

Flow propagation constraints:

$$u_{mapk}^{rs}(t) = h_{mp}^{rs}(k)\delta_{mapk}^{rs}(t) \quad \forall r,s,m,a,p,k,t \tag{8.56}$$

$$\sum_t \delta_{mapk}^{rs}(t) = 1 \quad \forall r,s,p,m,a \in p,k \tag{8.57}$$

$$\delta_{mapk}^{rs}(t) = \{0,1\} \quad \forall r,s,m,a,p,k,t \tag{8.58}$$

Nonnegativity constraint:

$$h_{mp}^{rs}(k) \geq 0 \quad \forall r,s,p,m,k \tag{8.59}$$

Definitional constraints:

$$\sum_k \sum_p h_{mp}^{rs}(k) = q_m^{rs} \quad \forall r,s,m \tag{8.60}$$

$$u_{ma}(t) = \sum_{rs} \sum_p \sum_k h_{mp}^{rs}(k)\delta_{mapk}^{rs}(t) \quad \forall m,a,t \tag{8.61}$$

$$c_{mp}^{rs}(k) = \sum_a \sum_t c_{ma}(t)\delta_{mapk}^{rs}(t) \quad \forall r,s,m,p,k \tag{8.62}$$

Equation (8.55) expresses the time-independent O-D demand in terms of route flows among all modes, and conserves the fixed total number of trips between O-D pair $rs$. Equation (8.60) is definitional, and expresses time-independent O-D demand in terms of time-dependent route flows among all modes. The other constraints are essentially the same as those for the DUO variable demand/departure time/route choice model, except that mode choice is also considered.

#### 8.2.1.3 Equivalence analysis

***Theorem 8.5***: Under a certain flow propagation relationship $\delta_{mapk}^{rs}(t) = \delta_{mapk}^{rs*}(t)$, $\forall r,s,m,a,p,k,t$, DUO mode choice/departure time/route choice conditions (8.48)~(8.49) imply VIP (8.54) and vice versa.

***Proof of necessity***: We need to prove that under a certain flow propagation relationship $\left(\delta_{mapk}^{rs}(t) = \delta_{mapk}^{rs*}(t), \forall r,s,m,a,p,k,t\right)$, DUO mode choice/departure time/route choice conditions (8.48)~(8.49) imply VIP (8.54). We first rearrange equilibrium conditions (8.48) as follows:

$$\left[c_{mp}^{rs*}(k) - \pi_m^{rs}\right]\left[h_{mp}^{rs}(k) - h_{mp}^{rs*}(k)\right] \geq 0 \quad \forall r,s,m,p,k \tag{8.63}$$

By summing over $r,s,m,p,k$, and then making a substitution of $\sum_k \sum_p h_{mp}^{rs}(k) = q_m^{rs}$, $\forall r,s,m$, one obtains:

$$\sum_{rs}\sum_m\sum_p\sum_k c_{mp}^{rs*}(k)\left[h_{mp}^{rs}(k) - h_{mp}^{rs*}(k)\right] - \sum_{rs}\sum_m \pi_m^{rs}\left[q_m^{rs} - q_m^{rs*}\right] \geq 0 \tag{8.64}$$

By applying equation (8.62), one obtains:

$$\begin{aligned}\sum_{rs}\sum_m\sum_k\sum_p\left[\sum_a\sum_t c_{ma}^{*}(t)\delta_{mapk}^{rs*}(t)\right]\left[h_{mp}^{rs}(k) - h_{mp}^{rs*}(k)\right]\\ - \sum_{rs}\sum_m \pi_m^{rs}\left[q_m^{rs} - q_m^{rs*}\right] \geq 0\end{aligned} \tag{8.65}$$

By changing the order of the summation on the first term, using equation (8.61), it follows:

$$\sum_m\sum_a\sum_t c_{ma}^{*}(t)\left[u_{ma}(t) - u_{ma}^{*}(t)\right] - \sum_{rs}\sum_m \pi_m^{rs}\left[q_m^{rs} - q_m^{rs*}\right] \geq 0 \tag{8.66}$$

Since $q_{m_2}^{rs} = \bar{q}^{rs} - q_{m_1}^{rs}$, using equation (8.49), we have:

$$\sum_m\sum_a\sum_t c_{ma}^{*}(t)\left[u_{ma}(t) - u_{ma}^{*}(t)\right] - \sum_{rs}\left(D_{m_1}^{rs*}\right)^{-1}\left[q_{m_1}^{rs} - q_{m_1}^{rs*}\right] \geq 0 \tag{8.67}$$

The inequality (8.67) is identical to VIP (8.54).

***Proof of sufficiency:*** We next prove that the VIP (8.54) implies the mode choice/departure time/route choice equilibrium conditions (8.48)~(8.49). First we demonstrate that VIP (8.54) implies expression (8.48). Without loss of generality, we now define the feasible solution $\{h_{mp}^{rs}(k),\ q_m^{rs*}\}$ be the same as $\{h_{mp}^{rs*}(k),\ q_m^{rs*}\}$, except for two routes, $p_1^{rs}$ during interval $k_1$, and $p_2^{rs}$ during interval $k_2$, for mode $m_1$. We consider two situations that could arise at equilibrium:

(i) Both route flows are positive, i.e., $h_{m_1p_1}^{rs*}(k_1) > 0$ and $h_{m_1p_2}^{rs*}(k_2) > 0$. We switch a small amount of flow $\Delta_1$ from route $p_1^{rs}$ to $p_2^{rs}$ with $0 < \Delta_1 \leq h_{m_1p_1}^{rs*}(k_1)$. That is:

$$h_{m_1p_1}^{rs}(k_1) = h_{m_1p_1}^{rs*}(k_1) - \Delta_1 \text{ and} \tag{8.68}$$

$$h_{m_1p_2}^{rs}(k_2) = h_{m_1p_2}^{rs*}(k_2) + \Delta_1 \tag{8.69}$$

Substituting the feasible solution $\{h_{mp}^{rs}(k),\ q_m^{rs*}\}$ into equation (8.54) yields:

$$c_{m_1p_1}^{rs*}(k_1)\left[h_{m_1p_1}^{rs}(k_1) - h_{m_1p_1}^{rs*}(k_1)\right] + c_{m_1p_2}^{rs*}(k_2)\left[h_{m_1p_2}^{rs}(k_2) - h_{m_1p_2}^{rs*}(k_2)\right] \geq 0 \tag{8.70}$$

By applying equations (8.68) and (8.69) and dividing by $\Delta_1$, we have:

$$c_{m_1p_2}^{rs*}(k_2) \geq c_{m_1p_1}^{rs*}(k_1) \tag{8.71}$$

Similarly, by switching a small amount of flow $\Delta_2$ with $0 < \Delta_2 \leq h_{m_1p_2}^{rs*}(k_2)$ from route $p_2^{rs}$ to $p_1^{rs}$, we obtain:

$$c_{m_1p_1}^{rs*}(k_1) \geq c_{m_1p_2}^{rs*}(k_2) \tag{8.72}$$

Equations (8.71) and (8.72) together imply:

$$c_{m_1p_2}^{rs*}(k_2) = c_{m_1p_1}^{rs*}(k_1) \tag{8.73}$$

We repeat this procedure to verify for each O-D pair and mode that all used routes with positive flow have the same actual route travel time.

(ii) One route flow is positive and the other route flow is nil. We arbitrarily assume, without loss of generality, $h_{m_1p_1}^{rs*}(k_1) > 0$ and $h_{m_1p_2}^{rs*}(k_2) = 0$. We switch a small amount of flow $\Delta_1$ from route $p_1^{rs}$ to $p_2^{rs}$ with $0 < \Delta_1 \leq h_{m_1p_1}^{rs*}(k_1)$. By the same argument shown in (i), we have $c_{m_1p_2}^{rs*}(k_2) \geq c_{m_1p_1}^{rs*}(k_1)$. We repeat this procedure to verify for each O-D pair and mode that all unused routes with zero flow have higher or equal actual route travel times.

By cases (i) and (ii), using expression (8.1), it follows that VIP (8.54) implies expression (8.48).

We now conclude the proof by showing that VIP (8.51) implies equation (8.49). As before, we construct a feasible solution $\{h_{mp}^{rs}(k), q_m^{rs}\}$, that differs from $\{h_{mp}^{rs*}(k), q_m^{rs*}\}$ only for the flows of the two used routes, $p_1$ of mode $m_1$, and $p_2$ of mode $m_2$, for which $h_{m_1p_1}^{rs}(k) = h_{m_1p_1}^{rs*}(k) - \Delta$, $h_{m_2p_2}^{rs}(k) = h_{m_2p_2}^{rs*}(k) + \Delta$, where $0 < |\Delta| \leq \min\left(h_{m_1p_1}^{rs*}(k),\ h_{m_2p_2}^{rs*}(k)\right)$. Note that $q_{m_1}^{rs} = q_{m_1}^{rs*} - \Delta$, and $q_{m_2}^{rs} = q_{m_2}^{rs*} + \Delta$. Applying VIP (8.54) yields $c_{m_1p_1}^{rs}(k)(-\Delta) + c_{m_2p_2}^{rs}(k)\Delta + \left(D_m^{rs*}\right)^{-1}\Delta \geq 0$, which after dividing by $\Delta$, and using equation (8.48), results in $-\pi_{m_1}^{rs} + \pi_{m_2}^{rs} + \left(D_{m_1}^{rs*}\right)^{-1} \geq 0$, for $\Delta > 0$, and $-\pi_{m_1}^{rs} + \pi_{m_2}^{rs} + \left(D_{m_1}^{rs*}\right)^{-1} \leq 0$, for $\Delta < 0$; hence, equation (8.49) is satisfied. This completes the proof.

### 8.2.2 Nested Diagonalization Method

In this section, we first show the network representation for the DUO mode choice/departure time/route choice problem. Then, the nested diagonalization method is formally stated.

#### 8.2.2.1 Time-space network

For the DUO mode choice/departure time/route choice model, the corresponding time-space network is shown in Figure 8.4, in which the auto and transit modes share a basic network, but are separated by links.

#### 8.2.2.2 Solution algorithm

We adopt the nested diagonalization method to solve the DUO mode choice/departure time/route choice model as follows.

*Nested Diagonalization Method*

**Step 0: Initialization.**

Step 0.1: Let $m'$=0. Set $\tau_a^0(t) = NINT\left[c_{a_0}(t)\right], \forall a, t$.

Step 0.2: Let $n$=1. Find an initial feasible solution $\left\{\left(q_m^{rs}\right)^1, u_{ma}^1(t)\right\}$. Compute the associated link travel times $\left\{c_{ma}^1(t)\right\}$.

**Step 1: *First Loop* Operation.**

Let $m' = m'+1$. Update the estimated actual link travel times by

$$\tau_a^{m'}(t) = NINT\left[(1-\gamma)\tau_a^{m'-1}(t) + \gamma c_a^n(t)\right] \quad \forall a, t \tag{8.74}$$

where $0 < \gamma \leq 1$.

Construct the corresponding feasible time-space network based on the estimated actual link travel times.

**Step 2: *Second Loop* Operation.**

Step 2.1: Let $n$=1. Compute and reset the initial feasible solution $\left\{\left(q_m^{rs}\right)^n, u_{ma}^n(t)\right\}$, based on the time-space network constructed by the estimated actual link travel times $\left\{\tau_{ma}^{m'}(t)\right\}$.

Step 2.2: Fix the inflows for all time-space links other than on the subject time-space link at the current level, yielding the following optimization problem.

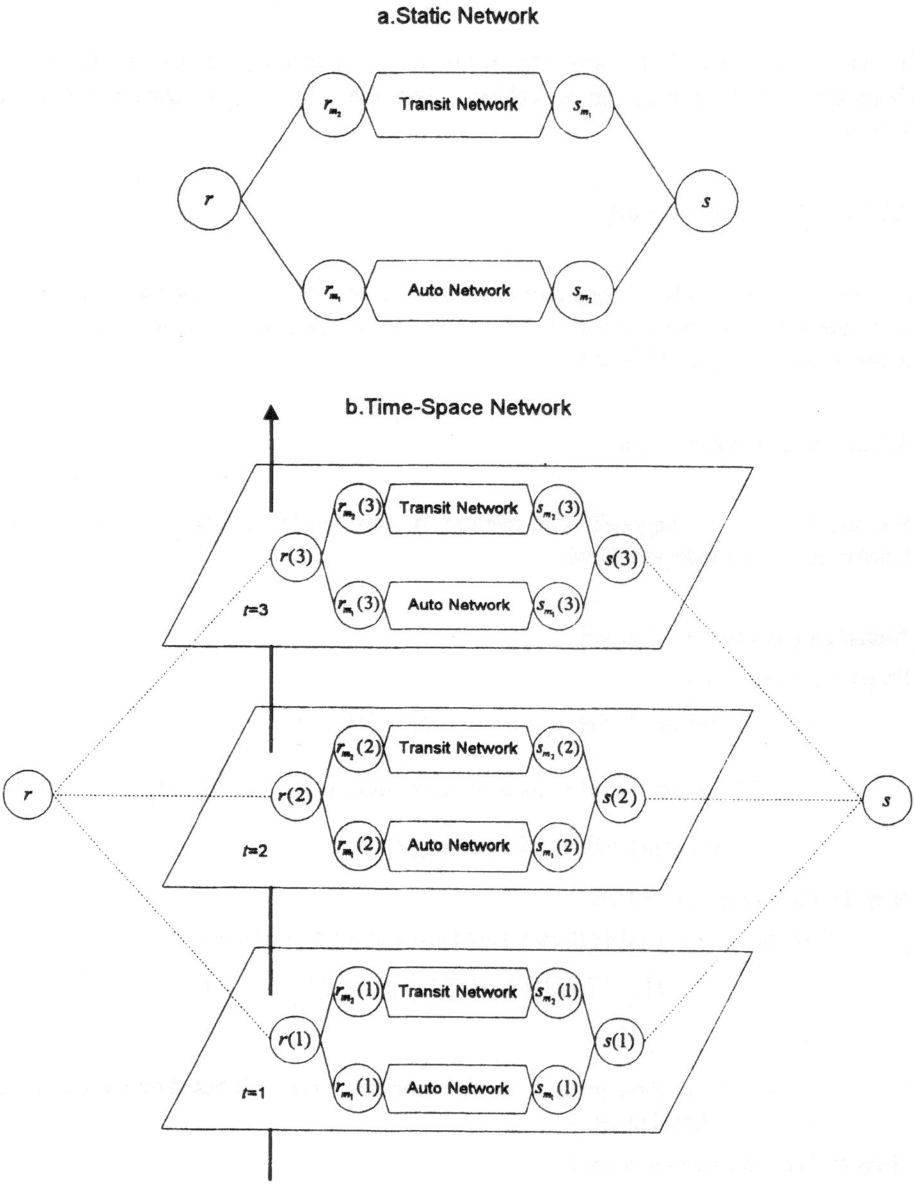


Figure 8.4: Time-Space Network for the Dynamic User-Optimal Mode Choice/Departure Time/Route Choice Model

$$\min\ z(\mathbf{u},\mathbf{q})=\sum_a\sum_t\int_0^{u_{m_1a}^{n+1}(t)} c_{m_1a}\left(u_{m_1a}^n(1),\cdots,u_{m_1a}^n(t-1),\omega,\mathbf{u}_{m_2}^n\right)d\omega$$
$$+\sum_a\sum_t\int_0^{u_{m_2a}^{n+1}(t)} c_{m_2a}\left(\mathbf{u}_{m_1}^n,u_{m_2a}^n(1),\cdots,u_{m_2a}^n(t-1),\omega\right)d\omega \quad (8.75)$$
$$-\sum_{rs}\sum_k\int_0^{q_{m_1}^{rs\ n+1}}\left(D_{m_1}^{rs}(\omega)\right)^{-1}d\omega$$

Flow conservation constraint:

$$\sum_m\sum_k\sum_p h_{mp}^{rs}(k)=\bar{q}^{rs} \quad \forall r,s \quad (8.76)$$

Nonnegativity constraint:

$$h_{mp}^{rs}(k)\ge 0 \quad \forall r,s,m,p,k \quad (8.77)$$

Definitional constraints:

$$u_{mapk}^{rs}(t)=h_{mp}^{rs}(k)\bar{\delta}_{mapk}^{rs}(t) \quad \forall r,s,m,a,p,k,t \quad (8.78)$$

$$\bar{\delta}_{mapk}^{rs}(t)=\{0,1\} \quad \forall r,s,m,a,p,k,t \quad (8.79)$$

$$\sum_k\sum_p h_{mp}^{rs}(k)=q_m^{rs} \quad \forall r,s,m \quad (8.80)$$

$$u_{ma}(t)=\sum_{rs}\sum_p\sum_k h_{mp}^{rs}(k)\bar{\delta}_{mapk}^{rs}(t) \quad \forall m,a,t \quad (8.81)$$

$$c_{mp}^{rs}(k)=\sum_a\sum_t c_{ma}(t)\bar{\delta}_{mapt}^{rs}(t) \quad \forall r,s,m,p,k \quad (8.82)$$

**Step 3: *Third Loop* Operation.**

Solve for the solution, $\left\{q_m^{rs}(k)^{n+1},\ u_{ma}^{n+1}(t)\right\}$, in optimization problem (8.75)~(8.82) by the partial linearization technique, also known as the Evans' method. Compute the resulting link travel times $\left\{c_{ma}^{n+1}(t)\right\}$.

**Step 4: Convergence Check for the *Second Loop* Operation.**

If $u_a^{n+1}(t)\approx u_a^n(t),\forall a,t$ and $\left(q_m^{rs}\right)^{n+1}\approx\left(q_m^{rs}\right)^n,\forall r,s,m$, go to Step 5; otherwise, set $n=n+1$, go to Step 2.2.

**Step 5: Convergence Check for the *First Loop* Operation.**

If $\tau_{ma}^{m'}(t)\approx c_{ma}^{n+1}(t),\forall m,a,t$, stop; the current solution is optimal. Otherwise, set $n=n+1$, and go to Step 1.

Analogous to the discussion in Section 5.3, the nested projection algorithm can be obtained by the following modification.

Step 2.2: Fix the inflows for all time-space links other than on the subject time-space link at the current level, yielding the following optimization problem.

$$\min\ z(\mathbf{u},\mathbf{q}) = \frac{1}{2}\left(\mathbf{u}^{n+1}\right)^T \mathbf{G}_1 \mathbf{u}^{n+1} + \left(\rho_1 \mathbf{c}\left(\mathbf{u}^n\right) - \mathbf{G}_1 \mathbf{u}^n\right)\mathbf{u}^{n+1}$$
$$-\frac{1}{2}\left(\mathbf{q}^{n+1}\right)^T \mathbf{G}_2 \mathbf{q}^{n+1} - \left(\rho_2 \left(\mathbf{D}(\mathbf{q}^n)\right)^{-1} - \mathbf{G}_2 \mathbf{q}^n\right)\mathbf{q}^{n+1} \quad (8.83)$$

Subject to: Equations (8.76)~(8.82)

where matrices $\mathbf{G}_1$ and $\mathbf{G}_2$ are symmetric and positive definite, and $\rho_1$ and $\rho_2$ are contraction operators.

### 8.2.2.3 Single transit link with constant travel time

In many circumstances, a single transit link, such as a subway, is operated on a closed system with constant travel time $\bar{\pi}^{rs}_{m_2}$. By an appropriate network representation, this specific DUO mode choice/departure time/route choice problem can be treated as the DUO route choice problem. In Figure 8.5(a), we first assume that the network consists of a basic auto network and a single transit link with constant travel time. In Figure 8.5(b), the corresponding time-space network contains three types of artificial time-space links. The first type of artificial time-space links connects time-independent origins $r$ to the time-independent destination $s$. The associated travel time functions are set as $c^{rs}_{m_2} = \bar{\pi}^{rs}_{m_2} + \left(D^{rs}_{m_1}\right)^{-1}$. The second type of artificial time-space links directly joins each time-independent origin $r$ to the auto-specific time-dependent origin $r_{m_1}(k)$, where the corresponding travel time function is defined as $c^{rr_m}(k) = 0$. The third type of artificial time-space links directly joins each auto-specific time-dependent destination $s_{m_1}(k)$ to the time-independent destination $s$, and the corresponding travel time function is defined as $c^{s_m s}(k) = 0$. The following theorem states that this specific DUO mode choice/departure time/route choice problem is equivalent to, in the mathematical sense, the DUO route choice problem.

***Theorem* 8.6**: The DUO mode choice/route choice problem is equivalent to the DUO route choice problem with the time-space network shown in Figure 8.5(b).

***Proof***: For the DUO route choice problem with time-space network shown in Figure 8.5(b), the corresponding VIP is as follows:

$$\sum_a \sum_t c^*_{m_1 a}(t)\left[u_{m_1 a}(t) - u^*_{m_1 a}(t)\right] + \sum_{rs}\left[\bar{\pi}^{rs}_{m_2} + \left(D^{rs*}_{m_1}\right)^{-1}\right]\left[q^{rs}_{m_2} - q^{rs*}_{m_2}\right] \geq 0 \quad \forall(\mathbf{u},\mathbf{q}) \in \Omega^* \tag{8.84}$$

By rearranging the second term, using $q^{rs}_{m_2} = \bar{q}^{rs} - q^{rs}_{m_1}$, one obtains:

$$\begin{aligned}&\sum_a \sum_t c^*_{m_1 a}(t)\left[u_{m_1 a}(t) - u^*_{m_1 a}(t)\right] + \sum_{rs} \bar{\pi}^{rs}_{m_2}\left(q^{rs}_{m_2} - q^{rs*}_{m_2}\right) \\ &\quad - \sum_{rs}\left(D^{rs*}_{m_1}\right)^{-1}\left[q^{rs}_{m_1} - q^{rs*}_{m_1}\right] \geq 0 \quad \forall(\mathbf{u},\mathbf{q}) \in \Omega^*\end{aligned} \tag{8.85}$$

Since the second term is equivalent to $\sum_a \sum_t c^*_{m_2 a}(t)\left[u_{m_2 a}(t) - u^*_{m_2 a}(t)\right]$, we have:

$$\sum_m \sum_a \sum_t c^*_{ma}(t)\left[u_{ma}(t) - u^*_{ma}(t)\right] - \sum_{rs}\left(D^{rs*}_{m_1}\right)^{-1}\left[q^{rs}_{m_1} - q^{rs*}_{m_1}\right] \geq 0 \quad \forall(\mathbf{u},\mathbf{q}) \in \Omega^* \tag{8.86}$$

The inequality (8.86) is the exact VIP for the DUO mode choice/route choice problem as shown in expression (8.54). The steps shown above are essentially reversible. This completes the proof.

When mode choice decisions are described by the following logit formula:

$$q^{rs}_{m_1} = \bar{q}^{rs}\frac{1}{1 + e^{\theta\left(\pi^{rs}_{m_1} - \bar{\pi}^{rs}_{m_2} - \varphi^{rs}\right)}} \quad \forall r,s \tag{8.87}$$

and we further assume that the dispersion parameter is set as $\theta = 1$, and the auto preference parameters $\varphi^{rs} = 0, \forall r,s$, then the nested diagonalization method can still be adopted for solutions with the following modifications.

Step 2.2: Fix the inflows for all physical links other than on the subject time-space link at the current level, yielding the following optimization problem.

$$\begin{aligned}\min z(\mathbf{u},\mathbf{q}) = &\sum_a \sum_t \int_0^{u^{n+1}_{m_1 a}(t)} c_{m_1 a}\left(u^n_{m_1 a}(1), \cdots, u^n_{m_1 a}(t-1), \omega, \mathbf{u}^n_{m_2}\right)d\omega \\ &+ \sum_{rs}\left(q^{rs}_{m_2}\right)^{n+1}\bar{\pi}^{rs\,n+1}_{m_2} - \sum_{rs}\int_0^{q^{rs\,n+1}_{m_1}} \ln\frac{\bar{q}^{rs} - \omega}{\omega}\,d\omega\end{aligned} \tag{8.88}$$

Subject to: Equations (8.76)~(8.82)

**Step 4: Convergence Check for the *Second Loop* Operation.**

If the following convergence criteria are met, go to Step 5; otherwise, set $n=n+1$, and go to Step 2.2.

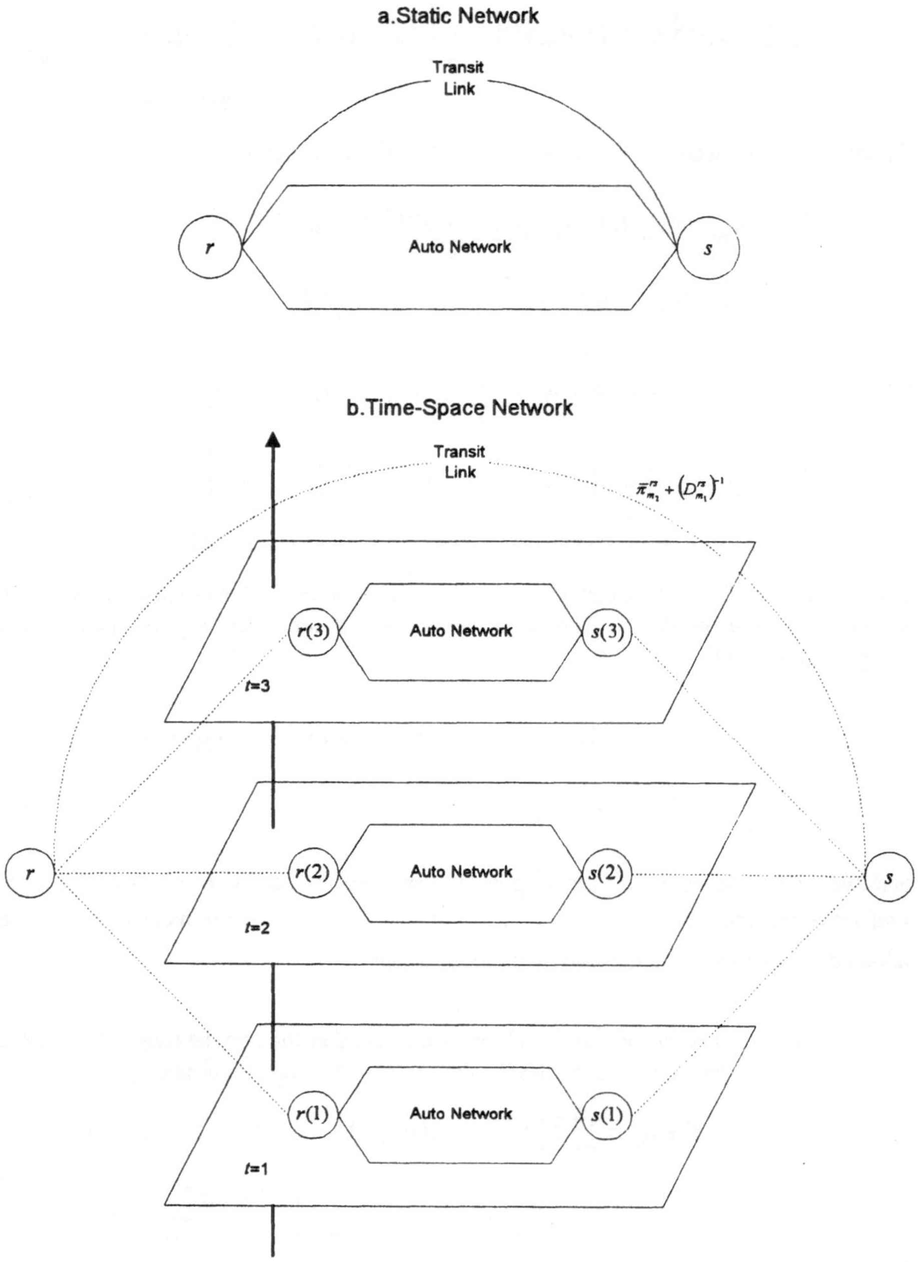


Figure 8.5: Time-Space Network for the Dynamic User-Optimal Mode Choice/Departure Time/Route Choice Model (Single Transit Link with Constant Travel Time)

$$u_a^{n+1}(t) \approx u_a^n(t) \quad \forall a,t \text{ and} \tag{8.89}$$

$$\frac{q_{m_1}^{rs\ n+1}}{\bar{q}^{rs}} \approx \frac{1}{1+e^{\left(c_{m_1}^{rs\ n+1} - \bar{\pi}_{m_1}^{rs\ n+1}\right)}} \quad \forall r,s \tag{8.90}$$

## 8.2.3 Numerical Example

### 8.2.3.1 Input data

A simple network shown in Figure 8.6 is used for testing. The test network consists of 8 links (of which links 1→5 and 2→5 denote the transit links) and 5 nodes, in which nodes 1 and 2 are origin nodes, node 5 is the destination, and nodes 3 and 4 are intermediate nodes.

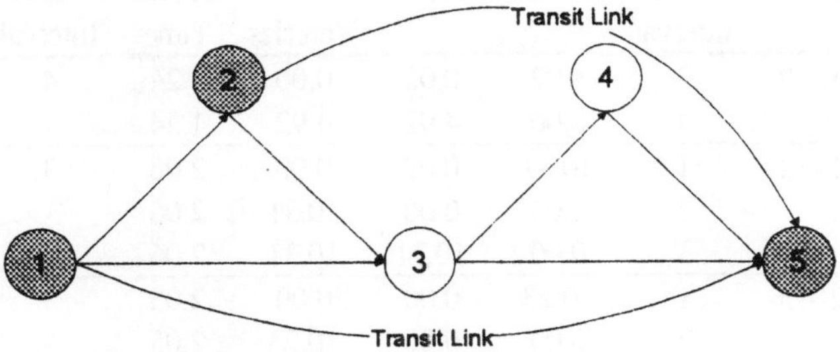


Figure 8.6: Test Network 2

The adopted dynamic travel time functions are arbitrarily constructed as follows:

$$c_{m_1a}(t) = 1 + 0.01\left(u_{m_1a}(t)\right)^2 + 0.01\left(x_{m_1a}(t)\right)^2 \quad \forall a,t \tag{8.91}$$

$$c_{m_2a} = 5 + \left(D^{rs}\right)^{-1} \quad \forall a,t \tag{8.92}$$

The time-independent origin-destination (O-D) demands are assumed in Table 8.5:

Table 8.5: Time-Independent Maximal O-D Demands for Test Network 2

| O-D Pair | Time Interval $k$=1~3 |
|---|---|
| 1-5 | 30 |
| 2-5 | 20 |

The mode choice functions are assumed to be the following logit formulas:

$$q_{m_1}^{rs} = \bar{q}^{rs} \frac{1}{1+e^{\left(\pi_{m_1}^{rs} - \bar{\pi}_{m_2}^{rs}\right)}} \qquad \forall r,s \tag{8.93}$$

$$q_{m_2}^{rs} = \bar{q}^{rs} \frac{1}{1+e^{\left(\bar{\pi}_{m_2}^{rs} - \pi_{m_1}^{rs}\right)}} \qquad \forall r,s \tag{8.94}$$

#### 8.2.3.2 Test results

The resulting flow pattern and time-independent O-D demands among all modes for the DUO mode choice/departure time/route choice model with the given data are summarized in Tables 8.6 and 8.7, respectively.

Table 8.6: Resulting Flow Pattern for Test Network 2

| Link | Entering Time Interval | Inflow | Exit Flow | Number of Vehicles | Link Travel Time | Exiting Time Interval |
|---|---|---|---|---|---|---|
| 1→2 | 3 | 4.92 | 0.00 | 0.00 | 1.24 | 4 |
| | 4 | 0.00 | 4.92 | 4.92 | 1.24 | - |
| 1→3 | 1 | 10.31 | 0.00 | 0.00 | 2.06 | 3 |
| | 2 | 0.00 | 0.00 | 10.31 | 2.06 | - |
| | 3 | 0.00 | 10.31 | 10.31 | 2.06 | - |
| 2→3 | 1 | 10.23 | 0.00 | 0.00 | 2.05 | 3 |
| | 2 | 0.00 | 0.00 | 10.23 | 2.05 | - |
| | 3 | 0.00 | 10.23 | 10.23 | 2.05 | - |
| | 4 | 4.92 | 0.00 | 0.00 | 1.24 | 5 |
| | 5 | 0.00 | 4.92 | 4.92 | 1.24 | - |
| 3→4 | 3 | 6.74 | 0.00 | 0.00 | 1.45 | 4 |
| | 4 | 0.00 | 6.74 | 6.74 | 1.45 | - |
| | 5 | 4.92 | 0.00 | 0.00 | 1.24 | 6 |
| | 6 | 0.00 | 4.92 | 4.92 | 1.24 | - |
| 3→5 | 3 | 13.80 | 0.00 | 0.00 | 2.91 | 6 |
| | 4 | 0.00 | 0.00 | 13.80 | 2.91 | - |
| | 5 | 0.00 | 0.00 | 13.80 | 2.91 | - |
| | 6 | 0.00 | 13.80 | 13.80 | 2.91 | - |
| 4→5 | 4 | 6.74 | 0.00 | 0.00 | 1.45 | 5 |
| | 5 | 0.00 | 6.74 | 6.74 | 1.45 | - |
| | 6 | 4.92 | 0.00 | 0.00 | 1.24 | 7 |
| | 7 | 0.00 | 4.92 | 4.92 | 1.24 | - |

Table 8.7: Resulting Time-Independent Mode-Specific O-D Demands for Test Network 2

| O-D Pair | Time Interval $k$=1~3 |
|---|---|
| 1-5 | 15.23/(14.77)* |
| 2-5 | 10.23/(9.77) |

* "Numbers" in brackets denote transit O-D demands

The corresponding actual route travel times are computed and summarized in Table 8.8.

Table 8.8: Actual Route Travel Times for Test Network 2

| Route | Time Interval | | |
|---|---|---|---|
| | $k$=1 | $k$=2 | $k$=3 |
| 1→2→3→4→5 | NA | – | 4.97/(4.92) |
| 1→3→4→5 | 4.97/(6.72)* | – | – |
| 1→3→5 | 4.97/(3.58) | – | – |
| 1→5 (Transit Demand) | 4.97/(14.77) | – | – |
| 2→3→4→5 | 4.95/(0.01) | – | – |
| 2→3→5 | 4.95/(10.22) | – | – |
| 2→5 (Transit Demand) | 4.95/(9.77) | – | – |

* "Numbers" in brackets refer to route flows

## 8.3 Notes

Florian and Spiess (1983) showed that the convergence of the iterative method proposed for problem (8.7)~(8.15) is ensured by sufficient conditions that are not always satisfied for all instances of their static model. On the other hand, the sufficient conditions that ensure the uniqueness of the solution are less stringent. Their experience shows that applications of this method, in practice, have always resulted in satisfactory convergence.

When the flow propagation relationships are prespecified, we can use a similar approach to derive sufficient conditions for the DUO mode choice/route choice model. We first reformulate link-based VIP (8.7) as a route-based VIP as follows.

$$\sum_{rs}\sum_{m}\sum_{p}\sum_{k} c_{mp}^{rs*}(k)\left[h_{mp}^{rs}(k) - h_{mp}^{rs*}(k)\right] - \sum_{rs}\sum_{k}\left(D_{m_1}^{rs*}(k)\right)^{-1}\left[q_{m_1}^{rs}(k) - q_{m_1}^{rs*}(k)\right] \geq 0 \qquad (8.95)$$

At equilibrium, any used route has the minimal travel time $\pi_m^{rs}(k)$. When we assign the time-dependent O-D demand $q_m^{rs}(k)$, to any route with the minimal travel time and by using the fact that $q_{m_1}^{rs}(k)+q_{m_2}^{rs}(k)=\bar{q}^{rs}(k)$, it follows that:

$$\sum_{rs}\sum_{k}\left[\pi_{m_1}^{rs}(k)-\pi_{m_2}^{rs}(k)\right]\left[q_{m_1}^{rs}(k)-q_{m_1}^{rs*}(k)\right]$$
$$-\sum_{rs}\sum_{k}\left(D_{m_1}^{rs*}(k)\right)^{-1}\left[q_{m_1}^{rs}(k)-q_{m_1}^{rs*}(k)\right]\geq 0 \tag{8.96}$$

By manipulation, we have:

$$\sum_{rs}\sum_{k}\left[\pi_{m_1}^{rs}(k)-\pi_{m_2}^{rs}(k)-\left(D_{m_1}^{rs*}(k)\right)^{-1}\right]\left[q_{m_1}^{rs}(k)-q_{m_1}^{rs*}(k)\right]\geq 0 \tag{8.97}$$

Let $F\left(q_{m_1}^{rs}(k)\right)=\pi_{m_1}^{rs}(k)-\pi_{m_2}^{rs}(k)-\left(D_{m_1}^{rs*}(k)\right)^{-1}$. Then, the above inequality may be restated as:

$$\sum_{rs}\sum_{k}F\left(q_{m_1}^{rs}(k)\right)\left(q_{m_1}^{rs}(k)-q_{m_1}^{rs*}(k)\right)\geq 0 \tag{8.98}$$

To justify the monotonicity of $F\left(q_{m_1}^{rs}(k)\right)$, one of the following conditions must hold:

i) $\left(\pi_{m_1}^{rs}(k)-\pi_{m_2}^{rs}(k)-\left(D_{m_1}^{rs*}(k)\right)^{-1}\right)$ is strictly monotone.

ii) $\left(\pi_{m_2}^{rs}(k)-\left(D_{m_1}^{rs*}(k)\right)^{-1}\right)$ is strictly monotone, since $\pi_{m_1}^{rs}(k)$ is monotone.

iii) $\left(\pi_{m_1}^{rs}(k)-\pi_{m_2}^{rs}(k)\right)$ is monotome since $-\left(D_{m_1}^{rs*}(k)\right)^{-1}$ is strictly monotone.

iv) $\pi_{m_2}^{rs}(k)$ is monotone since $\left(\pi_{m_1}^{rs}(k)-\left(D_{m_1}^{rs*}(k)\right)^{-1}\right)$ is strictly monotone.

Conditions (ii), (iii), (iv) are stronger than (i), and therefore imply it. Generally speaking, the monotonicity of $F\left(q_{m_1}^{rs}(k)\right)$ is likely to be satisfied in practice, since the inverse mode choice functions are quite steep; that is, a change in $q_{m_1}^{rs}(k)$ induces a much smaller change in $\left(\pi_{m_1}^{rs}(k)-\pi_{m_2}^{rs}(k)\right)$ than in the corresponding $\left(D_{m_1}^{rs*}(k)\right)^{-1}$.

# Chapter 9

# Dynamic User-Optimal Singly Constrained O-D Choice Models

Dynamic user-optimal singly constrained O-D choice models assume that the total number of trips either leaving each origin, or arriving at each destination, is predetermined. From the computational point of view, both types of situations can be covered within a unified framework by simply reversing the direction of the traffic assignment procedure. As an illustration, we only consider DUO singly constrained models with trips fixed at origins throughout this chapter.

DUO singly constrained models are appropriate for analyzing travelers' behavior on non-work trips because only the total number of trips leaving each origin is prespecified. During a national holiday or a weekend, a great number of people are inclined to leave home to enjoy their leisure time; but the decision as to where to go usually involves many objective factors and subjective preferences such as the attractiveness of destinations or required in-vehicle travel times. In many circumstances, the departure time of travelers is not necessarily restricted, and the choice of departure time is strongly influenced in such a way that the necessary *en route* travel time is also minimized. This issue has been a subject of contention. Some researches do not agree with such a hypothesis. The purpose of including this statement is to show that the mathematical formulation for such a hypothesis could be made. Consequently, two types of problems can be identified, i.e., the DUO singly constrained O-D choice/route choice problem, and the DUO singly constrained O-D choice/departure time/route choice problem.

In the following sections, the DUO singly constrained O-D choice/route choice model, including the equilibrium conditions and model formulation, the nested diagonalization method and a numerical example, is first considered in Section 9.1. A similar discussion for the DUO singly constrained O-D choice/departure time/route choice model is provided in Section 9.2. Finally, concluding notes are given in Section 9.3.

# 9.1 Dynamic User-Optimal Singly Constrained O-D Choice/Route Choice Model

In this section, we discuss the equilibrium conditions and model formulation, the nested diagonalization method and a numerical example.

## 9.1.1 Equilibrium Conditions and Model Formulation

The dynamic user-optimal conditions are defined to characterize the travelers' travel behavior for choosing the preferred destination using the minimal travel time route. Due to inherent link interactions, the variational inequality approach is adopted to formulate the DUO O-D choice/route choice model. The equivalence between the dynamic user-optimal conditions and the variational inequality formulation is then stated by a theorem and verified by a proof.

### 9.1.1.1 Dynamic user-optimal conditions

When the trip decision for choosing a destination and route choice is considered, the corresponding dynamic user-optimal conditions may be characterized by equilibrium conditions on both the route choice behavior and time-dependent O-D demands.

For route choice behavior, the conditions state for each O-D pair that the actual route travel times experienced by travelers departing from the same origin during the same interval are equal and minimal, or no traveler would be better off by unilaterally changing his/her route. In contrast, the actual route travel time of any unused route for each O-D pair is greater than or equal to the minimal actual route travel time. In other words, at equilibrium, if the flow departing from origin $r$ during interval $k$ over route $p$ toward destination $s$ is positive, i.e., $h_p^{rs*}(k) > 0$, then the corresponding actual route travel time is minimal. On the other hand, if no flow occurs on route $p$, i.e., $h_p^{rs*}(k) = 0$, then the corresponding actual route travel time is at least as great as the minimal actual route travel time. These equilibrium conditions can be mathematically expressed as follows:

$$c_p^{rs*}(k)\begin{cases} = \pi^{rs}(k) & \text{if } h_p^{rs*}(k) > 0 \\ \geq \pi^{rs}(k) & \text{if } h_p^{rs*}(k) = 0 \end{cases} \quad \forall r, s, p, k \tag{9.1}$$

For the time-dependent O-D demands, the O-D demand function is assumed to be a strictly decreasing function of the difference between the minimal time-dependent O-D route travel time and the time-dependent constant, $\bar{c}^r(k)$. Thus, the inverse O-D demand function has a strictly monotone mapping. The general form can be expressed mathematically as follows:

$$\pi^{rs}(k) - \bar{c}^r(k) = \left(D^{rs*}(k)\right)^{-1} \quad \forall r, s, k \tag{9.2}$$

The inverse O-D demand function may be expressed as follows:

$$\left(D^{rs}(k)\right)^{-1} = -\frac{1}{\theta}\ln\left(\frac{q^{rs}(k)}{\overline{q}^{r}(k)-q^{rs}(k)}\sum_{s'\neq s} e^{-\theta\left(\pi^{rs'}(k)-\bar{c}^{r}(k)-M^{s'}(k)\right)}\right)+M^{s}(k) \qquad (9.3)$$

$$\forall r,s,k$$

where $\theta > 0$ denotes a logit parameter and, $M^{s}(k) \geq 0$ denotes the attractiveness associated with destination *s* during interval *k*. Note that when equation (9.3) is inserted into equation (9.2), the following logit model results:

$$q^{rs}(k) = \overline{q}^{r}(k)\frac{e^{-\theta\left(\pi^{rs}(k)-M^{s}(k)\right)}}{\sum_{s'} e^{-\theta\left(\pi^{rs'}(k)-M^{s'}(k)\right)}} \qquad \forall r,s,k \qquad (9.4)$$

### 9.1.1.2 Variational inequality formulation

The DUO singly constrained O-D choice/route choice problem can be formulated using the variational inequality approach.

***Theorem 9.1***: The DUO singly constrained O-D choice/route choice problem is equivalent to finding a vector $(\mathbf{u}^*, \mathbf{q}^*) \in \Omega$ such that the following VIP holds:

$$\mathbf{c}^*(\mathbf{u}-\mathbf{u}^*) - \mathbf{D}^{-1*}(\mathbf{q}-\mathbf{q}^*) \geq 0 \quad \forall(\mathbf{u},\mathbf{q}) \in \Omega^* \qquad (9.5)$$

or, alternatively, in expanded form:

$$\sum_{a}\sum_{t} c_a^*(t)\left[u_a(t)-u_a^*(t)\right] - \sum_{rs}\sum_{k}\left(D^{rs*}(k)\right)^{-1}\left[q^{rs}(k)-q^{rs*}(k)\right] \geq 0 \qquad (9.6)$$

$$\forall(\mathbf{u},\mathbf{q}) \in \Omega^*$$

where $\Omega^*$ is a subset of $\Omega$ with $\delta_{apk}^{rs}(t)$ being realized at equilibrium, i.e., $\delta_{apk}^{rs}(t) = \delta_{apk}^{rs*}(t), \forall r,s,a,p,k,t$. The symbol $\Omega$ denotes the feasible region that is delineated by the following constraints:

Flow conservation constraint:

$$\sum_{s}\sum_{p} h_p^{rs}(k) = \overline{q}^{r}(k) \quad \forall r,k \qquad (9.7)$$

Flow propagation constraints:

$$u_{apk}^{rs}(t) = h_p^{rs}(k)\delta_{apk}^{rs}(t) \quad \forall r,s,a,p,k,t \qquad (9.8)$$

$$\sum_{t} \delta_{apk}^{rs}(t) = 1 \quad \forall r,s,p,a \in p,k \qquad (9.9)$$

$$\delta_{apk}^{rs}(t) = \{0,1\} \quad \forall r, s, a, p, k, t \tag{9.10}$$

Nonnegativity constraint:

$$h_p^{rs}(k) \geq 0 \quad \forall r, s, p, k \tag{9.11}$$

Definitional constraints:

$$\sum_p h_p^{rs}(k) = q^{rs}(k) \quad \forall r, s, k \tag{9.12}$$

$$u_a(t) = \sum_{rs} \sum_p \sum_k h_p^{rs}(k) \delta_{apk}^{rs}(t) \quad \forall a, t \tag{9.13}$$

$$c_p^{rs}(k) = \sum_a \sum_t c_a(t) \delta_{apk}^{rs}(t) \quad \forall r, s, p, k \tag{9.14}$$

The feasible region for the DUO singly constrained O-D choice/route choice model is similar to that for the DUO variable demand/route choice model, except for equation (9.7). Equation (9.7) expresses the time-dependent trip departures in terms of route flows and states that summing the route flows over all possible destinations $s$ and routes $p$ must be equal to the trips departing from origin $r$ during interval $k$.

#### 9.1.1.3 Equivalence analysis

***Theorem 9.2***: Under a certain flow propagation relationship $\left(\delta_{apk}^{rs}(t) = \delta_{apk}^{rs*}(t)\right)$, DUO singly constrained O-D choice/route choice conditions (9.1)–(9.2) imply VIP (9.6) and vice versa.

***Proof of necessity***: We need to prove that under a certain flow propagation relationship $\left(\delta_{apk}^{rs}(t) = \delta_{apk}^{rs*}(t)\right)$, DUO singly constrained O-D choice/route choice conditions (9.1)–(9.2) imply VIP (9.6). We first rearrange equilibrium conditions (9.1) as follows:

$$\left[c_p^{rs*}(k) - \pi^{rs}(k)\right]\left[h_p^{rs}(k) - h_p^{rs*}(k)\right] \geq 0 \quad \forall r, s, p, k \tag{9.15}$$

By summing over $r,s,p,k$ and then making a substitution of $\sum_p h_p^{rs}(k) = q^{rs}(k)$, we have:

$$\begin{aligned} &\sum_{rs} \sum_k \sum_p c_p^{rs*}(k)\left[h_p^{rs}(k) - h_p^{rs*}(k)\right] \\ &\quad - \sum_{rs} \sum_k \pi^{rs}(k)\left[q^{rs}(k) - q^{rs*}(k)\right] \geq 0 \end{aligned} \tag{9.16}$$

By using equation (9.2) and making a substitution of $\sum_s q^{rs}(k) = \bar{q}^r(k)$, it follows that:

$$\sum_{rs}\sum_{k}\sum_{p} c_p^{rs*}(k)\left[h_p^{rs}(k)-h_p^{rs*}(k)\right] \\ -\sum_{rs}\sum_{k}\left(D^{rs*}(k)\right)^{-1}\left[q^{rs}(k)-q^{rs*}(k)\right]\geq 0 \tag{9.17}$$

By applying equations (9.13)~(9.14), one obtains:

$$\sum_{a}\sum_{t} c_a^*(t)\left[u_a(t)-u_a^*(t)\right]-\sum_{rs}\sum_{k}\left(D^{rs*}(k)\right)^{-1}\left[q^{rs}(k)-q^{rs*}(k)\right]\geq 0 \tag{9.18}$$

Equation (9.18) is identical to VIP (9.6).

***Proof of sufficiency***: We next show that VIP (9.6) implies DUO singly constrained O-D choice/route choice conditions (9.1)~(9.2). The proof that VIP (9.6) implies expression (9.1) is given in Section 4.1.3.

We conclude the proof by showing that VIP (9.6) implies expression (9.2). We define the feasible solution $\left\{h_p^{rs}(k),\ q^{rs}(k)\right\}$ to be the same as $\left\{h_p^{rs*}(k),\ q^{rs*}(k)\right\}$ except for the two route flows $h_{p_1}^{rs_1}(k)$ and $h_{p_2}^{rs_2}(k)$, for which $h_{p_1}^{rs_1}(k)=h_{p_1}^{rs_1^*}(k)-\Delta$, and $h_{p_2}^{rs_2}(k)=h_{p_2}^{rs_2^*}(k)+\Delta$, where $0<|\Delta|\leq \min\left\{h_{p_1}^{rs_1}(k),h_{p_2}^{rs_2^*}(k)\right\}$. Note also that $q^{rs_1}(k)=q^{rs_1^*}(k)-\Delta$ and $q^{rs_2}(k)=q^{rs_2^*}(k)+\Delta$. Substituting this feasible solution into VIP (9.6), one obtains:

$$-\Delta\left[c_{p_1}^{rs_1^*}(k)-c_{p_2}^{rs_2^*}(k)\right]+\Delta\left[\left(D^{rs_1}(k)\right)^{-1}-\left(D^{rs_2}(k)\right)^{-1}\right]\geq 0 \tag{9.19}$$

Since $h_{p_1}^{rs_1^*}(k)$ and $h_{p_2}^{rs_2^*}(k)$ are positive, by expression (9.1), we have:

$$-\Delta\left[\pi^{rs_1}(k)-\pi^{rs_2}(k)\right]+\Delta\left[\left(D^{rs_1}(k)\right)^{-1}-\left(D^{rs_2}(k)\right)^{-1}\right]\geq 0 \tag{9.20}$$

We can now set the value of $\Delta$ either positive or negative. It follows that:

$$\left[\pi^{rs_1}(k)-\left(D^{rs_1}(k)\right)^{-1}\right]=\left[\pi^{rs_2}(k)-\left(D^{rs_2}(k)\right)^{-1}\right] \tag{9.21}$$

We repeat the above procedure for all O-D pair $rs_i$ and introduce the constant $\bar{c}^r(k)$. It then follows that:

$$\left[\pi^{rs_1}(k)-\left(D^{rs_1}(k)\right)^{-1}\right]=\left[\pi^{rs_2}(k)-\left(D^{rs_2}(k)\right)^{-1}\right]=\cdots=\cdots=\bar{c}^r(k) \tag{9.22}$$

Equation (9.28) implies expression (9.2). This completes the proof.

### 9.1.2 Nested Diagonalization Method

In this section, we first show the network representation for the DUO singly constrained O-

D choice/route choice problem. Then, the nested diagonalization method is formally stated.

#### 9.1.2.1 Time-space network

The network representation for the DUO singly constrained O-D choice/route choice model is shown in Figure 9.1. In Figure 9.1(a), we hypothesize a basic network. In Figure 9.1(b), the corresponding time-space network is constructed by reproducing the static network over time dimension.

#### 9.1.2.2 Solution algorithm

We adopt the nested diagonalization method as follows to solve the DUO singly constrained O-D choice/route choice model. Two types of link interactions are embedded in DUO singly constrained O-D choice/route choice model. One is due to estimated actual link travel times, and the other is from inflows on the time-space links other than the subject time-space link. However, once these two types of link interactions are temporarily fixed at current levels, the original problem reduces to a quadratic programming problem, which can be solved by the partial linearization method, also known as the Evans method.

*Nested Diagonalization Method*

**Step 0: Initialization.**

Step 0.1: Let $m$=0. Set $\tau_a^0(t) = NINT\left[c_{a_0}(t)\right], \forall a, t$.

Step 0.2: Let $n$=1. Find an initial feasible solution $\left\{q^{rs}(k)^1, u_a^1(t)\right\}$. Compute the associated link travel times $\left\{c_a^1(t)\right\}$.

**Step 1: *First Loop* Operation.**

Let $m$=$m$+1. Update the estimated actual link travel times by

$$\tau_a^m(t) = NINT\left[(1-\gamma)\tau_a^{m-1}(t) + \gamma c_a^n(t)\right] \quad \forall a, t \tag{9.23}$$

where $0 < \gamma \leq 1$.

Construct the corresponding feasible time-space network based on the estimated actual link travel times.

**Step 2: *Second Loop* Operation.**

Step 2.1: Let $n$=1. Compute and reset the initial feasible solution $\left\{q^{rs}(k)^n, u_a^n(t)\right\}$, based on the time-space network constructed by the estimated actual link travel times $\left\{\tau_a^m(t)\right\}$.

Step 2.2: Fix the inflows for all time-space links other than on the subject time-space link at the current level, yielding the following optimization problem.

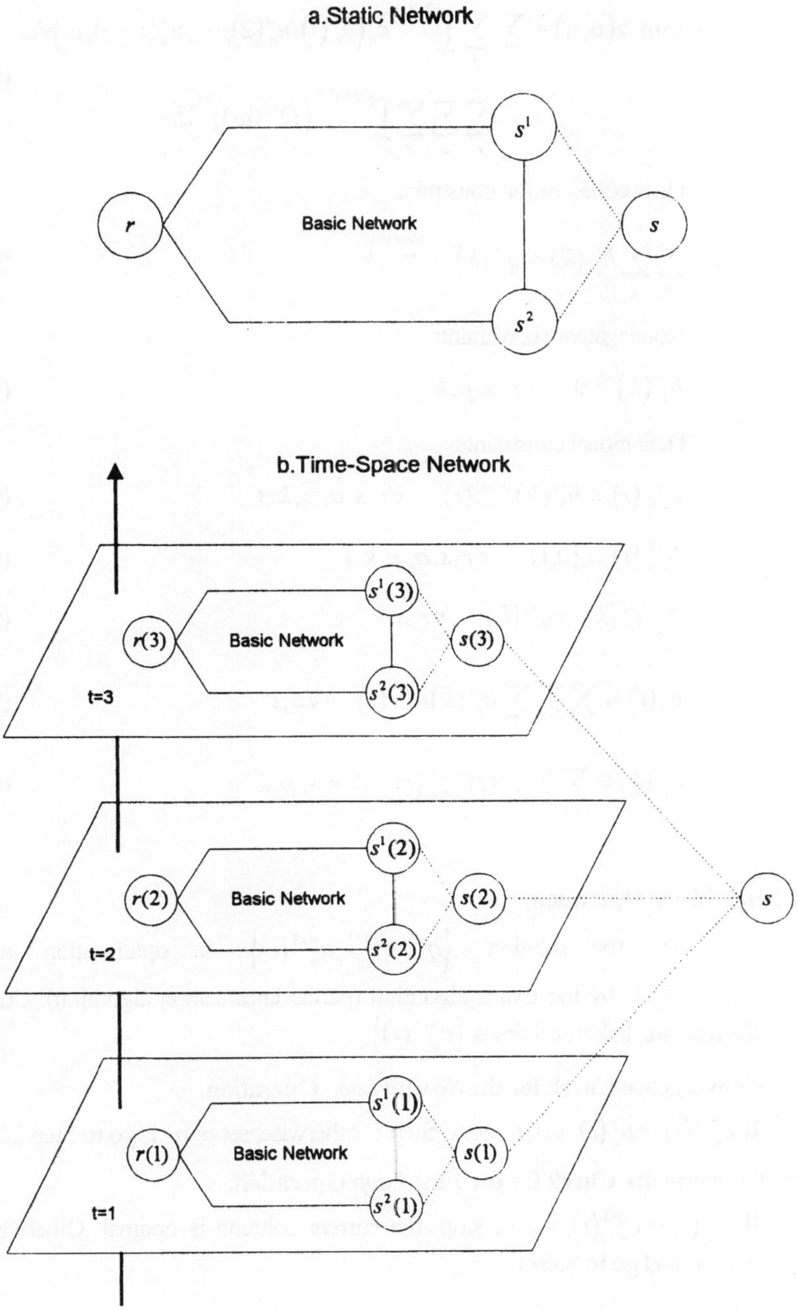


Figure 9.1: Time-Space Network for the Dynamic User-Optimal Singly Constrained O-D Choice/Route Choice Model

$$\min\ z(\mathbf{u},\mathbf{q}) = \sum_a \sum_t \int_0^{u_a^{n+1}(t)} c_a\left(u_a^n(1), u_a^n(2), \cdots, u_a^n(t-1), \omega\right) d\omega \\ - \sum_r \sum_s \sum_k \int_0^{q^{rs}(k)^{n+1}} \left(D^{rs}(\omega)\right)^{-1} d\omega \qquad (9.24)$$

Flow conservation constraint:

$$\sum_s \sum_p h_p^{rs}(k) = \bar{q}^r(k) \quad \forall r,k \qquad (9.25)$$

Nonnegativity constraint:

$$h_p^{rs}(k) \geq 0 \quad \forall r,s,p,k \qquad (9.26)$$

Definitional constraints:

$$u_{apk}^{rs}(t) = h_p^{rs}(k)\bar{\delta}_{apk}^{rs}(t) \quad \forall r,s,a,p,k,t \qquad (9.27)$$

$$\bar{\delta}_{apk}^{rs}(t) = \{0,1\} \quad \forall r,s,a,p,k,t \qquad (9.28)$$

$$\sum_p h_p^{rs}(k) = q^{rs}(k) \quad \forall r,s,k \qquad (9.29)$$

$$u_a(t) = \sum_{rs} \sum_p \sum_k h_p^{rs}(k)\bar{\delta}_{apk}^{rs}(t) \quad \forall a,t \qquad (9.30)$$

$$c_p^{rs}(k) = \sum_a \sum_t c_a(t)\bar{\delta}_{apk}^{rs}(t) \quad \forall r,s,p,k \qquad (9.31)$$

**Step 3: *Third Loop* Operation.**

Solve for the solution, $\left\{q^{rs}(k)^{n+1}, u_a^{n+1}(t)\right\}$, in optimization problem (9.24)~(9.31) by the Evans algorithm (partial linearization algorithm). Compute the resulting link travel times $\left\{c_a^{n+1}(t)\right\}$.

**Step 4: Convergence Check for the *Second Loop* Operation.**

If $u_a^{n+1}(t) \approx u_a^n(t), \forall a,t$, go to Step 5; otherwise, set $n=n+1$, go to Step 2.2.

**Step 5: Convergence Check for the *First Loop* Operation.**

If $\tau_a^m(t) \approx c_a^{n+1}(t), \forall a,t$, stop; the current solution is optimal. Otherwise, set $m=m+1$, and go to Step 1.

Analogous to the discussion in Section 5.3, the nested projection algorithm can be obtained by the following modification.

Step 2.2: Fix the inflows for all time-space links other than on the subject time-space link at the current level, yielding the following optimization problem.

$$\min\ Z = \frac{1}{2}\left(\mathbf{u}^{n+1}\right)^T \mathbf{G}_1 \mathbf{u}^{n+1} + \left(\rho_1 \mathbf{c}\left(\mathbf{u}^n\right) - \mathbf{G}_1 \mathbf{u}^n\right)\mathbf{u}^{n+1}$$
$$- \frac{1}{2}\left(\mathbf{q}^{n+1}\right)^T \mathbf{G}_2 \mathbf{q}^{n+1} - \left(\rho_2 \left(\mathbf{D}(\mathbf{q}^n)\right)^{-1} - \mathbf{G}_2 \mathbf{q}^n\right)\mathbf{q}^{n+1} \quad (9.32)$$

Subject to: Equations (9.31)~(9.37).

where matrices $\mathbf{G_1}$, $\mathbf{G_2}$ are symmetric and positive definite, and symbols $\rho_1$, $\rho_2$ are contraction operators.

### 9.1.3 Numerical Example

#### 9.1.3.1 Input data

The simple network shown in Figure 9.2 is used for testing. The test network consists of 6 links and 5 nodes in which node 1 is the origin, nodes 4 and 5 are destinations, and nodes 2 and 3 are intermediate nodes.

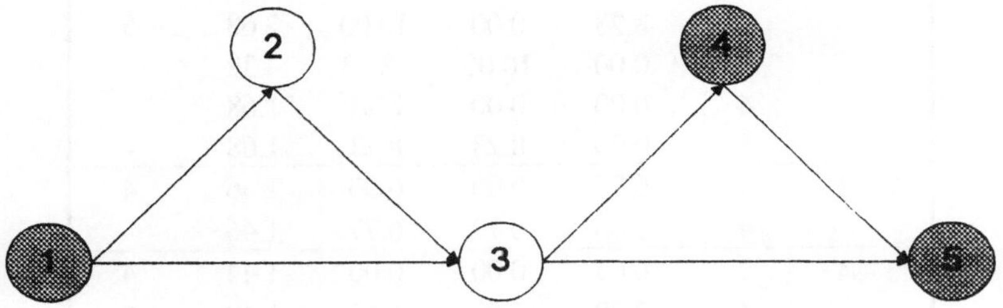


Figure 9.2: Test Network 1

The adopted dynamic travel time functions are arbitrarily constructed as follows:

$$c_a(t) = 2 + 0.01\left(u_a(t)\right)^2 + 0.01\left(x_a(t)\right)^2 \qquad a = (3,5), \forall t \qquad (9.33)$$

$$c_a(t) = 1 + 0.01\left(u_a(t)\right)^2 + 0.01\left(x_a(t)\right)^2 \qquad \forall a \neq (3,5), t \qquad (9.34)$$

The time-dependent O-D demand function is assumed as follows:

$$q^{rs}(k) = \bar{q}^r(k) \frac{e^{-\pi^{rs}(k)}}{\sum_{s'} e^{-\pi^{rs'}(k)}} \qquad \forall r,s,k \qquad (9.35)$$

The trips departing from origin 1 during two time intervals are summarized in Table 9.1:

Table 9.1: Time-Dependent O-D Demands

| O-D Pair | Time Interval $k$=1 | $k$=2 |
|---|---|---|
| 1-4 | 10 | 15 |
| 1-5 | | |

#### 9.1.3.2 Test results

The results for the DUO singly constrained O-D choice/route choice model are summarized in Table 9.2.

Table 9.2: Results for Test Network 1

| Link | Entering Time Interval | Inflow | Exit Flow | Number of Vehicles | Link Travel Time | Exiting Time Interval |
|---|---|---|---|---|---|---|
| 1→2 | 2 | 6.77 | 0.00 | 0.00 | 1.46 | 3 |
| | 3 | 0.00 | 6.77 | 6.77 | 1.46 | - |
| 1→3 | 1 | 10.00 | 0.00 | 0.00 | 2.00 | 3 |
| | 2 | 8.23 | 0.00 | 10.00 | 2.68 | 5 |
| | 3 | 0.00 | 10.00 | 18.23 | 4.32 | - |
| | 4 | 0.00 | 0.00 | 8.23 | 1.68 | - |
| | 5 | 0.00 | 8.23 | 8.23 | 1.68 | - |
| 2→3 | 3 | 6.77 | 0.00 | 0.00 | 1.46 | 4 |
| | 4 | 0.00 | 6.77 | 6.77 | 1.46 | - |
| 3→4 | 3 | 6.63 | 0.00 | 0.00 | 1.44 | 4 |
| | 4 | 2.28 | 6.63 | 6.63 | 1.49 | 5 |
| | 5 | 8.23 | 2.28 | 2.28 | 1.73 | 7 |
| | 6 | 0.00 | 0.00 | 8.23 | 1.68 | - |
| | 7 | 0.00 | 8.23 | 8.23 | 1.68 | - |
| 3→5 | 3 | 3.37 | 0.00 | 0.00 | 2.11 | 5 |
| | 4 | 4.48 | 0.00 | 3.37 | 2.31 | 6 |
| | 5 | 0.00 | 3.37 | 7.86 | 2.62 | - |
| | 6 | 0.00 | 4.48 | 4.48 | 2.20 | - |

The rationale of the proposed model and associated solution algorithm can be verified by checking if the resulting actual route travel times satisfy the dynamic user-optimal conditions. For example, consider route 1→3→5 departing during interval 1; hen, the corresponding actual route travel time can be obtained by summing the actual link travel time on link 1→3 during interval 1 and the actual link travel time on link 3→5 during interval $1+c_{1\to 3}(1)$ as follows:

$$\begin{aligned} c_{1\to3\to5}(1) &= c_{1\to3}(1) + c_{3\to5}\left(1 + c_{1\to3}(1)\right) \\ &= 2.00 + c_{3\to5}(3.00) = 2.00 + 2.11 = 4.11 \end{aligned} \tag{9.36}$$

The remaining used actual route times are also computed and summarized in Table 9.3. Note that the trips departing from the same origin and interval have approximately the same route travel time.

Table 9.3: Actual Route Travel Times

| Departure Interval | Route | | | |
|---|---|---|---|---|
| | 1→3→4 | 1→2→3→4 | 1→3→5 | 1→2→3→5 |
| $k$=1 | 3.44 | NA | 4.11 | NA |
| Average | 3.44 | | 4.11 | |
| $k$=2 | 4.41 | 4.41 | NA | 5.23 |
| Average | 4.41 | | 5.23 | |

NA: not applicable because the corresponding route is not used.

The computed equilibrium O-D trips are summarized in Table 9.4. Of the 10 unit trips departing from origin 1 during interval 1, 6.63 unit trips are distributed to destination 4, and 3.38 unit trips to destination 5. Similarly, of the 15 unit trips departing from origin 1 during interval 2, 10.15 unit trips are distributed to destination 4, and 4.85 unit trips to destination 5. In this numerical example, the time-dependent O-D demands can also be computed by the logit formula. The results obtained using the logit formula are included in the fifth column of Table 9.4. By comparing the third and the fifth columns, it can be seen that the equilibrium O-D demands obtained from the computer code and those obtained directly using the logit model are very similar, implying that the computed equilibrium O-D trips are convincing.

Table 9.4: Time-Dependent O-D Demands

| Interval | O-D Pair | Time-Dependent O-D Demand Computed by Computer Code | Actual Route Travel Time | Time-Dependent O-D Demand Computed by Logit Formula* |
|---|---|---|---|---|
| $k$=1 | (1,4) | 6.63 | 3.44 | 6.63 |
| | (1,5) | 3.37 | 4.11 | 3.37 |
| $k$=2 | (1,4) | 10.52 | 4.41 | 10.43 |
| | (1,5) | 4.48 | 5.23 | 4.57 |

* denotes the demands were computed by the logit formula

$$q^{rs}(k) = \bar{q}^{r}(k)\frac{e^{-\pi^{rs}(k)}}{\sum_{s'} e^{-\pi^{rs'}(k)}}$$

## 9.2 Dynamic User-Optimal Singly Constrained O-D Choice/Departure Time/Route Choice Model

In this section, we discuss the equilibrium conditions and model formulation, the nested diagonalization method and a numerical example for the dynamic user-optimal singly constrained O-D choice/departure time/route choice model.

### 9.2.1 Equilibrium Conditions and Model Formulation

The dynamic user-optimal conditions are defined to characterize the travelers' behavior for choosing the preferred destination and using the minimal travel time route. Due to inherent link interactions, the variational inequality approach is adopted to formulate the DUO O-D choice/departure time/route choice model. The equivalence between the dynamic user-optimal conditions and the variational inequality formulation is then stated by a theorem and verified by a proof.

#### 9.2.1.1 Dynamic user-optimal equilibrium conditions

When the trip decision for choosing a destination and route choice is considered, the corresponding dynamic user-optimal conditions may be characterized by equilibrium conditions on both route choice behavior and time-independent O-D demands. The term *time-independent O-D demands* refers to the characteristics of O-D demands within defined *windows* of time. That is, the O-D demands are not completely independent of time of day; however, they are time-independent within certain periods within a day.

For route choice behavior, the conditions state for each O-D pair that the actual route travel times experienced by travelers departing from the same origin, regardless the departure times, are equal and minimal; or, no traveler would be better off by unilaterally changing his/her route. On the contrary, the actual route travel time of any unused route for each O-D pair is greater than or equal to the minimal actual route travel time. In other words, at equilibrium, if the flow departing from origin $r$ during interval $k$ over route $p$ toward destination $s$ is positive, i.e., $h_p^{rs*}(k) > 0$, then the corresponding actual route travel time is minimal. On the other hand, if no flow occurs on route $p$, i.e., $h_p^{rs*}(k) = 0$, then the corresponding actual route travel time is at least as great as the minimal actual route travel time. These equilibrium conditions can be mathematically expressed as follows:

$$c_p^{rs*}(k)\begin{cases} = \pi^{rs} & \text{if } h_p^{rs*}(k) > 0 \\ \geq \pi^{rs} & \text{if } h_p^{rs*}(k) = 0 \end{cases} \quad \forall r, s, p, k \tag{9.37}$$

For the time-independent O-D demands, the O-D demand function is assumed to be a strictly decreasing function of the difference between the minimal time-independent O-D route travel time and the time-independent constant, $\bar{c}^r$. Thus, the inverse O-D demand function has strictly monotone mapping. The general form can be expressed mathematically as follows:

$$\pi^{rs} - \bar{c}^{r} = \left(D^{rs*}\right)^{-1} \quad \forall r,s \tag{9.38}$$

The well known inverse O-D demand function is as follows:

$$\left(D^{rs}\right)^{-1} = -\frac{1}{\theta}\ln\left(\frac{q^{rs}}{\bar{q}^{r} - q^{rs}}\sum_{s'\neq s} e^{-\theta\left(\pi^{rs'} - \bar{c}^{r} - M^{s'}\right)}\right) + M^{s} \quad \forall r,s \tag{9.39}$$

where $\theta > 0$ denotes a logit parameter, and $M^{s} \geq 0$ denotes the attractiveness associated with destination $s$. Note that when equation (9.45) is inserted into equation (9.44), the following logit model results:

$$q^{rs} = \bar{q}^{r}\,\frac{e^{-\theta\left(\pi^{rs} - M^{s}\right)}}{\sum_{s'} e^{-\theta\left(\pi^{rs'} - M^{s'}\right)}} \quad \forall r,s \tag{9.40}$$

#### 9.2.1.2 Variational inequality formulation

The DUO singly constrained O-D choice/departure time/route choice problem can be formulated using the variational inequality approach.

***Theorem 9.3***: The DUO singly constrained O-D choice/departure time/route choice problem is equivalent to finding a solution $(\mathbf{u}^*, \mathbf{q}^*) \in \Omega$ such that the following VIP holds:

$$\mathbf{c}^*(\mathbf{u} - \mathbf{u}^*) - \mathbf{D}^{-1*}(\mathbf{q} - \mathbf{q}^*) \geq 0 \quad \forall(\mathbf{u}, \mathbf{q}) \in \Omega^* \tag{9.41}$$

or, alternatively, in expanded form:

$$\sum_{a}\sum_{t} c_a^*(t)\left[u_a(t) - u_a^*(t)\right] - \sum_{rs}\left(D^{rs*}\right)^{-1}\left[q^{rs} - q^{rs*}\right] \geq 0 \qquad \forall(\mathbf{u}, \mathbf{q}) \in \Omega^* \tag{9.42}$$

where $\Omega^*$ is a subset of $\Omega$ with $\delta_{apk}^{rs}(t)$ being realized at equilibrium, i.e., $\delta_{apk}^{rs}(t) = \delta_{apk}^{rs*}(t), \forall r,s,a,p,k,t$. The symbol $\Omega$ denotes the feasible region that is delineated by the following constraints:

Flow conservation constraint:

$$\sum_{s}\sum_{k}\sum_{p} h_p^{rs}(k) = \bar{q}^{r} \quad \forall r \tag{9.43}$$

Flow propagation constraints:

$$u_{apk}^{rs}(t) = h_p^{rs}(k)\delta_{apk}^{rs}(t) \quad \forall r,s,a,p,k,t \tag{9.44}$$

$$\sum_t \delta_{apk}^{rs}(t) = 1 \quad \forall r,s,p,a \in p,k \tag{9.45}$$

$$\delta_{apk}^{rs}(t) = \{0,1\} \quad \forall r,s,a,p,k,t \tag{9.46}$$

Nonnegativity constraint:

$$h_p^{rs}(k) \geq 0 \quad \forall r,s,p,k \tag{9.47}$$

Definitional constraints:

$$\sum_k \sum_p h_p^{rs}(k) = q^{rs} \quad \forall r,s \tag{9.48}$$

$$u_a(t) = \sum_{rs} \sum_p \sum_k h_p^{rs}(k) \delta_{apk}^{rs}(t) \quad \forall a,t \tag{9.49}$$

$$c_p^{rs}(k) = \sum_a \sum_t c_a(t) \delta_{apk}^{rs}(t) \quad \forall r,s,p,k \tag{9.50}$$

The feasible region for the DUO singly constrained O-D choice/departure time/route choice model is similar to that for the DUO variable demand/route choice model, except for equation (9.43). Equation (9.43) expresses the time-independent trip departures in turns of route flows, and states that summing the route flows over all possible destinations $s$ and routes $p$ during the entire analysis period must be equal to the trips departing from origin $r$.

#### 9.2.1.3 Equivalence analysis

***Theorem 9.6***: Under a certain flow propagation relationship $\left(\delta_{apk}^{rs}(t) = \delta_{apk}^{rs*}(t)\right)$, DUO singly constrained O-D choice/departure time/route choice conditions (9.37)~(9.38) imply VIP (9.42) and vice versa.

***Proof of necessity***: We need to prove that under a certain flow propagation relationship $\left(\delta_{apk}^{rs}(t) = \delta_{apk}^{rs*}(t)\right)$, DUO singly constrained O-D choice/departure time/route choice conditions (9.37)~(9.38) imply VIP (9.42). We first rearrange equilibrium conditions (9.37) as follows:

$$\left[c_p^{rs*}(k) - \pi^{rs}\right]\left[h_p^{rs}(k) - h_p^{rs*}(k)\right] \geq 0 \quad \forall r,s,p,k \tag{9.51}$$

By summing over $r,s,p,k$ and then making a substitution of $\sum_k \sum_p h_p^{rs}(k) = q^{rs}$, we have:

$$\sum_{rs} \sum_k \sum_p c_p^{rs*}(k)\left[h_p^{rs}(k) - h_p^{rs*}(k)\right] - \sum_{rs} \pi^{rs}\left[q^{rs} - q^{rs*}\right] \geq 0 \tag{9.52}$$

By using equation (9.44) and making a substitution of $\sum_s q^{rs} = \bar{q}^r$, it follows:

$$\sum_{rs}\sum_{k}\sum_{p} c_p^{rs*}(k)\left[h_p^{rs}(k)-h_p^{rs*}(k)\right]-\sum_{rs}\left(D^{rs*}\right)^{-1}\left[q^{rs}-q^{rs*}\right]\geq 0 \qquad (9.53)$$

By applying equations (9.55)~(9.56), we obtain:

$$\sum_{a}\sum_{t} c_a^*(t)\left[u_a(t)-u_a^*(t)\right]-\sum_{rs}\left(D^{rs*}\right)^{-1}\left[q^{rs}-q^{rs*}\right]\geq 0 \qquad (9.54)$$

Equation (9.54) is identical to VIP (9.42).

***Proof of sufficiency***: We next show that VIP (9.42) implies DUO singly constrained O-D choice/departure time /route choice conditions (9.37)~(9.38). The proof that VIP (9.42) implies expression (9.37) is given in Section 6.1.3.

We conclude the proof by showing that VIP (9.48) implies expression (9.44). We define now a feasible solution $\left\{h_p^{rs}(k),q^{rs}\right\}$ to be the same as $\left\{h_p^{rs*}(k),q^{rs*}\right\}$, except for the two route flows, $h_{p_1}^{rs_1}(k_1)$ during time interval $k_1$, and $h_{p_2}^{rs_2}(k_2)$ during time interval $k_2$, for which $h_{p_1}^{rs_1}(k_1)=h_{p_1}^{rs_1^*}(k_1)-\Delta$, $h_{p_2}^{rs_2}(k_2)=h_{p_2}^{rs_2^*}(k_2)+\Delta$, where $0<|\Delta|\leq \min\left\{h_{p_1}^{rs_1^*}(k_1),h_{p_2}^{rs_2^*}(k_2)\right\}$. Note also that $q^{rs_1}=q^{rs_1^*}-\Delta$ and $q^{rs_2}=q^{rs_2^*}+\Delta$. Substituting this feasible solution into VIP (9.48), one obtains:

$$-\Delta\left[c_{p_1}^{rs_1^*}(k_1)-c_{p_2}^{rs_2^*}(k_2)\right]+\Delta\left[\left(D^{rs_1}\right)^{-1}-\left(D^{rs_2}\right)^{-1}\right]\geq 0 \qquad (9.55)$$

Since $h_{p_1}^{rs_1^*}(k_1)$ and $h_{p_2}^{rs_2^*}(k_2)$ are positive, by expression (9.43), we have:

$$-\Delta\left[\pi^{rs_1}-\pi^{rs_2}\right]+\Delta\left[\left(D^{rs_1}\right)^{-1}-\left(D^{rs_2}\right)^{-1}\right]\geq 0 \qquad (9.56)$$

We can now set the value of $\Delta$ either positive or negative. It follows:

$$\left[\pi^{rs_1}-\left(D^{rs_1}\right)^{-1}\right]=\left[\pi^{rs_2}-\left(D^{rs_2}\right)^{-1}\right] \qquad (9.57)$$

We repeat the above procedure for all O-D pairs $rs_i$ and introduce the constant $\bar{c}^r$. It follows:

$$\left[\pi^{rs_1}-\left(D^{rs_1}\right)^{-1}\right]=\left[\pi^{rs_2}-\left(D^{rs_2}\right)^{-1}\right]=\cdots=\cdots=\bar{c}^r \qquad (9.58)$$

Equation (9.58) implies expression (9.38). This completes the proof.

### 9.2.2 Nested Diagonalization Method

In this section, we first show the network representation for the DUO singly constrained O-D choice/departure time/route choice problem. Then, the nested diagonalization method is formally stated.

### 9.2.2.1 Time-space network

The network representation for the DUO singly constrained O-D choice/departure time/route choice model is shown in Figure 9.3. In Figure 9.3(a), we first assume a basic network. In Figure 9.3(b), the corresponding time-space network is constructed by reproducing the static network over time dimension.

### 9.2.2.2 Solution algorithm

The following nested diagonalization algorithm is proposed to solve the singly constrained DUO O-D choice/departure time/route choice problem.

*Nested Diagonalization Method*

**Step 0: Initialization.**

Step 0.1: Let $m$=0. Set $\tau_a^0(t) = NINT\left[c_{a_0}(t)\right], \forall a, t$.

Step 0.2: Let $n$=1. Find an initial feasible solution $\left\{\left(q^{rs}\right)^1, u_a^1(t)\right\}$. Compute the associated link travel times $\left\{c_a^1(t)\right\}$.

**Step 1: *First Loop* Operation.**

Let $m$=$m$+1. Update the estimated actual link travel times by

$$\tau_a^m(t) = NINT\left[(1-\gamma)\tau_a^{m-1}(t) + \gamma c_a^n(t)\right] \quad \forall a, t \tag{9.59}$$

where $0 < \gamma \le 1$.

Construct the corresponding feasible time-space network based on the estimated actual link travel times.

**Step 2: *Second Loop* Operation.**

Step 2.1: Let $n$=1. Compute and reset the initial feasible solution $\left\{\left(q^{rs}\right)^n, u_a^n(t)\right\}$, based on the time-space network constructed by the estimated actual link travel times $\left\{\tau_a^m(t)\right\}$.

Step 2.2: Fix the inflows for all time-space links other than on the subject time-space link at the current level, yielding the following optimization problem.

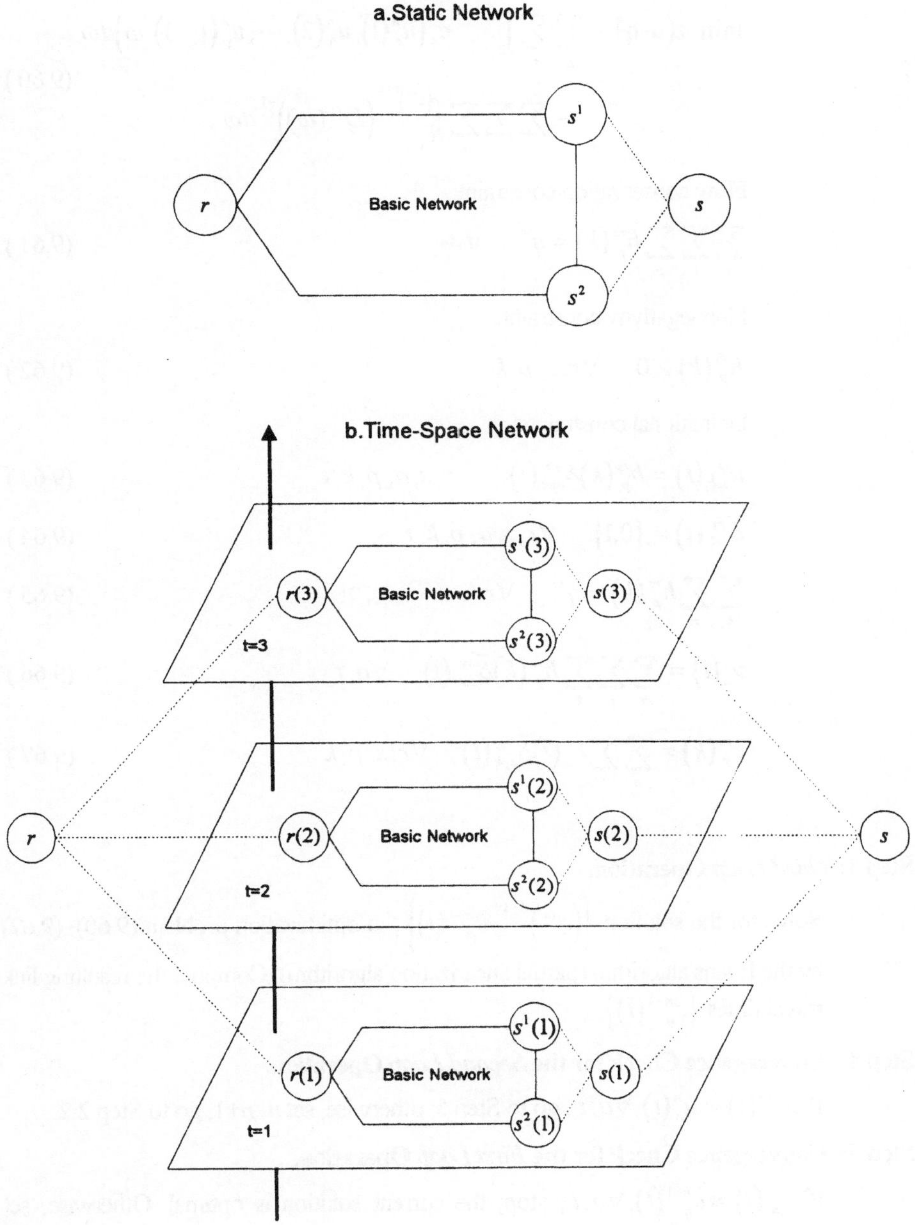


Figure 9.3: Time-Space Network for the Dynamic User-Optimal Singly Constrained O-D Choice/Departure Time/Route Choice Model

$$\min\ z(\mathbf{u},\mathbf{q}) = \sum_a \sum_t \int_0^{u_a^{n+1}(t)} c_a\left(u_a^n(1), u_a^n(2), \cdots, u_a^n(t-1), \omega\right) d\omega - \sum_r \sum_s \sum_k \int_0^{\left(q^{rs}\right)^{n+1}} \left(D^{rs}(\omega)\right)^{-1} d\omega \tag{9.60}$$

Flow conservation constraint:

$$\sum_s \sum_k \sum_p h_p^{rs}(k) = \bar{q}^r \quad \forall r \tag{9.61}$$

Nonnegativity constraint:

$$h_p^{rs}(k) \geq 0 \quad \forall r, s, p, k \tag{9.62}$$

Definitional constraints:

$$u_{apk}^{rs}(t) = h_p^{rs}(k) \bar{\delta}_{apk}^{rs}(t) \quad \forall r, s, a, p, k, t \tag{9.63}$$

$$\bar{\delta}_{apk}^{rs}(t) = \{0,1\} \quad \forall r, s, a, p, k, t \tag{9.64}$$

$$\sum_k \sum_p h_p^{rs}(k) = q^{rs} \quad \forall r, s \tag{9.65}$$

$$u_a(t) = \sum_{rs} \sum_p \sum_k h_p^{rs}(k) \bar{\delta}_{apk}^{rs}(t) \quad \forall a, t \tag{9.66}$$

$$c_p^{rs}(k) = \sum_a \sum_t c_a(t) \bar{\delta}_{apk}^{rs}(t) \quad \forall r, s, p, k \tag{9.67}$$

**Step 3: *Third Loop* Operation.**

Solve for the solution $\left\{\left(q^{rs}\right)^{n+1}, u_a^{n+1}(t)\right\}$, in optimization problem (9.60)~(9.67) by the Evans algorithm (partial linearization algorithm). Compute the resulting link travel times $\left\{c_a^{n+1}(t)\right\}$.

**Step 4: Convergence Check for the *Second Loop* Operation.**

If $u_a^{n+1}(t) \approx u_a^n(t), \forall a, t$, go to Step 5; otherwise, set $n$=$n$+1, go to Step 2.2.

**Step 5: Convergence Check for the *First Loop* Operation.**

If $\tau_a^m(t) \approx c_a^{n+1}(t), \forall a, t$, stop; the current solution is optimal. Otherwise, set $n$=$n$+1, and go to Step 1.

Analogous to the discussion in Section 5.3, the nested projection algorithm can be obtained by the following modification.

Step 2.2: Fix the inflows for all time-space links other than on the subject time-space link at the current level, yielding the following optimization problem.

$$\min\ z = \frac{1}{2}\left(\mathbf{u}^{n+1}\right)^T \mathbf{G}_1 \mathbf{u}^{n+1} + \left(\rho_1 \mathbf{c}\left(\mathbf{u}^n\right) - \mathbf{G}_1 \mathbf{u}^n\right)\mathbf{u}^{n+1}$$
$$- \frac{1}{2}\left(\mathbf{q}^{n+1}\right)^T \mathbf{G}_2 \mathbf{q}^{n+1} - \left(\rho_2 \left(\mathbf{D}(\mathbf{q}^n)\right)^{-1} - \mathbf{G}_2 \mathbf{q}^n\right)\mathbf{q}^{n+1} \tag{9.68}$$

Subject to: Equations (9.67)~(9.37).

where matrices $\mathbf{G}_1$, $\mathbf{G}_2$ are symmetric and positive definite, and symbols $\rho_1$, $\rho_2$ are contraction operators.

### 9.2.3 Numerical Example

#### 9.2.3.1 Input data

A simple network shown in Figure 9.4 is used for testing. The test network consists of 6 links and 5 nodes in which node 1 is the origin, nodes 4 and 5 are destinations, and nodes 2 and 3 are intermediate nodes.

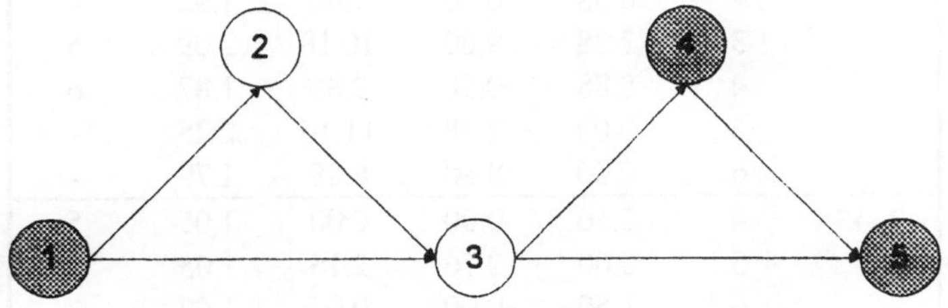


Figure 9.4: Test Network 2

The adopted dynamic travel time functions are arbitrarily constructed as follows:

$$c_a(t) = 2 + 0.01\left(u_a(t)\right)^2 + 0.01\left(x_a(t)\right)^2 \quad a = (3,5), \forall t \tag{9.69}$$

$$c_a(t) = 1 + 0.01\left(u_a(t)\right)^2 + 0.01\left(x_a(t)\right)^2 \quad \forall a \neq (3,5), t \tag{9.70}$$

The time-independent O-D demand function is assumed as follows:

$$q^{rs} = \bar{q}^r \frac{e^{-\pi^{rs}}}{\sum_{s'} e^{-\pi^{rs'}}} \quad \forall r, s \tag{9.71}$$

The trips departing from origin 1 in eight time intervals are summarized in Table 9.5:

Table 9.5: Time-Independent O-D Demands

| O-D Pair | Interval |
|---|---|
| | $k$=1~8 |
| 1-4 | 25 |
| 1-5 | |

### 9.2.3.2 Test results

The obtained results are summarized in Table 9.6.

Table 9.6: Results for Test Network 2

| Link | Entering Time Interval | Inflow | Exit Flow | Number of Vehicles | Link Travel Time | Exiting Time Interval |
|---|---|---|---|---|---|---|
| 1→2 | 3 | 2.16 | 0.00 | 0.00 | 1.05 | 4 |
| | 4 | 0.00 | 2.16 | 2.16 | 1.05 | - |
| | 5 | 1.50 | 0.00 | 0.00 | 1.02 | 6 |
| | 6 | 0.00 | 1.50 | 1.50 | 1.02 | - |
| 1→3 | 1 | 9.60 | 0.00 | 0.00 | 1.92 | 3 |
| | 2 | 0.58 | 0.00 | 9.60 | 1.92 | 4 |
| | 3 | 2.28 | 9.60 | 10.18 | 2.09 | 5 |
| | 4 | 8.88 | 0.58 | 2.86 | 1.87 | 6 |
| | 5 | 0.00 | 2.28 | 11.16 | 2.25 | - |
| | 6 | 0.00 | 8.88 | 8.88 | 1.79 | - |
| 2→3 | 4 | 2.16 | 0.00 | 0.00 | 1.05 | 5 |
| | 5 | 0.00 | 2.16 | 2.16 | 1.05 | - |
| | 6 | 1.50 | 0.00 | 0.00 | 1.02 | 7 |
| | 7 | 0.00 | 1.50 | 1.50 | 1.02 | - |
| 3→4 | 3 | 6.05 | 0.00 | 0.00 | 1.37 | 4 |
| | 4 | 0.40 | 6.05 | 6.05 | 1.37 | 5 |
| | 5 | 4.44 | 0.40 | 0.40 | 1.20 | 6 |
| | 6 | 4.69 | 4.44 | 4.44 | 1.42 | 7 |
| | 7 | 1.50 | 4.69 | 4.69 | 1.24 | 8 |
| | 8 | 0.00 | 1.50 | 1.50 | 1.02 | - |
| 3→5 | 3 | 3.55 | 0.00 | 0.00 | 2.13 | 5 |
| | 4 | 0.18 | 0.00 | 3.55 | 2.13 | 6 |
| | 5 | 0.00 | 3.55 | 3.73 | 2.14 | - |
| | 6 | 4.19 | 0.18 | 0.18 | 2.18 | 8 |
| | 7 | 0.00 | 0.00 | 4.19 | 2.18 | - |
| | 8 | 0.00 | 4.19 | 4.19 | 2.18 | - |

Table 9.7: Actual Route Travel Times

| Departure Interval | Route | | | | |
|---|---|---|---|---|---|
| | 1→3→4 | 1→2→3→4 | 1→3→5 | 1→3→4→5 | 1→2→3→4→5 |
| $k$=1 | 3.29 | NA | 4.05 | NA | NA |
| $k$=2 | 3.29 | NA | 4.05 | NA | NA |
| $k$=3 | 3.29 | 3.30 | NA | NA | NA |
| $k$=4 | 3.29 | NA | 4.05 | NA | NA |
| $k$=5 | NA | 3.28 | NA | NA | NA |
| Average | 3.29 | | 4.05 | | |

NA: not applicable because the corresponding route is not used.

Table 9.8: Time-Independent O-D Demands

| O-D Pair $(r,s)$ | Time-Independent O-D Demand Computed by Computer Code | Actual Route Travel Time | Time-Independent O-D Demand Computed by the Logit Formula* |
|---|---|---|---|
| (1,4) | 17.08 | 3.29 | 17.03 |
| (1,5) | 7.92 | 4.05 | 7.97 |

* denotes the demands were computed by the logit formula

$$q^{rs} = \bar{q}^{r} \frac{e^{-\pi^{rs}}}{\sum_{s'} e^{-\pi^{rs'}}}$$

## 9.3 Notes

Within the nested diagonalization method, we use the Evans method (1976) instead of the FW method in the *third loop* operation. The Evans method may be regarded as a generalization of the FW (1956) linearization method. Given a current solution to the *main problem*, all travel times are fixed at their current values. O-D demands of a *subproblem* are then solved directly from the available optimality conditions; whereas subproblem route choices are identified by minimizing the linearized portion of the objective function that pertains to travel times resulting from all trips being assigned to the minimum travel time route. A new main problem solution is obtained by minimizing the objective function with respect to a step size $\lambda$, $0 \leq \lambda \leq 1$, where $\lambda$ is a weight for combining the main and subproblem solutions. This procedure is called "bisection" in which a gradient search method is applied.

DUO doubly constrained distribution models are not explored in this chapter because trips constrained on both ends are too stringent for dynamic models, and quite difficult to solve.

# Chapter 10

# Dynamic System-Optimal Route Choice Model

The dynamic system-optimal (DSO) route choice problem requires that travelers comply with a certain driving behavior so that the system-optimal flow pattern, in terms of some predetermined performance measures, can be obtained. This problem is *normative* in the sense that the transportation system operator has control over the travelers, who must follow the system's instruction. This assumption is rather restrictive because it is extremely difficult to impose such constraints on travelers, especially in an open transportation system network. As a consequence, travelers who are instructed to take longer routes may unilaterally change routes so as to decrease their travel times, which implies the dynamic system-optimal flow pattern is not stable, and therefore unlikely to sustain itself. Nevertheless, the dynamic system-optimal problem is important because the system-optimal travel time may serve as a yardstick by which different flow patterns can be measured. Indeed, the system-optimal flow pattern is desired by many types of traffic control policies such as tolls and signal controls in order to enhance the efficiency of road network utilization.

In this chapter, we first perform a marginal link travel time function analysis in Section 10.1. Then, we present the equilibrium conditions and the model formulation for the DSO route choice problem in Section 10.2. After that, the nested diagonalization method is proposed in Section 10.3 and a numerical example is provided in Section 10.4. Network travel times for the DSO and DUO route choice problems are compared in Section 10.5. A static DSO counterpart is approximated in Section 10.6. Braess's paradox is studied in Section 10.7. A two-stage procedure is proposed for computing tolls in Section 10.8, and finally, some concluding remarks are given in Section 10.9.

## 10.1 Marginal Link Travel Time Function Analysis

In this section, we are not concerned with evaluating the appropriateness of each kind of performance measure, such as minimal network travel time or maximal network capacity; rather, a unified analysis framework is attempted. Therefore, without loss of generality, a dynamic marginal travel time function is first derived.

The dynamic marginal link travel time function $\hat{c}_a(t)$ can be obtained by taking the derivative of the total network travel time with respect to link inflows, as follows:

$$\hat{c}_a(t) = \frac{\partial\left(\sum_{a'}\sum_{t'} c_{a'}(t')u_{a'}(t')\right)}{\partial u_a(t)} = c_a(t)\frac{\partial u_a(t)}{\partial u_a(t)} + \sum_{a'}\sum_{t'} u_{a'}(t')\frac{\partial c_{a'}(t')}{\partial u_a(t)} \quad (10.1)$$
$$= c_a(t) + \sum_{a'}\sum_{t'} u_{a'}(t')\frac{\partial c_{a'}(t')}{\partial u_a(t)}$$

This dynamic marginal link travel time function can be interpreted as the effect of an additional traveler on link $a$ during interval $t$ to the total travel time on all links during all intervals. It is the sum of two components, $c_a(t)$ which is the travel time experienced by that additional traveler when the total link inflow rate is $u_a(t)$, and $\sum_{a'}\sum_{t'} u_{a'}(t')\frac{\partial c_{a'}(t')}{\partial u_a(t)}$ which is the additional travel time that this traveler inflicts on each of the $u_{a'}(t')$ travelers using another link $a'$ during another interval $t'$, where general link $a'$ includes our specific link $a$ and $t'$ includes $t$. Strictly speaking, the effect is the result of increasing the inflows $u_a(t)$ by one unit. When the topological link interaction is not considered, equation (10.1) can be simplified to:

$$\hat{c}_a(t) = c_a(t) + \sum_{t'} u_a(t')\frac{\partial c_a(t')}{\partial u_a(t)} \quad (10.2)$$

The relationships among inflow, exit flow and number of vehicles derived in Section 3.7 are restated as follows.

$$c_a(t) = c\left(u_a(1), u_a(2), \cdots, u_a(t)\right) \quad (10.3)$$

Equations (10.2) and (10.3) together imply

$$\hat{c}_a(t) = \hat{c}\left(u_a(1), u_a(2), \cdots, u_a(t)\right) \quad (10.4)$$

Note that when the FIFO restraint is imposed, then the inflow entering the subject link before a specific interval $t$ would have exited the link already. This means that the inflows that need to be considered in the link travel time function are only those entering the subject link between intervals $t$ and $t'$, where $t'$ is the current interval, i.e.,

$$\hat{c}_a(t') = \hat{c}\left(u_a(t), u_a(t+1), \cdots, u_a(t')\right)$$
$$t \le t', t + \tau_a(t) \ge t', t - 1 + \tau_a(t-1) < t' \quad (10.5)$$

In the case where the dynamic link travel time is defined as a function of the number of vehicles on that time-space link, $c_a(t) = c(x_a(t))$, then by the relationship between the inflows and the number of vehicles on that link, equation (10.5) is reduced to

$$\hat{c}_a(t') = \hat{c}(u_a(t), u_a(t+1), \cdots, u_a(t'-1))$$
$$t \le t', t + \tau_a(t) \ge t', t - 1 + \tau_a(t-1) < t' \tag{10.6}$$

## 10.2 Equilibrium Conditions and Model Formulation

In this section, a dynamic system-optimal (or equilibrium) conditions are first defined to characterize the travelers' driving behavior for using routes with the minimal marginal travel time. The marginal route travel times can be computed by adding up marginal link travel times in consideration of the flow propagation requirement along that route. Then, the optimization formulation is presented and the optimality conditions are derived for the DSO route choice problem.

### 10.2.1 Dynamic system-optimal conditions

Given O-D demands that are fixed and time-dependent, the dynamic system-optimal conditions for each O-D pair state that the marginal route travel times for travelers departing during the same interval are equal and minimal; or, no traveler could decrease his/her marginal route travel time by unilaterally changing his/her route. In contrast, the marginal route travel time of any unused route for each O-D pair is greater than or equal to the minimal marginal route travel time. In other words, at system-optimality, if the flow departing from origin $r$ during interval $k$ over route $p$ toward destination $s$ is positive, i.e., $h_p^{rs*}(k) > 0$, then the corresponding marginal route travel time $\hat{c}_p^{rs*}(k)$ is minimal. However, if no flow occurs on route $p$, i.e., $h_p^{rs*}(k) = 0$, then the corresponding marginal route travel time is at least as great as the minimal marginal route travel time $\hat{\pi}^{rs}(k)$. These equilibrium conditions can be mathematically expressed as follows:

$$\hat{c}_p^{rs*}(k)\begin{cases} = \hat{\pi}^{rs}(k) & \text{if } h_p^{rs*}(k) > 0 \\ \ge \hat{\pi}^{rs}(k) & \text{if } h_p^{rs*}(k) = 0 \end{cases} \quad \forall r,s,p,k \tag{10.7}$$

where

$$\hat{c}_p^{rs}(k) = \sum_a \sum_t \hat{c}_a(t)\delta_{apk}^{rs}(t) \quad \forall r,s,p,k \tag{10.8}$$

Equation (10.8) expresses the marginal route travel time in terms of marginal link travel times through the use of indicator variables.

### 10.2.2 Optimization formulation

The dynamic system-optimal (DSO) route choice problem can be formulated as an optimization model as follows:

$$\min \quad \mathbf{cu} \qquad \forall \mathbf{u} \in \Omega \tag{10.9}$$

Or, alternatively, in expanded form:

$$\min \quad \sum_{a}\sum_{t} c_a(t)u_a(t) \qquad \forall \mathbf{u} \in \Omega \tag{10.10}$$

where $\Omega$ denotes the feasible region that is delineated by flow conservation, flow propagation, nonnegativity, and definitional constraints.

Flow conservation constraint:

$$\sum_{p} h_p^{rs}(k) = \bar{q}^{rs}(k) \qquad \forall r,s,k \tag{10.11}$$

Flow propagation constraints:

$$u_{apk}^{rs}(t) = h_p^{rs}(k)\delta_{apk}^{rs}(t) \qquad \forall r,s,a,p,k,t \tag{10.12}$$

$$\sum_{t} \delta_{apk}^{rs}(t) = 1 \qquad \forall r,s,p,a \in p,k \tag{10.13}$$

$$\delta_{apk}^{rs}(t) = \{0,1\} \qquad \forall r,s,a,p,k,t \tag{10.14}$$

Nonnegativity constraint:

$$h_p^{rs}(k) \geq 0 \qquad \forall r,s,p,k \tag{10.15}$$

Definitional constraints:

$$u_a(t) = \sum_{rs}\sum_{p}\sum_{k} h_p^{rs}(k)\delta_{apk}^{rs}(t) \qquad \forall a,t \tag{10.16}$$

$$c_p^{rs}(k) = \sum_{a}\sum_{t} c_a(t)\delta_{apk}^{rs}(t) \qquad \forall r,s,p,k \tag{10.17}$$

The above constraints are the same as that for the DUO route choice model.

### 10.2.3 First-order conditions

***Theorem 10.1***: Under a certain flow propagation relationship $\left(\delta_{apk}^{rs}(t) = \delta_{apk}^{rs*}(t)\right)$, the first-order (or necessary) conditions for dynamic system-optimal route choice problem (10.10) are identical to equilibrium conditions (10.7).

***Derivation of the first-order conditions***: Under a certain flow propagation relationship $\left(\delta_{apk}^{rs}(t)=\delta_{apk}^{rs*}(t)\right)$, dynamic system-optimal route choice problem (10.10) can be simplified as:

$$\min \quad \sum_a \sum_t c_a(t) u_a(t) \tag{10.18}$$

Subject to:

Flow conservation constraint:

$$\sum_p h_p^{rs}(k)=\bar{q}^{rs}(k) \quad \forall r,s,k \tag{10.19}$$

Nonnegativity constraint:

$$h_p^{rs}(k)\geq 0 \quad \forall r,s,p,k \tag{10.20}$$

Definitional constraints:

$$u_{apk}^{rs}(t)=h_p^{rs}(k)\delta_{apk}^{rs*}(t) \quad \forall r,s,a,p,k,t \tag{10.21}$$

$$\delta_{apk}^{rs*}(t)=\{0,1\} \quad \forall r,s,a,p,k,t \tag{10.22}$$

$$u_a(t)=\sum_{rs}\sum_p\sum_k h_p^{rs}(k)\delta_{apk}^{rs*}(t) \quad \forall a,t \tag{10.23}$$

$$c_p^{rs}(k)=\sum_a\sum_t c_a(t)\delta_{apk}^{rs*}(t) \quad \forall r,s,p,k \tag{10.24}$$

The Lagrangian associated with this problem can be written as:

$$L(\mathbf{h},\hat{\boldsymbol{\pi}})=z(\mathbf{u}(\mathbf{h}))+\sum_{rs}\sum_k \hat{\pi}^{rs}(k)\left[\bar{q}^{rs}(k)-\sum_p h_p^{rs}(k)\right] \tag{10.25}$$

The first order conditions for this problem are as follows:

$$h_p^{rs*}(k)\frac{\partial L(\mathbf{h},\hat{\boldsymbol{\pi}})}{\partial h_p^{rs}(k)}=0 \quad \forall r,s,p,k \tag{10.26}$$

$$\frac{\partial L(\mathbf{h},\hat{\boldsymbol{\pi}})}{\partial h_p^{rs}(k)}\geq 0 \quad \forall r,s,p,k \tag{10.27}$$

$$\frac{\partial L(\mathbf{h},\hat{\boldsymbol{\pi}})}{\partial \hat{\pi}^{rs}(k)}=0 \quad \forall r,s,k \tag{10.28}$$

$$h_p^{rs}(k)\geq 0 \quad \forall r,s,p,k \tag{10.29}$$

The derivative of the Lagrangian with respect to a path-flow variable, $h_p^{rs}(k)$, is:

$$\frac{\partial L(\mathbf{h},\hat{\pi})}{\partial h_p^{rs}(k)} = \frac{\partial z(\mathbf{u}(\mathbf{h}))}{\partial h_p^{rs}(k)} + \frac{\partial}{\partial h_p^{rs}(k)}\left\{\sum_{rs}\sum_{k}\hat{\pi}^{rs}(k)\left[\bar{q}^{rs}(k) - \sum_{p} h_p^{rs}(k)\right]\right\} \tag{10.30}$$

The first term on the right-hand side of equation (10.30) can be written out as follows:

$$\begin{aligned}
\frac{\partial z(\mathbf{u}(\mathbf{h}))}{\partial h_p^{rs}(k)} &= \sum_a\sum_t \frac{\partial z(\mathbf{u}(\mathbf{h}))}{\partial u_a(t)}\frac{\partial u_a(t)}{\partial h_p^{rs}(k)} = \sum_a\sum_t \frac{\partial z(\mathbf{u}(\mathbf{h}))}{\partial u_a(t)}\delta_{apk}^{rs*}(t) \\
&= \sum_a\sum_t \left[\frac{\partial}{\partial u_a(t)}\sum_{a'}\sum_{t'} c_{a'}(t')u_{a'}(t')\right]\delta_{apk}^{rs*}(t) \\
&= \sum_a\sum_t \left[c_a(t) + \sum_{a'}\sum_{t'} u_{a'}(t')\frac{\partial c_{a'}(t')}{\partial u_a(t)}\right]\delta_{apk}^{rs*}(t)
\end{aligned} \tag{10.31}$$

By equations (10.1) and (10.8), we have:

$$\frac{\partial z(\mathbf{u}(\mathbf{h}))}{\partial h_p^{rs}(k)} = \sum_a\sum_t \hat{c}_a(t)\delta_{apk}^{rs*}(t) = \hat{c}_p^{rs}(k) \quad \forall r,s,p,k \tag{10.32}$$

The second term on the right-hand side of equation (10.30) can be simplified to:

$$\frac{\partial}{\partial h_p^{rs}(k)}\left\{\sum_{rs}\sum_{k}\hat{\pi}^{rs}(k)\left[\bar{q}^{rs}(k) - \sum_{p} h_p^{rs}(k)\right]\right\} = -\hat{\pi}^{rs}(k) \quad \forall r,s,p,k \tag{10.33}$$

Consequently, Lagrangian (10.30) becomes:

$$\frac{\partial L(\mathbf{h},\hat{\pi})}{\partial h_p^{rs}(k)} = \hat{c}_p^{rs}(k) - \hat{\pi}^{rs}(k) \tag{10.34}$$

Moreover, the derivative of the Lagrangian with respect to a dual variable, $\hat{\pi}^{rs}(k)$, is:

$$\frac{\partial L(\mathbf{h},\hat{\pi})}{\partial \hat{\pi}^{rs}(k)} = \bar{q}^{rs}(k) - \sum_p h_p^{rs}(k) = 0 \quad \forall r,s,k \tag{10.35}$$

By equations (10.34) and (10.35), the first-order conditions can now be expressed as follows:

$$\left[\hat{c}_p^{rs}(k) - \hat{\pi}^{rs}(k)\right]h_p^{rs*}(k) = 0 \quad \forall r,s,p,k \tag{10.36}$$

$$\hat{c}_p^{rs}(k) - \hat{\pi}^{rs}(k) \geq 0 \quad \forall r,s,p,k \tag{10.37}$$

$$\sum_p h_p^{rs}(k) = \bar{q}^{rs}(k) \quad \forall r,s,k \tag{10.38}$$

$$h_p^{rs}(k) \geq 0 \quad \forall r,s,p,k \tag{10.39}$$

We now show that the first-order conditions for the DSO route choice problem are identical to equilibrium conditions (10.7). By equations (10.37) and (10.39), we have:

$$\left[\hat{c}_p^{rs}(k) - \hat{\pi}^{rs}(k)\right] h_p^{rs}(k) \geq 0 \quad \forall r,s,p,k \tag{10.40}$$

Subtracting equation (10.36) from equation (10.40) yields:

$$\left[\hat{c}_p^{rs}(k) - \hat{\pi}^{rs}(k)\right]\left[h_p^{rs}(k) - h_p^{rs*}(k)\right] \geq 0 \quad \forall r,s,p,k \tag{10.41}$$

To preserve the inequality relation, marginal route travel time $\hat{c}_p^{rs*}(k)$ must equal minimal route travel time $\hat{\pi}^{rs}(k)$ if $h_p^{rs*}(k) > 0$, while greater than or equal to minimal route travel time $\hat{\pi}^{rs}(k)$ if $h_p^{rs*}(k) = 0$. These conditions imply equilibrium conditions (10.7). This completes the proof.

## 10.3 Nested Diagonalization Method

By taking into account two link interactions, i.e., actual link travel times $\tau_a(t)$ and the interference among link inflows which appear at different time intervals, we propose the nested diagonalization method to solve the DSO route choice model as follows.

***Algorithm***

**Step 0: Initialization.**

Step 0.1: Let $m$=0. Set $\tau_a^0(t) = NINT\left[c_{a_0}(t)\right], \forall a,t$.

Step 0.2: Let $n$=1. Find an initial feasible solution $\left\{u_a^1(t)\right\}$. Compute the associated link travel times $\left\{c_a^1(t)\right\}$.

**Step 1: *First Loop* Operation.**

Let $m$=$m$+1. Update the estimated actual link travel times by

$$\tau_a^m(t) = NINT\left[(1-\gamma)\tau_a^{m-1}(t) + \gamma c_a^n(t)\right] \quad \forall a,t \tag{10.42}$$

where $0 < \gamma \leq 1$.

Construct the corresponding feasible time-space network based on the estimated actual link travel times.

**Step 2: *Second Loop* Operation.**

Step 2.1: Let $n$=1. Compute and reset the initial feasible solution $\left\{u_a^n(t)\right\}$, based on the time-space network constructed by the estimated actual link travel times $\left\{\tau_a^m(t)\right\}$.

Step 2.2: Fix the inflows for all time-space links other than that on the subject time-space link at the current level, yielding the following optimization problem.

$$\min\ z(\mathbf{u}) = \sum_a \sum_t c_a\left(u_a^n(1), u_a^n(2), \cdots, u_a^n(t-1), u_a^{n+1}(t)\right) u_a^{n+1}(t)$$

(10.43)

Flow conservation constraint:

$$\sum_p h_p^{rs}(k) = \bar{q}^{rs}(k) \quad \forall r,s,k \tag{10.44}$$

Nonnegativity constraint:

$$h_p^{rs}(k) \geq 0 \quad \forall r,s,p,k \tag{10.45}$$

Definitional constraints:

$$u_{apk}^{rs}(t) = h_p^{rs}(k)\bar{\delta}_{apk}^{rs}(t) \quad \forall r,s,a,p,k,t \tag{10.46}$$

$$\bar{\delta}_{apk}^{rs}(t) = \{0,1\} \quad \forall r,s,a,p,k,t \tag{10.47}$$

$$u_a(t) = \sum_{rs}\sum_p\sum_k h_p^{rs}(k)\bar{\delta}_{apk}^{rs}(t) \quad \forall a,t \tag{10.48}$$

$$c_p^{rs}(k) = \sum_a\sum_t c_a(t)\bar{\delta}_{apk}^{rs}(t) \quad \forall r,s,p,k \tag{10.49}$$

**Step 3: *Third Loop* Operation.**

Solve for the solution $\{u_a^{n+1}(t)\}$ in the optimization problem (10.43)~(10.49) by the FW method. Compute the resulting link travel times $\{c_a^{n+1}(t)\}$.

**Step 4: Convergence Check for the *Second Loop* Operation.**

If $u_a^{n+1}(t) \approx u_a^n(t), \forall a,t$, go to Step 5; otherwise, set $n$=$n$+1, and go to Step 2.2.

**Step 5: Convergence Check for the *First Loop* Operation.**

If $\tau_a^m(t) \approx c_a^{n+1}(t), \forall a,t$, stop; the current solution is optimal. Otherwise, set $n$=$n$+1, and go to Step 1.

Note the FW method is applied in Step 3 to the corresponding time-space network. Therefore, after searching for the shortest *marginal* route from time-dependent origin $r(k)$ toward time-independent destination $s$, the AON assignment is performed in a backward manner. If there is a branching from a "superorigin" having a route travel time longer than the preset entire analysis period within the solution procedure, this route must be discarded. Superorigin refers to the sum of the time-dependent origins in the time-space representation of the transportation network (Figure 10.1); it could also be called the "time-independent origin". Analogous to the discussion shown in Section 5.3, the nested projection algorithm can be obtained by the following modification.

Step 2.2: Fix the inflows for all time-space links other than on the subject time-space link at the current level, yielding the following optimization problem.

$$\min\ z(\mathbf{u}) = \frac{1}{2}\left(\mathbf{u}^{n+1}\right)^T \mathbf{G}\mathbf{u}^{n+1} + \left(\rho\hat{\mathbf{c}}\left(\mathbf{u}^n\right) - \mathbf{G}\mathbf{u}^n\right)\mathbf{u}^{n+1} \quad (10.50)$$

Subject to: (10.44)~(10.49)

where matrix **G** is symmetric and positive definite and $\rho$ is a contraction operator.

# 10.4 Numerical Example

## 10.4.1 Input Data

A simple network shown in Figure 10.1 is used for testing. The test network consists of 6 links and 5 nodes in which node 1 is the origin, node 5 is the destination, and nodes 2, 3 and 4 are intermediate nodes.

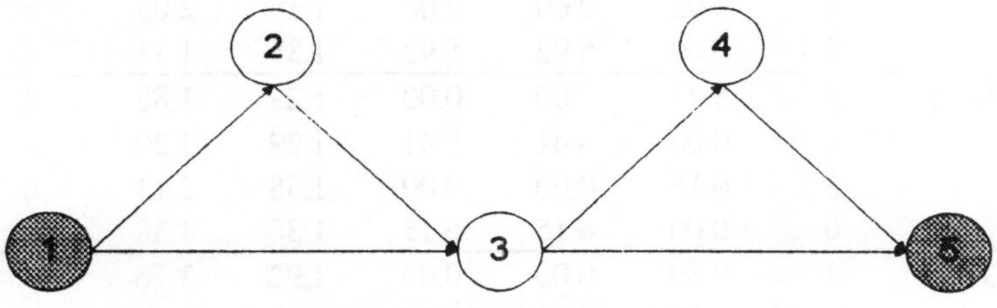


Figure 10.1: Test Network 1

The adopted dynamic travel time function is arbitrarily constructed as follows:

$$c_a(t) = 1 + 0.01\left(u_a(t)\right)^2 + 0.01\left(x_a(t)\right)^2 \quad \forall a, t \quad (10.51)$$

where $u_a(t)$ denotes the inflow rate on link $a$ during interval $t$, and $x_a(t)$ indicates the number of vehicles on link $a$ during interval $t$. The time-dependent origin-destination (O-D) demands are assumed in Table 10.1:

Table 10.1: Time-Dependent O-D Demands

| O-D | Time Interval | |
|---|---|---|
| Pair | $k$=1 | $k$=2 |
| 1-5 | 15 | 15 |

## 10.4.2 Test Results

The DSO route choice model with the above given data was solved by a computer program coded with $C^{++}$. The results are summarized in Table 10.2.

Table 10.2: Results for Test Network 1

| Link | Entering Time Interval | Inflow | Exit Flow | Number of Vehicles | Average Link Travel Time | Marginal Link Travel Time | Exiting Time Interval |
|---|---|---|---|---|---|---|---|
| 1→2 | 1 | 5.41 | 0.00 | 0.00 | 1.29 | 1.88 | 2 |
| | 2 | 5.92 | 5.41 | 5.41 | 1.64 | 2.34 | 4 |
| | 3 | 0.00 | 0.00 | 5.92 | 1.35 | 1.35 | - |
| | 4 | 0.00 | 5.92 | 5.92 | 1.35 | 1.35 | - |
| 1→3 | 1 | 9.59 | 0.00 | 0.00 | 1.92 | 3.76 | 3 |
| | 2 | 9.08 | 0.00 | 9.59 | 2.74 | 4.39 | 5 |
| | 3 | 0.00 | 9.59 | 18.67 | 4.49 | 4.49 | - |
| | 4 | 0.00 | 0.00 | 9.08 | 1.83 | 1.83 | - |
| | 5 | 0.00 | 9.08 | 9.08 | 1.83 | 1.83 | - |
| 2→3 | 2 | 5.41 | 0.00 | 0.00 | 1.29 | 1.88 | 3 |
| | 3 | 0.00 | 5.41 | 5.41 | 1.29 | 1.29 | - |
| | 4 | 5.92 | 0.00 | 0.00 | 1.35 | 2.05 | 5 |
| | 5 | 0.00 | 5.92 | 5.92 | 1.35 | 1.35 | - |
| 3→4 | 3 | 5.41 | 0.00 | 0.00 | 1.29 | 1.88 | 4 |
| | 4 | 0.00 | 5.41 | 5.41 | 1.29 | 1.29 | - |
| | 5 | 6.15 | 0.00 | 0.00 | 1.38 | 2.13 | 6 |
| | 6 | 0.00 | 6.15 | 6.15 | 1.38 | 1.38 | - |
| 3→5 | 3 | 9.59 | 0.00 | 0.00 | 1.92 | 3.76 | 5 |
| | 4 | 0.00 | 0.00 | 9.59 | 1.92 | 1.92 | - |
| | 5 | 8.85 | 9.59 | 9.59 | 2.70 | 4.27 | 8 |
| | 6~7 | 0.00 | 0.00 | 8.85 | 1.78 | 1.78 | - |
| | 8 | 0.00 | 8.85 | 8.85 | 1.78 | 1.78 | - |
| 4→5 | 4 | 5.41 | 0.00 | 0.00 | 1.29 | 1.88 | 5 |
| | 5 | 0.00 | 5.41 | 5.41 | 1.29 | 1.29 | - |
| | 6 | 6.15 | 0.00 | 0.00 | 1.38 | 2.13 | 7 |
| | 7 | 0.00 | 6.15 | 6.15 | 1.38 | 1.38 | - |

The correctness of the results can be verified by checking if the marginal route travel times satisfy the dynamic system-optimal conditions. If we consider route 1→3→5 departing during interval 1 as an example, we can see that the corresponding marginal route travel can be obtained by summing the marginal link travel time on link 1→3 during interval 1 and the marginal link travel time on link 3→5 during time interval $1+c_{1\to3}(1)$ as follows:

$$\begin{aligned}\hat{c}_{1\to3\to5}(1) &= \hat{c}_{1\to3}(1) + \hat{c}_{3\to5}\left(1 + c_{1\to3}(1)\right) \\ &= 3.76 + \hat{c}_{3\to5}(2.92) \approx 3.76 + \hat{c}_{3\to5}(3) = 3.76 + 3.76 \qquad (10.52) \\ &= 7.52\end{aligned}$$

The remaining marginal route times are also computed and summarized in Table 10.3. The trips departing the same origin during the same time interval have approximately the same marginal route travel time. The minor difference is basically due to the round-off error accrued from actual link travel times.

Table 10.3: Average and Marginal Route Travel Times

| Route | Time Interval $k$=1 | | | | Time Interval $k$=2 | | | |
|---|---|---|---|---|---|---|---|---|
| | Average Travel Time | Marginal Travel Time | Route* Flow | Total Travel Time | Average Travel Time | Marginal Travel Time | Route* Flow | Total Travel Time |
| 1→2→3→4→5 | 5.16 | 7.52 | 5.40 | 27.86 | 5.75 | 8.65 | 5.90 | 33.98 |
| 1→2→3→5 | 4.50 | 7.52 | – | – | 5.69 | 8.66 | – | – |
| 1→3→4→5 | 4.50 | 7.52 | – | – | 5.50 | 8.65 | 0.30 | 1.65 |
| 1→3→5 | 3.84 | 7.52 | 9.60 | 36.86 | 5.44 | 8.66 | 8.80 | 47.87 |
| Total | – | – | 15.00 | 64.73 | – | – | 15.00 | 83.50 |

*represents one possible route flow pattern because equilibrated route flow patterns are not unique.

## 10.5 Comparison of Network Travel Times for the DSO and DUO Route Choice Problems

In Section 10.1, we claim that an adaptation of the marginal travel time is desired in order to minimize the total network travel time. In other words, the total network travel time of the DSO route choice model should be lower than that for the DUO route choice model. To verify this, we once again use the hypothesized network data for test network 1. By re-solving the two models with the assumed input data, we can obtain the actual route travel times as shown in Tables 10.4 and 10.5, respectively.

Table 10.4: DUO Results for Test Network 1

| Link | Entering Time Interval | Inflow | Exit Flow | Number of Vehicles | Link Travel Time | Exiting Time Interval |
|---|---|---|---|---|---|---|
| 1→2 | 1 | 3.7 | 0.0 | 0.0 | 1.14 | 2 |
| | 2 | 6.5 | 3.7 | 3.7 | 1.56 | 4 |
| | 3 | 0.0 | 0.0 | 6.5 | 1.43 | - |
| | 4 | 0.0 | 6.5 | 6.5 | 1.43 | - |
| 1→3 | 1 | 11.3 | 0.0 | 0.0 | 2.27 | 3 |
| | 2 | 8.5 | 0.0 | 11.3 | 2.99 | 5 |
| | 3 | 0.0 | 11.3 | 19.8 | 4.90 | - |
| | 4 | 0.0 | 0.0 | 8.5 | 1.72 | - |
| | 5 | 0.0 | 8.5 | 8.5 | 1.72 | - |
| 2→3 | 2 | 3.7 | 0.0 | 0.0 | 1.14 | 3 |
| | 3 | 0.0 | 3.7 | 3.7 | 1.14 | - |
| | 4 | 6.5 | 0.0 | 0.0 | 1.43 | 5 |
| | 5 | 0.0 | 6.5 | 6.5 | 1.43 | - |

Table 10.4: DUO Results for Test Network 1 (continued)

| Link | Entering Time Interval | Inflow | Exit Flow | Number of Vehicles | Link Travel Time | Exiting Time Interval |
|---|---|---|---|---|---|---|
| 3→4 | 3 | 3.7 | 0.0 | 0.0 | 1.14 | 4 |
| | 4 | 0.0 | 3.7 | 3.7 | 1.14 | - |
| | 5 | 6.9 | 0.0 | 0.0 | 1.47 | 6 |
| | 6 | 0.0 | 6.9 | 6.9 | 1.47 | - |
| 3→5 | 3 | 11.3 | 0.0 | 0.0 | 2.27 | 5 |
| | 4 | 0.0 | 0.0 | 11.3 | 2.27 | - |
| | 5 | 8.1 | 11.3 | 11.3 | 2.94 | 8 |
| | 6~7 | 0.0 | 0.0 | 8.1 | 1.66 | - |
| | 8 | 0.0 | 8.1 | 8.1 | 1.66 | - |
| 4→5 | 4 | 3.7 | 0.0 | 0.0 | 1.14 | 5 |
| | 5 | 0.0 | 3.7 | 3.7 | 1.14 | - |
| | 6 | 6.9 | 0.0 | 0.0 | 1.47 | 7 |
| | 7 | 0.0 | 6.9 | 6.9 | 1.47 | - |

Table 10.5: Route Travel Times for the DUO Route Choice Model

| Route | Time Interval $k$=1 | | | | Time Interval $k$=2 | | | |
|---|---|---|---|---|---|---|---|---|
| | Average Travel Time | Marginal Travel Time | Route* Flow | Total Travel Time | Average Travel Time | Marginal Travel Time | Route* Flow | Total Travel Time |
| 1→2→3→4→5 | 4.56 | – | 3.70 | 16.87 | 5.93 | – | 6.50 | 38.55 |
| 1→2→3→5 | 4.55 | – | – | – | 5.93 | – | – | – |
| 1→3→4→5 | 4.55 | – | – | – | 5.93 | – | 0.40 | 2.37 |
| 1→3→5 | 4.54 | – | 11.30 | 51.30 | 5.93 | – | 8.10 | 48.03 |
| Total | – | – | 15.00 | 68.17 | – | – | 15.00 | 88.86 |

*represents one possible route flow pattern because equilibrated route flow patterns are not unique.

By comparing Tables 10.3 and 10.5, we can observe that the total network travel times for the DSO route choice problem are always less than or equal to that for the DUO route choice problem. This result is also consistent with our intuition.

## 10.6 Static Counterpart Approximation

The major difference between the dynamic system-optimal route choice model and its static counterpart is the inclusion of the flow propagation constraint. The flow propagation constraint requires that an inflow cannot exit the time-space link unless the actual link travel time elapses. If we force the actual link travel times to equal zero $(\tau_a = 0)$, then our DSO route choice model reduces to its static counterpart. To verify this, we re-solve the model with all actual link travel times set equal to zero. The results for the static counterpart approximation are summarized in Table 10.6.

Table 10.6: Static Counterpart Approximation for Test Network 1

| Link | Entering Time Interval | Inflow | Exit Flow | Number of Vehicles | Average Link Travel Time | Marginal Link Travel Time | Exiting Time Interval |
|---|---|---|---|---|---|---|---|
| 1→2 | 1 | 5.4 | 5.4 | 0.0 | 1.29 | 1.88 | 1 |
| | 2 | 5.4 | 5.4 | 0.0 | 1.29 | 1.88 | 2 |
| 1→3 | 1 | 9.6 | 9.6 | 0.0 | 1.92 | 3.76 | 1 |
| | 2 | 9.6 | 9.6 | 0.0 | 1.92 | 3.76 | 2 |
| 2→3 | 1 | 5.4 | 5.4 | 0.0 | 1.29 | 1.88 | 1 |
| | 2 | 5.4 | 5.4 | 0.0 | 1.29 | 1.88 | 2 |
| 3→4 | 1 | 5.4 | 5.4 | 0.0 | 1.29 | 1.88 | 1 |
| | 2 | 5.4 | 5.4 | 0.0 | 1.29 | 1.88 | 2 |
| 3→5 | 1 | 9.6 | 9.6 | 0.0 | 1.92 | 3.76 | 1 |
| | 2 | 9.6 | 9.6 | 0.0 | 1.92 | 3.76 | 2 |
| 4→5 | 1 | 5.4 | 5.4 | 0.0 | 1.29 | 1.88 | 1 |
| | 2 | 5.4 | 5.4 | 0.0 | 1.29 | 1.88 | 2 |

The associated route travel times are computed and shown in Table 10.7. For the sake of comparison, the manually computed solutions are also included in brackets. Clearly, the code-computed static approximation and the manually computed approximation are very similar. The minor difference, as mentioned before, is attributed to the round-off arithmetic operations.

Table 10.7: Route Travel Times for the Static Counterpart Approximation

| Route | Time Interval $k$=1 | | | | Time Interval $k$=2 | | | |
|---|---|---|---|---|---|---|---|---|
| | Average Travel Time | Marginal Travel Time | Route* Flow | Total Travel Time | Average Travel Time | Marginal Travel Time | Route* Flow | Total Travel Time |
| 1→2→3 →4→5 | 5.16 (5.17) | 7.52 (7.52) | 5.40 (5.41) | 27.86 (27.99) | 5.16 (5.17) | 7.52 (7.52) | 5.40 (5.41) | 27.86 (27.99) |
| 1→2→3 →5 | 4.50 (4.51) | 7.52 (7.52) | – | – | 4.50 (4.51) | 7.52 (7.52) | – | – |
| 1→3→4 →5 | 4.50 (4.51) | 7.52 (7.52) | – | – | 4.50 (4.51) | 7.52 (7.52) | – | – |
| 1→3→5 | 3.84 (3.84) | 7.52 (7.52) | 9.60 (9.59) | 36.86 (36.80) | 3.84 (3.84) | 7.52 (7.52) | 9.60 (9.59) | 36.86 (36.80) |
| Total | – | – | 15.00 | 64.73 (64.79) | – | – | 15.00 | 64.73 (64.79) |

Remark: *represents one possible route flow pattern because equilibrated route flow patterns are not unique; "numbers" in brackets are manually computed solutions.

## 10.7 Braess's Paradox

Braess's paradox illustrates a counter-intuitive result: additions (deletions) of links to a network may make congestion and delays worse (better) (Sheffi, 1985). The increase (decrease) in travel time is rooted in the essence of the user equilibrium, where each motorist minimizes his or her own travel time. The individual choice of route is carried out without consideration of the effect of this action on other network users. There is no reason, therefore, to expect the total travel time to decrease (increase). By means of an example we show that Braess's paradox may also appear in the DUO route choice model.

Consider the simple 2*2 grid networks shown in Figure 10.2; Figure 10.2(a) consists of 4 links and 4 nodes, and Figure 10.2(b) consists of 5 links and 4 nodes (the additional link is directed from node 3 to node 2, denoted as 3→2). Nodes 1 and 4, respectively, are the origin and the destination, whereas nodes 2 and 3 are intermediate.

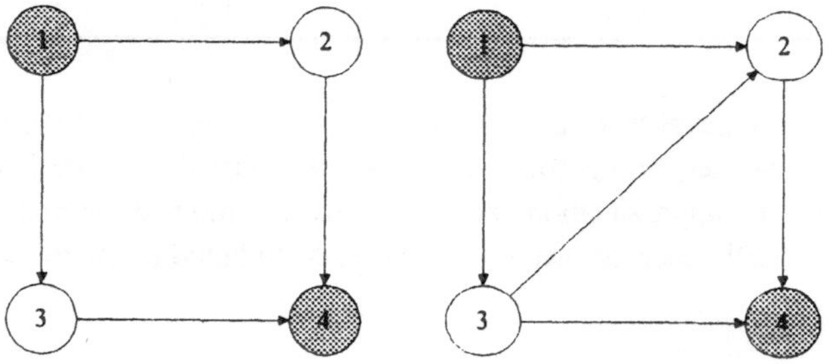


(a) Before Improvement (b) After Improvement

Figure 10.2: Test Network 2

The link travel time functions are redefined as follows:

$$c_a(t) = 8 + 0.01(u_a(t))^2 + 0.01(x_a(t))^2 \quad a = 1 \to 2, 3 \to 4, \forall t \quad (10.53)$$

$$c_a(t) = 1 + 0.01(u_a(t))^2 + 0.01(x_a(t))^2 \quad a = 1 \to 3, 2 \to 4, 3 \to 2, \forall t \quad (10.54)$$

The time-dependent O-D demands are assumed in Table 10.8.

Table 10.8: Time Dependent O-D Demands for Test Network 2

| O-D | Time Interval | |
|---|---|---|
| Pair | $k$=1 | $k$=2 |
| 1-4 | 25 | 25 |

With the above data, the results for network 2 are summarized in Tables 10.9 and 10.10, respectively.

Table 10.9: Braess's Paradox Results for Test Network 2(a)

| Link | Entering Time Interval | Inflow | Exit Flow | Number of Vehicles | Link Travel Time | Exiting Time Interval |
|---|---|---|---|---|---|---|
| 1→2 | 1 | 12.5 | 0.0 | 0.0 | 9.56 | 11 |
| | 2 | 12.5 | 0.0 | 12.5 | 11.13 | 13 |
| | 3~10 | 0.0 | 0.0 | 25.0 | 14.25 | - |
| | 11 | 0.0 | 12.5 | 25.0 | 14.25 | - |
| | 12 | 0.0 | 0.0 | 12.5 | 9.56 | - |
| | 13 | 0.0 | 12.5 | 12.5 | 9.56 | - |
| 1→3 | 1 | 12.5 | 0.0 | 0.0 | 2.56 | 4 |
| | 2 | 12.5 | 0.0 | 12.5 | 4.12 | 6 |
| | 3 | 0.0 | 0.0 | 25.0 | 7.25 | - |
| | 4 | 0.0 | 12.5 | 25.0 | 7.25 | - |
| | 5 | 0.0 | 0.0 | 12.5 | 2.56 | - |
| | 6 | 0.0 | 12.5 | 12.5 | 2.56 | - |
| 2→4 | 11 | 12.5 | 0.0 | 0.0 | 2.56 | 14 |
| | 12 | 0.0 | 0.0 | 12.5 | 2.56 | - |
| | 13 | 12.5 | 0.0 | 12.5 | 4.13 | 17 |
| | 14 | 0.0 | 12.5 | 25.0 | 7.25 | - |
| | 15~16 | 0.0 | 0.0 | 12.5 | 2.56 | - |
| | 17 | 0.0 | 12.5 | 12.5 | 2.56 | - |
| 3→4 | 4 | 12.5 | 0.0 | 0.0 | 9.56 | 14 |
| | 5 | 0.0 | 0.0 | 12.5 | 9.56 | - |
| | 6 | 12.5 | 0.0 | 12.5 | 11.12 | 17 |
| | 7~13 | 0.0 | 0.0 | 25.0 | 14.25 | - |
| | 14 | 0.0 | 12.5 | 25.0 | 14.25 | - |
| | 15~16 | 0.0 | 0.0 | 12.5 | 9.56 | - |
| | 17 | 0.0 | 12.5 | 12.5 | 9.56 | - |

Table 10.10: Braess's Paradox Results for Test Network 2(b)

| Link | Entering Time Interval | Inflow | Exit Flow | Number of Vehicles | Link Travel Time | Exiting Time Interval |
|---|---|---|---|---|---|---|
| 1→2 | 1 | 4.9 | 0.0 | 0.0 | 8.24 | 9 |
| | 2 | 11.5 | 0.0 | 4.9 | 9.57 | 12 |
| | 3~8 | 0.0 | 0.0 | 16.5 | 10.71 | - |
| | 9 | 0.0 | 4.9 | 16.5 | 10.71 | - |
| | 10 | 0.0 | 0.0 | 11.5 | 9.32 | - |
| | 11 | 0.0 | 0.0 | 11.5 | 9.32 | - |
| | 12 | 0.0 | 11.5 | 11.5 | 9.32 | - |
| 1→3 | 1 | 20.1 | 0.0 | 0.0 | 5.02 | 6 |
| | 2 | 13.5 | 0.0 | 20.1 | 6.84 | 9 |
| | 3~5 | 0.0 | 0.0 | 33.5 | 12.25 | - |
| | 6 | 0.0 | 20.1 | 33.5 | 12.25 | - |
| | 7 | 0.0 | 0.0 | 13.5 | 2.82 | - |
| | 8 | 0.0 | 0.0 | 13.5 | 2.82 | - |
| | 9 | 0.0 | 13.5 | 13.5 | 2.82 | - |
| 2→4 | 9 | 20.0 | 0.0 | 0.0 | 4.98 | 14 |
| | 10~11 | 0.0 | 0.0 | 20.0 | 4.98 | - |
| | 12 | 13.4 | 0.0 | 20.0 | 6.77 | 19 |
| | 13 | 0.0 | 0.0 | 33.3 | 12.10 | - |
| | 14 | 0.0 | 20.0 | 33.3 | 12.10 | - |
| | 15~18 | 0.0 | 0.0 | 13.4 | 2.78 | - |
| | 19 | 0.0 | 13.4 | 13.4 | 2.78 | - |
| 3→2 | 6 | 15.0 | 0.0 | 0.0 | 3.25 | 9 |
| | 7~8 | 0.0 | 0.0 | 15.0 | 3.25 | - |
| | 9 | 1.8 | 15.0 | 15.0 | 3.29 | 13 |
| | 10~11 | 0.0 | 0.0 | 1.8 | 1.03 | - |
| | 12 | 0.0 | 1.8 | 1.8 | 1.03 | - |
| 3→4 | 6 | 5.0 | 0.0 | 0.0 | 8.25 | 14 |
| | 7~8 | 0.0 | 0.0 | 5.0 | 8.25 | - |
| | 9 | 11.6 | 0.0 | 5.0 | 9.61 | 19 |
| | 10~13 | 0.0 | 0.0 | 16.7 | 10.79 | - |
| | 14 | 0.0 | 5.0 | 16.7 | 10.79 | - |
| | 15 | 0.0 | 0.0 | 11.6 | 9.36 | - |
| | 16~18 | 0.0 | 0.0 | 11.6 | 9.36 | - |
| | 19 | 0.0 | 11.6 | 11.6 | 9.36 | - |

The actual route travel times for network 2 are computed and summarized in Table 10.11.

Table 10.11: Actual Route Travel Times (Braess's Paradox)

| Results for Test Network 2(a) | | |
|---|---|---|
| Route | Time Interval | |
| | *k*=1 | *k*=2 |
| 1→2→4 | 12.12 | 15.26 |
| 1→3→4 | 12.12 | 15.24 |
| **Results for Test Network 2(b)** | | |
| Route | Time Interval | |
| | *k*=1 | *k*=2 |
| 1→2→4 | 13.22 | 16.34 |
| 1→3→4 | 13.27 | 16.45 |
| 1→3→2→4 | 13.25 | 16.9 |

The actual route travel times for network 2(b) are higher than those for network 2(a). The addition of a link in the test network resulted in longer actual route travel times, which implies a Braess's Paradox.

## 10.8 Toll Policies

One could expect the system-optimizing flow pattern to be achieved when a central authority dictates the links/routes to be selected so as to achieve system optimum. Tolls, either in the form of route tolls or link tolls, can be imposed in order to make the user-optimal solution satisfy the system-optimal conditions. Thus, tolls serve as a mechanism for modifying the travel time as perceived by the individual travelers. The amount of each toll can be computed by means of solving a two-stage problem in which the first stage problem is designed to achieve the system-optimal flow pattern, and the second stage problem is to calculate the necessary tolls to maintain the system-optimal flow pattern subject to the user-optimizing driving behavior.

We first consider the link toll policy by way of a two-stage approach. In the first stage, the dynamic system-optimal flow pattern is computed from optimization problem (10.10). In the second stage, link tolls are calculated by

$$\gamma_a^*(t) = \hat{c}_a^*(t) - c_a^*(t) \quad \forall a, t \tag{10.55}$$

where $c_a^*(t)$ is the average travel time on link $a$ during interval $t$ associated with system-optimal flow pattern $\mathbf{u}^*$ which is obtained from optimization problem (10.10). Hence, to determine the link toll policy, compute the system-optimal solution. Once the system-optimal solution is established, one then substitutes that flow pattern $\mathbf{u}^*$ into equation (10.55) to compute the link toll $\gamma_a^*(t)$ for all links $a$ and all intervals $t$.

Note that the marginal link travel times are always greater than or equal to the associated average link travel times, because the dynamic travel time function is assumed to be nondecreasing. Consequently, the amount of toll charged for each used link is positive; no toll is required for unused links. Given network 1 data, the link tolls calculated by the proposed two-stage problem are summarized in Table 10.12.

Table 10.12: Link Toll Imposition

| Link | Entering Time Interval | Inflow | Number of Vehicles | Exit Flow | Average Link Travel Time | Marginal Link Travel Time | Link Toll | Link Toll Revenue |
|---|---|---|---|---|---|---|---|---|
| 1→2 | 1 | 5.4 | 0.0 | 0.0 | 1.29 | 1.88 | 0.59 | 3.19 |
| | 2 | 5.9 | 5.4 | 5.4 | 1.65 | 2.35 | 0.70 | 4.13 |
| | 3 | 0.0 | 0.0 | 5.9 | 1.35 | 1.35 | 0.00 | 0.00 |
| | 4 | 0.0 | 5.9 | 5.9 | 1.35 | 1.35 | 0.00 | 0.00 |
| 1→3 | 1 | 9.6 | 0.0 | 0.0 | 1.92 | 3.75 | 1.83 | 17.57 |
| | 2 | 9.1 | 0.0 | 9.6 | 2.74 | 4.39 | 1.65 | 15.02 |
| | 3 | 0.0 | 9.6 | 18.7 | 4.48 | 4.48 | 0.00 | 0.00 |
| | 4 | 0.0 | 0.0 | 9.1 | 1.82 | 1.82 | 0.00 | 0.00 |
| | 5 | 0.0 | 9.1 | 9.1 | 1.82 | 1.82 | 0.00 | 0.00 |
| 2→3 | 2 | 5.4 | 0.0 | 0.0 | 1.29 | 1.88 | 0.59 | 3.19 |
| | 3 | 0.0 | 5.4 | 5.4 | 1.29 | 1.29 | 0.00 | 0.00 |
| | 4 | 5.9 | 0.0 | 0.0 | 1.35 | 2.05 | 0.70 | 4.13 |
| | 5 | 0.0 | 5.9 | 5.9 | 1.35 | 1.35 | 0.00 | 0.00 |
| 3→4 | 3 | 5.4 | 0.0 | 0.0 | 1.29 | 1.88 | 0.59 | 3.19 |
| | 4 | 0.0 | 5.4 | 5.4 | 1.29 | 1.29 | 0.00 | 0.00 |
| | 5 | 6.2 | 0.0 | 0.0 | 1.38 | 2.14 | 0.76 | 4.71 |
| | 6 | 0.0 | 6.2 | 6.2 | 1.38 | 1.38 | 0.00 | 0.00 |
| 3→5 | 3 | 9.6 | 0.0 | 0.0 | 1.92 | 3.75 | 1.83 | 17.57 |
| | 4 | 0.0 | 0.0 | 9.6 | 1.92 | 1.92 | 0.00 | 0.00 |
| | 5 | 8.8 | 9.6 | 9.6 | 2.70 | 4.26 | 1.56 | 13.73 |
| | 6 | 0.0 | 0.0 | 8.8 | 1.78 | 1.78 | 0.00 | 0.00 |
| | 7 | 0.0 | 0.0 | 8.8 | 1.78 | 1.78 | 0.00 | 0.00 |
| | 8 | 0.0 | 8.8 | 8.8 | 1.78 | 1.78 | 0.00 | 0.00 |
| 4→5 | 4 | 5.4 | 0.0 | 0.0 | 1.29 | 1.88 | 0.59 | 3.19 |
| | 5 | 0.0 | 5.4 | 5.4 | 1.29 | 1.29 | 0.00 | 0.00 |
| | 6 | 6.2 | 0.0 | 0.0 | 1.38 | 2.14 | 0.76 | 4.71 |
| | 7 | 0.0 | 6.2 | 6.2 | 1.38 | 1.38 | 0.00 | 0.00 |
| Total Network Toll Revenue | | | | | | | 94.31 | |

The amount of route tolls can be directly derived by summing up the link tolls along the route through the incidence variable, as follows:

$$\gamma_p^{rs*}(k) = \sum_a \sum_t \gamma_a^*(t)\delta_{apk}^{rs}(t) \quad \forall r, s, p, k \tag{10.56}$$

The amount of route toll required to achieve system-optimal flow pattern, subject to the user-optimal drivers' behavior, is computed and summarized in Table 10.13.

Table 10.13: Route Toll Imposition

| Route | Time Interval | Average Travel Time | Marginal Travel Time | Route* Flow | Route Toll | Route Toll Revenue |
|---|---|---|---|---|---|---|
| 1→2→3→4→5 | $k$=1 | 5.16 | 7.52 | 5.40 | 2.36 | 12.74 |
| 1→2→3→5 | | 4.50 | 7.51 | – | 3.01 | – |
| 1→3→4→5 | | 4.50 | 7.51 | – | 3.01 | – |
| 1→3→5 | | 3.84 | 7.50 | 9.60 | 3.66 | 35.14 |
| Total | | – | – | 15.00 | | 47.88 |
| 1→2→3→4→5 | $k$=2 | 5.76 | 8.68 | 5.90 | 2.92 | 17.23 |
| 1→2→3→5 | | 5.70 | 8.66 | – | 2.96 | – |
| 1→3→4→5 | | 5.50 | 8.67 | 0.30 | 3.17 | 0.95 |
| 1→3→5 | | 5.44 | 8.65 | 8.80 | 3.21 | 28.25 |
| Total | | – | – | 15.00 | | 46.66 |
| Total Network Toll Revenue | | | | 47.88+46.66=94.54 | | |

*represents one possible route flow pattern because route flow patterns are not unique.

# 10.9 Notes

Merchant and Nemhauser (1978a, 1978b) presented a discontinuous, nonlinear, nonconvex dynamic system-optimal route choice model (M-N model) for a many-to-one network and proposed a solution algorithm. Ho (1980) solved the M-N model by successively optimizing a sequence of linear problems, and derived sufficient condition for the optimal solution. Carey (1986) indicated that the M-N model's constraints satisfy the linear independence condition, and further reformulated (Carey, 1987) the M-N model as a convex, nonlinear programming problem and proved the existence and uniqueness of the optimal solution.

Luque and Friesz (1980) formulated the first dynamic system optimal (DSO) problem using optimal control theory, and later researchers such as Matsui (1987), Ran and Shimazaki (1989), Friesz et al (1989), and Wie et al (1990), also employed the optimal control theory to study the dynamic system-optimal and dynamic user-optimal problems.

In this chapter, we explored the DSO route choice problem using the optimization approach, though the variational inequality approach is equally applicable. Many important issues that are well known in static system-optimal models, such as Braess's paradox and toll impositions, are revisited in the dynamic sense. Other DSO travel choice models involving the variable demand, mode choice and O-D choice can also be formulated and solved accordingly without difficulty.

# Chapter 11

# Dynamic Signal Control Systems

In the standard traffic assignment problem, link capacities are determined exogenously and remain unchanged throughout the solution process. Even though such an assumption may have been necessary in the past to keep a traffic assignment problem manageable, it fails to reflect the real situation. In fact, the magnitude of link capacities are dependent on the prevailing conditions consisting of physical features of the roadway and traffic conditions. Traffic conditions are strongly influenced by the roadway characteristics. During short periods of time, however, the traffic conditions are mostly influenced by the traffic signal control system of the roadway. Thus, link capacities can be simplified as functions of traffic signals for dynamic analysis. To better represent the real situation, the effect of signal controls on the traffic flow pattern must be explicitly taken into account in the standard traffic assignment.

Theoretically, the purpose of traffic signal control policies is to optimize the overall network performance. However, such a system is too complex to be handled; as a result somewhat sub-optimal distributed traffic signal control policies, which only react to the traffic flow pattern approaching each isolated intersection, are often adopted.

In this chapter, a dynamic network signal control (DNSC) system is first explored in Section 11.1 with respect to its model formulation, solution algorithm, variational inequality sensitivity analysis and link capacity and queuing delay. In Section 11.2, the dynamic traffic-responsive signal control (DTSC) system is introduced and a method called iterative optimization and assignment (IOA) is proposed for solving the DTSC problem. A numerical example is also provided to illustrate the efficiency of the method,. Finally, concluding remarks are given in Section 11.3.

## 11.1 Dynamic Network Signal Control System

### 11.1.1 Model Formulation

The dynamic network-wide signal control (DNSC) system is stated in the following theorem.

***Theorem 11.1***: The DNSC system is equivalent to finding a vector $(\mathbf{u}^*, \mathbf{g}^*) \in \Omega$ such that the following optimization problem holds:

$$\min \sum_a \sum_t c_a(t) u_a(t) \quad \forall (\mathbf{u}, \mathbf{g}) \in \Omega \tag{11.1}$$

The symbol $\Omega$ denotes the feasible region that is delineated by the following constraints:

Cycle length conservation constraint:

$$\sum_m \left( g^{Im}(t) + l^{Im} \right) = C^I(t) \quad \forall I, t \tag{11.2}$$

Definitional constraints:

$$H_a^m(t) = \frac{v_a^m(t)}{S_a^m} \frac{C^I(t)}{g^{Im}(t)} \quad \forall I, m, a \in B(I), t \tag{11.3}$$

$$C^I(t) = C^I(\mathbf{v}) \quad \forall I, t \tag{11.4}$$

Boundary constraints:

$$H_a^m(t) < 1 \quad \forall m, a \in B(I), t \tag{11.5}$$

$$g^{Im}(t) \geq \underline{g}^{Im} \quad \forall I, m, t \tag{11.6}$$

$$\overline{C}^I \geq C^I(t) \geq \underline{C}^I \quad \forall I, t \tag{11.7}$$

$$c_{a_2}(t) v_a(t) \leq Q_a \quad \forall a, t \tag{11.8}$$

User equilibrium constraint:

$$\sum_a \sum_t c_a^*(t) \left[ u_a(t) - u_a^*(t) \right] \geq 0 \quad \forall \mathbf{u} \in \Omega(\mathbf{g}) \tag{11.9}$$

Expression (11.1) is an optimization formulation for the DNSC system. The symbol $\mathbf{c}$ denotes the vector $\{c_a(t)\}$, of which each element is defined as the link travel time function associated with intersection $I$ during phase $m$ and interval $t$. It is worth noting that, though not shown explicitly, the link travel time functions essentially contain both link inflows and green times as their arguments. To show this, consider the well-known FHWA travel time function, which can be decomposed into two components, i.e., $c_{a_1}$ denotes the free flow travel time on link $a$, and $c_{a_2}\left(t+c_{a_1}\right)$ denotes the downstream delay on link $a$ during interval $\left(t+c_{a_1}\right)$, as follows.

$$c_a(t) = c_{a_1} + c_{a_2}\left(t + c_{a_1}\right)$$

$$= c_{a_1} + 0.15 \times c_{a_1} \times \left( \frac{v_a\left(t + c_{a_1}\right)}{\dfrac{\sum_m S_a^m \times g^{Im}\left(t + c_{a_1}\right)}{C^I\left(t + c_{a_1}\right)}} \right)^4 \quad \forall a \in B(I) \tag{11.10}$$

Knowing that the inflow and exit flow variables can be transformed by each other, it is obvious that link travel time can be expressed as a function of both inflows and green times as follows.

$$c_a(t) = c_a(\mathbf{u}, \mathbf{g}) \quad \forall a \tag{11.11}$$

Equation (11.2) conserves the cycle length at intersections whereas at each intersection and interval, the sum of green times and loss times must be equal to the cycle length. Equation (11.3) defines the degree of saturation, $H_a^m(t)$, on link $a$ during phase $m$ and interval $t$. The degree of saturation can also be alternatively written as the ratio of exit flow to capacity on link $a$ at interaction $I$ during phase $m$ and interval $t$. Equation (11.4) determines the cycle length as a function of exit flows; by extension of Webster's (1976) formulation, the dynamic counterpart results in the following formula:

$$C^I(t) = \frac{5 + 4.5L}{1 - \sum_{a \in B(I)} \sum_m \dfrac{v_a^m(t)}{S_a^m}} \quad \forall I, t \tag{11.12}$$

where $L$ is the number of signal phases.

Equation (11.5) restricts the degree of saturation to be less than 1; implying that the exit flow can never be greater than the corresponding link capacity. Thus, oversaturation is not allowed. Equation (11.6) imposes the lower limit for the phase green times (usually set as 10 seconds) to ensure vehicles can safely pass through the intersection. In fact, the upper limit for the phase green times is also implicitly included through the use of equation (11.2). Expression (11.7) imposes both the upper and lower limits for the cycle length. Traditionally, the upper limit is set as 150 seconds to avoid too long of a queuing delay for individual drivers, and the lower limit is set as 40 seconds to reduce the total loss times. Inequality (11.8) imposes the maximum storage capacity $Q_a$ (vehicles) on link $a$ and expression (11.9) is itself a VIP which complies with the dynamic user-optimal conditions. The associated constraints include flow conservation, flow propagation, nonnegativity and variable definition, as follows:

Flow conservation constraint:

$$\sum_p h_p^{rs}(k) = \bar{q}^{rs}(k) \quad \forall r,s,k \tag{11.13}$$

Flow propagation constraints:

$$u_{apk}^{rs}(t) = h_p^{rs}(k)\delta_{apk}^{rs}(t) \quad \forall r,s,a,p,k,t \tag{11.14}$$

$$\sum_t \delta_{apk}^{rs}(t) = 1 \quad \forall r,s,p,a \in p,k \tag{11.15}$$

$$\delta_{apk}^{rs}(t) = \{0,1\} \quad \forall r,s,a,p,k,t \tag{11.16}$$

Nonnegativity constraint:

$$h_p^{rs}(k) \geq 0 \quad \forall r,s,p,k \tag{11.17}$$

Definitional Constraints:

$$u_a(t) = \sum_{rs}\sum_p\sum_k h_p^{rs}(k)\delta_{apk}^{rs}(t) \quad \forall a,t \tag{11.18}$$

$$c_p^{rs}(k) = \sum_a\sum_t c_a(t)\delta_{apk}^{rs}(t) \quad \forall r,s,p,k \tag{11.19}$$

### 11.1.2 Solution Algorithm

The feasible region delineated by expressions (11.2)~(11.9) for the DNSC system is essentially nonconvex, because expressions (11.3), (11.4), (11.8), (11.9) and the implicit reaction function $\mathbf{u}(\mathbf{g})$ are nonlinear. Nonconvexity portends existence of local optima; hence, the global optimum is difficult to find, even with the most computationally efficient procedures. In the following, we tentatively propose a solution algorithm based on the so-called (variational inequality) sensitivity analysis technique (Friesz et al., 1990) without numerical examples.

The algorithm begins with initial green times and the corresponding link capacities. The aforementioned nested diagonalization method is then applied with modifications for calculating the derivatives by the variational inequality sensitivity analysis. The derivative information is in turn adopted to formulate a linearized subproblem, to which an auxiliary (green time) solution is obtained. The direction pointing from the main problem solution to the subproblem solution indicates a descent search direction, along which the optimal move size is determined. The above procedure is repeated for the updated main problem solution until a convergence criterion is met. The formal steps of the algorithm can be described as follows.

***Algorithm***

**Step 0: Initialization.**

Step 0.1: Let $o$=1. Find initial feasible green times $\left\{g^{Im}(t)^1\right\}$.

Step 0.2: Compute the corresponding link capacities $\left\{CAP_a(t)^o\right\}$ as follows.

$$CAP_a^o(t) = \sum_m S_a^m \frac{g^{Im}(t)^o}{C^I(t)^o} \quad \forall a,t,I \tag{11.20}$$

Step 0.3: Let $n$=1. Find an initial feasible flow pattern $\left\{u_a^1(t)\right\}$. Compute the associated link travel times $\left\{c_a^1(t)\right\}$.

Step 0.4: Let $l$=0. Set $\tau_a^0(t) = 0, \forall a,t$.

**Step 1: *First Loop* Operation.**

Step 1.1: Let $l$=$l$+1. Update the estimated actual link travel times by

$$\tau_a^l(t) = NINT\left[(1-\gamma)\tau_a^{l-1}(t) + \gamma c_a^n(t)\right] \quad \forall a,t. \tag{11.21}$$

Step 1.2: Construct the corresponding feasible time-space network based on the estimated actual link travel times.

**Step 2: *Second Loop* Operation.**

Step 2.1: Let $n$=1. Compute and reset the initial feasible solution $\left\{u_a^n(t)\right\}$, based on the time-space network constructed by the estimated actual link travel times $\left\{\tau_a^l(t)\right\}$.

Step 2.2: Fix the inflows for each physical link other than on the subject time-space link at the current level, yielding the following optimization problem.

$$\min\ z(\mathbf{u}) = \sum_a \sum_t \int_0^{u_a^{n+1}(t)} c_a\left(u_a^n(1), u_a^n(2), \cdots, u_a^n(t-1), \omega\right) d\omega \tag{11.22}$$

s.t.

Flow conservation constraint:

$$\sum_p h_p^{rs}(k) = \bar{q}^{rs}(k) \quad \forall r,s \tag{11.23}$$

Nonnegativity constraint:

$$h_p^{rs}(k) \geq 0 \qquad \forall r,s,p,k \tag{11.24}$$

Definitional constraints:

$$u_{apk}^{rs}(t) = h_p^{rs}(k)\bar{\delta}_{apk}^{rs}(t) \quad \forall r,s,a,p,k,t \tag{11.25}$$

$$\bar{\delta}_{apk}^{rs}(t) = \{0,1\} \quad \forall r,s,a,p,k,t \tag{11.26}$$

$$u_a(t) = \sum_{rs}\sum_{p}\sum_{k} h_p^{rs}(k)\bar{\delta}_{apk}^{rs}(t) \quad \forall a,t \tag{11.27}$$

$$c_p^{rs}(k) = \sum_{a}\sum_{t} c_a(t)\bar{\delta}_{apk}^{rs}(t) \quad \forall r,s,p,k \tag{11.28}$$

**Step 3: *Third Loop* Operation.**

Solve for the solution $\{h_p^{rs}(k)\}$ and the associated inflow pattern $\{u_a^{n+1}(t)\}$ in the mathematical problem (11.22)~(11.28) by any path-based algorithm. Compute the resulting link travel times $\{c_a^{n+1}(t)\}$.

**Step 4: Convergence Check for the *Second Loop* Operation.**

If $u_a^{n+1}(t) \approx u_a^n(t), \forall a,t$, go to Step 5; otherwise, set $n=n+1$, and go to Step 2.2.

**Step 5: Convergence Check for the *First Loop* Operation.**

If $\tau_a^l(t) \approx c_a^{n+1}(t), \forall a,t$, go to Step 6; otherwise, set $n=n+1$, and go to Step 1.

**Step 6: Update the Phase Green Times.**

Step 6.1: Calculate the derivatives, $\frac{\partial v}{\partial g}$, using variational inequality sensitivity analysis (to be described in Sections 11.1.3 and 11.1.4).

Step 6.2: Formulate local linear approximations of the DNSC system using the derivative information and solve the resulting linear programming problem to obtain an auxiliary solution **g'**.

$$\min\ z(\mathbf{g}') = \sum_{I}\sum_{m}\sum_{a\in B(I)}\sum_{t} \frac{\partial\left(c_a(t)^{n+1} \times u_a(t)^{n+1}\right)}{\partial u_a(t)} \frac{\partial u_a^m(\mathbf{g}^o)}{\partial g^{Im}(t)} g'^{Im}(t) \tag{11.29}$$

s.t.

Cycle length conservation constraint:

$$\sum_{m}\left(g'^{Im}(t) + l^{Im}\right) = C^I(t) \quad \forall I,t \tag{11.30}$$

Definitional constraints:

$$H_a^m(t) = \frac{v_a^m(t)}{S_a^m}\frac{C^I(t)}{g'^{Im}(t)} \quad \forall I, m, a \in B(I), t \tag{11.31}$$

$$C^I(t) = C^I(\mathbf{v}) \quad \forall I, t \tag{11.32}$$

Boundary constraints:

$$H_a^m(t) < 1 \quad \forall m, a \in B(I), t \tag{11.33}$$

$$g'^{Im}(t) \geq \underline{g}^{Im} \quad \forall I, m, t \tag{11.34}$$

$$\overline{C}^I \geq C^I(t) \geq \underline{C}^I \quad \forall I, t \tag{11.35}$$

$$c_{a_2}(t) v_a(t) \leq Q_a \quad \forall a, t \tag{11.36}$$

Linear approximation constraints:

$$\frac{\partial v_a\left[\mathbf{g}^o\left(t + c_a(t)\right)\right]}{\partial g^{Im}\left(t + c_a(t)\right)} = \frac{\partial u_a\left(\mathbf{g}^o(t)\right)}{\partial g^{Im}\left(t + c_a(t)\right)} \quad \forall I, m, a \in B(I), t \tag{11.37}$$

$$v_a(\mathbf{g}) = v_a(\mathbf{g}^o) + \sum_m \frac{\partial v_a(\mathbf{g}^o)}{\partial g^{Im}(t)}\left(g'^{Im}(t) - \left(g^{Im}(t)\right)^o\right) \quad \forall I, a \in B(I), t \tag{11.38}$$

$$C^I(\mathbf{v}) = C^I(\mathbf{v}^{n+1}) + \sum_{a \in B(I)} \frac{\partial C^I(\mathbf{v}^{n+1})}{\partial v_a(t)} \frac{\partial v_a(t)^{n+1}}{\partial g^{Im}(t)} \quad \forall I, t \tag{11.39}$$

$$c_{a_1}(t) v_a(t) = c_{a_1}(t)^{n+1} v_a(t)^{n+1} + \sum_m \sum_{a \in B(I)} \left[\left(\frac{\partial c_{a_1}(t)^{n+1}}{\partial v_a(t)} \frac{\partial v_a(t)^{n+1}}{\partial g^{Im}(t)} v_a(t)^{n+1} + c_{a_1}(t)^{n+1} \frac{\partial v_a(t)^{n+1}}{\partial g^{Im}(t)}\right) \left(g'^{Im}(t) - \left(g^{Im}(t)\right)^o\right)\right] \quad \forall I, a \in B(I), t \tag{11.40}$$

$$v_a(t) C^I(\mathbf{v}) = v_a(t)^{n+1} C^I(\mathbf{v}^{n+1}) + \sum_m \sum_{a \in B(I)} \left[\left(\frac{\partial C^I(\mathbf{v}^{n+1})}{\partial v_a(t)} \frac{\partial v_a(t)^{n+1}}{\partial g^{Im}(t)} v_a(t)^{n+1} + C^I(\mathbf{v}^{n+1}) \frac{\partial v_a(t)^{n+1}}{\partial g^{Im}(t)}\right) \left(g'^{Im}(t) - \left(g^{Im}(t)\right)^o\right)\right] \quad \forall I, a \in B(I), t \tag{11.41}$$

Step 6.3: Update the phase green times

$$g^{Im}(t)^{o+1} = g^{Im}(t)^{o} + \tfrac{1}{o+1}\left(g'^{Im}(t)^{o} - g^{Im}(t)^{o}\right) \quad \forall I,m,t \qquad (11.42)$$

**Step 7: Convergence Check for the Phase Green Times.**

If $g^{Im}(t)^{o+1} \approx g^{Im}(t)^{o}$, or the maximum number of major iterations is reached, stop; the current solution is optimal. Otherwise, set $o$=$o$+1 and go to Step 0.2.

### 11.1.3 Variational Inequality Sensitivity Analysis

This section contains results summarized from Friesz et al. (1990), Tobin et al. (1988) and Tobin (1986), which are presented without proof. Now assume **g** and **u** represent the decision vectors. Note that **u** is implicitly determined by **g**, i.e., $\mathbf{u}(\mathbf{g})$. We wish to solve the following problem.

$$\min \quad \sum_a \sum_t c_a(t) u_a(t) \quad \forall(\mathbf{u},\mathbf{g}) \in \Omega \qquad (11.43)$$

where the feasible region $\Omega$ is defined by the following constraints:

$$\mathbf{L} \le \mathbf{g} \le \mathbf{U} \qquad (11.44)$$

$$\mathbf{c}(\mathbf{u}^*,\mathbf{g})(\mathbf{u} - \mathbf{u}^*) \ge 0, \quad \forall \mathbf{u} \in \Omega(\mathbf{g}) \qquad (11.45)$$

where $\hat{\mathbf{c}}: R^n \times R^q \to R^1$ is differentiable in $(\mathbf{u},\mathbf{g})$, $\mathbf{L} \in R^q_+$, $\mathbf{U} \in R^q_+$, and $\Omega(\mathbf{g})$ is the feasible set, possibly dependent on **g**, of **u**-variables.

$$\Omega(\mathbf{g}) = \left\{\mathbf{u} \in R^n : \mathbf{f}(\mathbf{u},\mathbf{g}) \ge 0, \mathbf{h}(\mathbf{u},\mathbf{g}) = 0\right\} \qquad (11.46)$$

We consider a finite dimensional problem with $\mathbf{u} = (u_1, \cdots, u_n)$; $\mathbf{g} = (g_1, \cdots, g_q)$; $f: R^n \times R^q \to R^m$, differentiable and concave in **u**; $h: R^n \times R^q \to R^p$, linear in **u**; and $\mathbf{c}: R^n \times R^q \to R^n$, differentiable and strictly monotone in **u**.

Let

$\mathbf{c}(\mathbf{u},\mathbf{g})$ be once differentiable in $(\mathbf{u},\mathbf{g})$.

$\mathbf{f}(\mathbf{u},\mathbf{g})$ be concave in **u** and twice continuously differentiable in $(\mathbf{u},\mathbf{g})$.

$\mathbf{h}(\mathbf{u},\mathbf{g})$ be linear affine in **u** and once continuously differentiable in **g**.

Now consider the following perturbed variational inequality, denoted as VI(**g**): find $\mathbf{u}^* \in \Omega(\mathbf{g})$ such that

$$\mathbf{c}(\mathbf{u}^*,\mathbf{g})(\mathbf{u} - \mathbf{u}^*) \ge 0, \quad \forall \mathbf{u} \in \Omega(\mathbf{g}) \qquad (11.47)$$

where

$$\Omega(\mathbf{g}) = \left\{\mathbf{u} \mid \mathbf{f}(\mathbf{u},\mathbf{g}) \ge 0, \mathbf{h}(\mathbf{u},\mathbf{g}) = 0\right\} \qquad (11.48)$$

***Theorem 11.2***: Consider the following conditions on the perturbed variational inequality VI($\bar{\mathbf{g}}$):

(a) The constraints $f_i(\mathbf{u},\bar{\mathbf{g}})$ are concave in $\mathbf{u}$, and $\mathbf{u}^* \in \Omega$, $\eta^* \in R^m$, $\mu^* \in R^p$ satisfy

$$\mathbf{c}(\mathbf{u}^*,\bar{\mathbf{g}}) - \nabla\mathbf{f}(\mathbf{u}^*,\bar{\mathbf{g}})^T \eta^* - \nabla\mathbf{h}(\mathbf{u}^*,\bar{\mathbf{g}})^T \mu^* = \mathbf{0} \tag{11.49}$$

$$\eta^{*T}\,\mathbf{f}(\mathbf{u}^*,\bar{\mathbf{g}}) = \mathbf{0} \tag{11.50}$$

$$\eta^* \geq 0 \tag{11.51}$$

(b) $\mathbf{c}(\mathbf{u}^*,\bar{\mathbf{g}})$ is such that

$$\mathbf{x}^T \nabla\mathbf{c}(\mathbf{u}^*,\bar{\mathbf{g}})\mathbf{x} > 0 \tag{11.52}$$

for all $\mathbf{x} \neq 0$ such that

$$\nabla f_i(\mathbf{u}^*,\bar{\mathbf{g}})\mathbf{x} \geq 0 \text{ for } i \text{ such that } g_i(\mathbf{u}^*,\bar{\mathbf{g}}) = 0, \tag{11.53}$$

$$\nabla f_i(\mathbf{u}^*,\bar{\mathbf{g}})\mathbf{x} = 0 \text{ for } i \text{ such that } \eta_i^* > 0, \tag{11.54}$$

$$\nabla h_i(\mathbf{u}^*,\bar{\mathbf{g}})\mathbf{x} = 0 \text{ for } i = 1,\ldots,p. \tag{11.55}$$

(b') $\nabla\mathbf{c}(\mathbf{u}^*,\bar{\mathbf{g}})$ is positive definite.

(c) The gradients, $\nabla f_i(\mathbf{u}^*,\bar{\mathbf{g}})$ for $i$ such that $f_i(\mathbf{u}^*,\bar{\mathbf{g}}) = 0$ and $\nabla h_i(\mathbf{u}^*,\bar{\mathbf{g}})$ for $i$=1,...,$p$ are linearly independent.

(d) The strict complementary slackness conditions

$$\eta_i^* > 0 \text{ when } f_i(\mathbf{u}^*,\bar{\mathbf{g}}) = 0 \tag{11.56}$$

are satisfied.

We then have the following:

(i) If (a) is satisfied, then $\mathbf{u}^*$ is a solution to VI($\bar{\mathbf{g}}$).

(ii) If, in addition, (b) or (b') is satisfied, then $\mathbf{u}^*$ is a locally unique solution to VI($\bar{\mathbf{g}}$).

(iii) If, in addition, (c) is satisfied, then $\eta^*$ and $\mu^*$ are unique, and for $\mathbf{g}$ in a neighborhood of $\bar{\mathbf{g}}$, there exists a unique, directionally differentiable, *Lipschitz* continuous function $\left[\mathbf{u}(\mathbf{g})^T, \eta(\mathbf{g})^T, \mu(\mathbf{g})^T\right]^T$, where $\mathbf{u}(\mathbf{g})$ is a locally unique solution to VI($\mathbf{g}$), and $\eta(\mathbf{g})$ and $\mu(\mathbf{g})$ are unique associated multipliers satisfying (a) and (b) above for VI($\mathbf{g}$), and with

$$\left[\mathbf{u}(\bar{\mathbf{g}})^T, \eta(\bar{\mathbf{g}})^T, \mu(\bar{\mathbf{g}})^T\right]^T = \left[\mathbf{u}^{*T}, \eta^{*T}, \mu^{*T}\right]^T \tag{11.57}$$

Additionally, in a neighborhood of $\bar{\mathbf{g}}$, the set of binding constraints is unchanged and the binding constraint gradients are linearly independent at $\mathbf{u}(\mathbf{g})$.

(iv) If, in addition, (d) is satisfied, then the function $\left[\mathbf{u}(\mathbf{g})^T, \eta(\mathbf{g})^T, \mu(\mathbf{g})^T\right]^T$ is differentiable in the neighborhood of $\bar{\mathbf{g}}$.

***Theorem 11.3***: If the vector $\mathbf{u}^*$ is a solution to the VI($\bar{\mathbf{g}}$) and the gradients $\nabla f_i\left(\mathbf{u}^*, \bar{\mathbf{g}}\right)$ for $i$ such that $f_i\left(\mathbf{u}^*, \bar{\mathbf{g}}\right) = 0$ and $\nabla h_i\left(\mathbf{u}^*, \bar{\mathbf{g}}\right)$ for $i$=1,...$p$ are linearly independent, then there exists $\eta^* \in R^m$, $\mu^* \in R^p$ such that expressions (11.49)~(11.51) are satisfied.

Let $\mathbf{u}^*$ be a solution to VI($\bar{\mathbf{g}}$) satisfying Theorem 11.3. Then for $\mathbf{g} = \bar{\mathbf{g}}$ and $(\mathbf{u}, \eta, \mu) = (\mathbf{u}^*, \eta^*, \mu^*)$ we have:

$$\mathbf{c}(\mathbf{u},\mathbf{g}) - \sum_{i=1}^{m} \eta_i \nabla f_i(\mathbf{u},\mathbf{g})^T - \sum_{i=1}^{p} \mu_i \nabla h_i(\mathbf{u},\mathbf{g})^T = \mathbf{0} \qquad (11.58)$$

$$\eta_i f_i(\mathbf{u},\mathbf{g}) = 0 \quad \text{for } i=1,2,\ldots,m \qquad (11.59)$$

$$h_i(\mathbf{u},\mathbf{g}) = 0 \quad \text{for } i=1,2,\ldots p \qquad (11.60)$$

Let the Jacobian matrix of the system (11.58)~(11.60) with respect to $\mathbf{x} = (\mathbf{u}, \eta, \mu)$ be denoted as $\mathbf{J}_\mathbf{x}^*$, and with respect to $\mathbf{g}$ as $\mathbf{J}_\mathbf{g}^*$. Then, we may state the following results:

***Corollary 11.1***: If the constraints $f_i\left(\mathbf{u}, \bar{\mathbf{g}}\right)$ are concave and the conditions (b) (or (b')), (c) and (d) in Theorem 11.2 are satisfied, then the inverse of $\mathbf{J}_\mathbf{x}^*$ exists and the derivatives of $(\mathbf{u}^*, \eta^*, \mu^*)$ with respect to $\mathbf{g}$ are given by

$$\nabla_\mathbf{g}\mathbf{x}(\bar{\mathbf{g}}) = \mathbf{J}_\mathbf{x}^{-1}(\bar{\mathbf{g}})\left[-\mathbf{J}_\mathbf{g}(\bar{\mathbf{g}})\right] \qquad (11.61)$$

Furthermore, the points of nondifferentiability of the implicit function $\mathbf{u}(\mathbf{g})$ defined by the variational inequality (11.43) form a set of measure zero, and, therefore, points exist in a small neighborhood of a nondifferential point at which this implicit function is differentiable. In particular, the following result holds:

***Theorem 11.4***: Suppose $\mathbf{u}^*$ solves the variational inequality (11.45) and (11.46) for $\mathbf{g} = \bar{\mathbf{g}}$ and the conditions of (a), (b) (or (b')) and (c) of Theorem 11.2 are satisfied. Then in any open neighborhood of $\bar{\mathbf{g}}$, there exists $\hat{\mathbf{g}}$, such that the implicit function $\mathbf{x}(\mathbf{g}) = \left[\mathbf{u}(\mathbf{g})^T, \eta(\mathbf{g})^T, \mu(\mathbf{g})^T\right]^T$ is differentiable at $\hat{\mathbf{g}}$.

Theorem 11.4 allows the development of algorithms for expressions (11.43)~(11.46) using the gradient to determine a descent direction. If at step $l$, $\mathbf{g}^l$ is such that $\mathbf{x}(\mathbf{g}^l)$ is not differentiable, then by a systematic search in an ε-neighborhood of $\mathbf{g}^l$, a point $\hat{\mathbf{g}}^l$ can be found where $\mathbf{x}(\hat{\mathbf{g}}^l)$ is differentiable. This may be accomplished by returning to $\mathbf{g}^{l-1}$ and perturbing the step length and/or direction. In practice, it is not likely that a step will land on a point at which the implicit function is non-differentiable. If it does, it is likely that a single change in the step length will be sufficient.

### 11.1.4 Sensitivity Analysis of the Restricted Variational Inequality

In this section, the perturbed network equilibrium network flow problem is of interest and can be written as the following perturbed variational inequality.

$$\mathbf{c}(\mathbf{u}^*,\varepsilon)(\mathbf{u}-\mathbf{u}^*)\geq 0 \quad \forall \mathbf{u}\in\Omega(\varepsilon) \tag{11.62}$$

where

$$\Omega(\varepsilon)=\left\{\mathbf{u}|\Lambda^1\,\mathbf{h}=\mathbf{u},\ \Lambda^2\,\mathbf{h}=\mathbf{q}(\varepsilon),\ \mathbf{h}\geq 0\right\} \tag{11.63}$$

where $\varepsilon$ is a vector of perturbation parameters. It is assumed that $\mathbf{c}(\mathbf{u},\varepsilon)$ is once continuously differentiable in $(\mathbf{u},\varepsilon)$, and $\mathbf{q}(\varepsilon)$ is once continuously differentiable in $\varepsilon$. The equilibrium route flows are generally not unique and are contained in the convex polytope

$$\Gamma^*(\varepsilon)=\left\{\mathbf{u}|\Lambda^1\,\mathbf{h}=\mathbf{u}^*,\Lambda^2\,\mathbf{h}=\mathbf{q}(\varepsilon),\mathbf{h}\geq 0\right\} \tag{11.64}$$

where u* solves expressions (11.62)~(11.63). Because for any vector $\varepsilon$, the set of route flow solutions in $\Gamma^*(\varepsilon)$ is a convex set, derivatives of a solution $\mathbf{h}^*$ with respect to the perturbation parameters do not exist. The nonuniqueness of **h** causes the conditions of Theorem 11.2(b) not to be met even though $\mathbf{u}^*$ is unique.

The approach taken here to calculate the derivatives of arc flows with respect to the perturbation parameters is to select one particular path flow solution, in particular an extreme point of $\Gamma^*(\varepsilon)$, and develop the derivatives for this particular solution point with respect to the perturbation parameters. Assuming that the strict complementary slackness conditions are satisfied and that the extreme point is nondegenerate, it is shown that if the perturbed problem is restricted to only those routes that are positive in the extreme point, it still solves the original perturbed problem for small perturbations. The calculation of the derivatives of arc flows with respect to the perturbation parameters through the use of the derivatives of route flows, do not depend on the nondegenerate extreme point chosen and therefore, are in fact the desired derivatives for the original problem. In the following, we reduce the network under consideration to that which contains only arcs which have positive flow in the solution, and consider only the routes on these arcs.

Given the solution of $\mathbf{u}^*(0)$, the first step is to choose a unique route flow vector $\mathbf{h}^*$ to associate with $\mathbf{u}^*(0)$. The only requirement in this choice is that $\mathbf{h}^*$ be a nondegenerate extreme point of $\Gamma^*(0)$, that is, an extreme point solution $\mathbf{h}^*$ in which the number of routes with positive flow is equal to the rank of $\left[(\Lambda^1)^T | (\Lambda^2)^T\right]$, which at most is equal to the number of arcs with positive flow plus the number of origin-destination pairs. A number of route generating methods can be used to generate $\mathbf{h}^*$, such as the Gradient Projection method, Aggregate Simplicial Decomposition method and Disaggregate Simplicial Decomposition method. In fact, the commonly used Convex Combination method can still be used with modifications. That is, the route flow information is saved from iteration to iteration until the algorithm terminates. The resulting equilibrium route flow can be used as $\mathbf{h}^*$, an extreme route flow.

Let $\mathbf{h}^*$ be a nondegenerate extreme point of $\Gamma^*(0)$. Since $\mathbf{h}^*$ is a solution to the perturbed variational inequality (11.62)~(11.63) at $\varepsilon = 0$, by Theorem 11.2(a) there exists a solution to the system

$$\mathbf{c}'(\mathbf{h}^*,0) - \eta - (\Lambda^2)^T \mu = \mathbf{0} \tag{11.65}$$

$$\eta^T \mathbf{h}^* = 0 \tag{11.66}$$

$$\Lambda^2 \mathbf{h}^* - \mathbf{q}(0) = 0 \tag{11.67}$$

$$\eta \geq 0 \tag{11.68}$$

Since we have restricted the network to those arcs with positive flow, any route $p$ is such that $c'_p(h^*,0) = \pi^{rs}$, and so, $\eta = 0$. Also, $\mathbf{h}^*$ is a nondegenerate extreme point of $\Gamma^*(0)$ that corresponds to a unique basis $\mathbf{B}^*$. For a neighborhood of $\varepsilon = 0$, $\mathbf{B}^*$ is a basis for $\Gamma^*(\varepsilon)$, because the right-hand side $\left[\mathbf{u}^*(\varepsilon)^T | \mathbf{q}(\varepsilon)^T\right]^T$ varies continuously with $\varepsilon$. Therefore, for $\varepsilon$ near zero, there exists an extreme point $\mathbf{h}^*(\varepsilon)$ of $\Gamma^*(\varepsilon)$ with the same positive route flows as those in $\mathbf{h}^*$. We then restrict the problem to only those route variables $h_p$ which are positive in $\mathbf{h}^*$ by denoting the set of routes as $\mathbf{P}^+$, the corresponding route flow vector as $\mathbf{h}^{*+}$, and the reduced origin-destination/route incidence matrix as $\Lambda^{2+}$. Other reduced vectors and matrices will be denoted similarly. Since all route flow variables are positive in this reduced system, and will remain so for perturbations in a neighborhood of 0, the nonnegativity constraints on $\mathbf{h}$ are not binding and may be eliminated without changing the solution in a neighborhood of 0. The system then reduces to

$$\mathbf{c}'(\mathbf{h}^*,0)-\left(\Lambda^{2+}\right)^T\mu=\mathbf{0} \tag{11.69}$$

$$\Lambda^{2+}\mathbf{h}^{*+}-\mathbf{q}(0)=0 \tag{11.70}$$

To see that Theorem 11.2(b) is satisfied for this restricted system, the reader is referred to Tobin et al (1988). It can easily be seen that the columns of $\left(\Lambda^2\right)^T$ are linearly independent and so $\mu$ is unique. If we denote this unique vector as $\mu^*$, the conditions for Theorem 11.3 are therefore satisfied by the system (11.69)~(11.70), and the derivatives of $\mathbf{h}^{*+}$ with respect to $\varepsilon$ may be calculated as follows: The Jacobian matrix of the system (11.69)~(11.70) with respect to $\left(\mathbf{h}^+,\mu\right)$ and evaluated at $\varepsilon=0$ is

$$\mathbf{J}_{h^+\mu}=\begin{bmatrix}\nabla\mathbf{c}'^+(\mathbf{h}^*,0) & -\left(\Lambda^2\right)^T\\ \Lambda^2 & 0\end{bmatrix} \tag{11.71}$$

Suppose

$$\left[\mathbf{J}_{h^+\mu}\right]^{-1}=\begin{bmatrix}\mathbf{B}_{11} & \mathbf{B}_{12}\\ \mathbf{B}_{21} & \mathbf{B}_{22}\end{bmatrix} \tag{11.72}$$

It is easily shown that

$$\mathbf{B}_{11}=\nabla\mathbf{c}'^+(\mathbf{h}^*,0)^{-1}\begin{bmatrix}\mathbf{I}-\left(\Lambda^{2+}\right)^T\left[\Lambda^{2+}\nabla\mathbf{c}'^+(\mathbf{h}^*,0)^{-1}\left(\Lambda^{2+}\right)^T\right]^{-1}\\ \cdot\Lambda^{2+}\nabla\mathbf{c}'^+(\mathbf{h}^*,0)^{-1}\end{bmatrix} \tag{11.73}$$

$$\mathbf{B}_{12}=\nabla\mathbf{c}'^+(\mathbf{h}^*,0)^{-1}\left(\Lambda^{2+}\right)^T\left[\Lambda^{2+}\left[\nabla\mathbf{c}'^+(\mathbf{h}^*,0)^{-1}\left(\Lambda^{2+}\right)^T\right]^{-1}\right] \tag{11.74}$$

$$\mathbf{B}_{21}=-\left[\Lambda^{2+}\left[\nabla\mathbf{c}'^+(\mathbf{h}^*,0)^{-1}\left(\Lambda^{2+}\right)^T\right]^{-1}\right]\Lambda^{2+}\nabla\mathbf{c}'^+(\mathbf{h}^*,0)^{-1} \tag{11.75}$$

$$\mathbf{B}_{22}=\left[\Lambda^{2+}\nabla\mathbf{c}'^+(\mathbf{h}^*,0)^{-1}\left(\Lambda^{2+}\right)^T\right]^{-1} \tag{11.76}$$

The Jacobian matrix of the system (11.69)~(11.70) with respect to $\varepsilon$ and evaluated at 0 is

$$\mathbf{J}_{\varepsilon}=\begin{bmatrix}\nabla_{\varepsilon}\mathbf{c}'^+(\mathbf{h}^*,0)\\ -\nabla_{\varepsilon}\mathbf{q}(0)\end{bmatrix} \tag{11.77}$$

Then

$$\begin{bmatrix}\nabla_{\varepsilon}\mathbf{h}^+\\ \nabla_{\varepsilon}\mu\end{bmatrix}=\begin{bmatrix}\mathbf{B}_{11} & \mathbf{B}_{12}\\ \mathbf{B}_{21} & \mathbf{B}_{22}\end{bmatrix}\begin{bmatrix}-\nabla_{\varepsilon}\mathbf{c}'^+(\mathbf{h}^*,0)\\ \nabla_{\varepsilon}\mathbf{q}(0)\end{bmatrix} \tag{11.78}$$

Therefore, the derivatives of route flows with respect to $\varepsilon$ at $\varepsilon = 0$ are

$$\nabla_{\varepsilon}\mathbf{h}^{+} = -\mathbf{B}_{11}\nabla_{\varepsilon}\mathbf{c}'^{+}(\mathbf{h}^{*},0) + \mathbf{B}_{12}\nabla_{\varepsilon}\mathbf{q}(0) \tag{11.79}$$

where $\mathbf{B}_{11}$ and $\mathbf{B}_{12}$ are given in equations (11.73) and (11.74).

By the relationships of arc flows and route flows, we know that $\nabla_{\varepsilon}\mathbf{u}^{+} = \Lambda^{1+}\nabla_{\varepsilon}\mathbf{h}^{+}$, and by knowing that $\nabla_{\varepsilon}\mathbf{c}'^{+}(\mathbf{h}^{*},0) = \left(\Lambda^{1+}\right)^{T}\nabla_{\varepsilon}\mathbf{c}(\mathbf{u}^{*},0)$, we can calculate the derivatives of arc flows with respect to $\varepsilon$ at $\varepsilon = 0$, as follows:

$$\nabla_{\varepsilon}\mathbf{u}^{+} = -\Lambda^{2+}\mathbf{B}_{11}\left(\Lambda^{2+}\right)^{T}\nabla_{\varepsilon}\mathbf{c}(\mathbf{u}^{*},0) + \Lambda^{2+}\mathbf{B}_{12}\nabla_{\varepsilon}\mathbf{q}(0) \tag{11.80}$$

Moreover, by observing that $\nabla_{h}\mathbf{c}'^{+}(\mathbf{h}^{*},0) = \left(\Lambda^{2+}\right)^{T}\nabla_{u}\mathbf{c}(\mathbf{u}^{*},0)\Lambda^{2+}$, we obtain

$$\nabla_{\varepsilon}\mathbf{u}^{+} = -\Lambda^{2+}\mathbf{B}'_{11}\left(\Lambda^{2+}\right)^{T}\nabla_{\varepsilon}\mathbf{c}(\mathbf{u}^{*},0) + \Lambda^{2+}\mathbf{B}'_{12}\nabla_{\varepsilon}\mathbf{q}(0) \tag{11.81}$$

where

$$\mathbf{B}'_{11} = \left[\left(\Lambda^{1+}\right)^{T}\nabla\mathbf{c}(\mathbf{h}^{*},0)\Lambda^{1+}\right]^{-1}$$
$$\left\{\mathbf{I} - \left(\Lambda^{2+}\right)^{T}\left[\Lambda^{2+}\left[\left(\Lambda^{1+}\right)^{T}\nabla\mathbf{c}(\mathbf{h}^{*},0)\Lambda^{1+}\right]^{-1}\left(\Lambda^{2+}\right)^{T}\right]^{-1}\Lambda^{2+}\left[\left(\Lambda^{1+}\right)^{T}\nabla\mathbf{c}(\mathbf{h}^{*},0)\Lambda^{1+}\right]^{-1}\right\} \tag{11.82}$$

$$\mathbf{B}'_{12} = \left[\left(\Lambda^{1+}\right)^{T}\nabla\mathbf{c}(\mathbf{u}^{*},0)\Lambda^{1+}\right]^{-1}\left(\Lambda^{2+}\right)^{T}\left[\Lambda^{2+}\left[\left(\Lambda^{1+}\right)^{T}\nabla\mathbf{c}(\mathbf{u}^{*},0)\Lambda^{1+}\right]^{-1}\left(\Lambda^{2+}\right)^{T}\right]^{-1} \tag{11.83}$$

### 11.1.5 Link Capacity and Queuing Delay

In saturated road networks, a queuing delay is apt to occur. The queuing delay is different from the signal delay because the former is due to limited capacity and should be determined from network equilibrium conditions, whereas the latter is due to the interruption of traffic by the traffic signal and could be determined by a formula developed from local traffic conditions (Yang and Yager, 1995). If we treat the signal delay and queuing delay separately, then the resulting travel time function becomes

$$c_{a}(t) = c_{a_1} + c_{a_2}\left(t + c_{a_1}\right) + c_{a_3}\left(t + c_{a_1}\right) \tag{11.84}$$

where $c_{a_3}\left(t + c_{a_1}\right)$ denotes the link queuing delay.

The queuing delay is due to limited capacity, and should be determined endogenously in a transportation network model. In past research, Thompson and Payne (1975) dealt with traffic assignment in transportation networks with capacity constraints and queuing, whereas the simplifying assumption that the cruise times on links are constant was made. They showed that at network equilibrium, the queuing delay at the exit of a link corresponds to the Lagrange multiplier associated with the link capacity constraint in simple linear programming problem. Smith (1987) extended the result of Thompson and Payne to a signal-controlled network and gave the equilibrium conditions of the interactions between flows, queues and green times when a control policy developed by himself was used. Yang and Yager (1995) further formulated the traffic assignment and signal control problem in saturated road networks as a bilevel model. The lower level problem represented a network equilibrium model involving queuing explicitly on saturated links, which predicts how drivers react to any given signal control pattern, and the upper level problem determined signal splits to optimize a system objective function, taking account of the drivers' route choice behavior in response to signal split changes. Then a sensitivity analysis was implemented for the queuing network equilibrium problem to obtain the derivatives of the equilibrium link flows and equilibrium queuing delays with respect to signal splits, and the derivative information was used to develop a gradient descent algorithm to solve the proposed bilevel traffic signal control problem.

The queuing delay can also be accommodated endogenously within our DNSC framework, but modifications are necessary:

1. Link capacity related constraints (11.3) and (11.5) must be moved into the feasible region $\Omega(\mathbf{g})$ shown in expression (11.9).
2. The travel time function must explicitly include free flow travel time, signal delay and queuing delay, as follows.

   $$c_a(t) = c_{a_1} + c_{a_2}\left(t + c_{a_1}\right) + c_{a_3}\left(t + c_{a_1}\right) \quad \forall a \in B(I) \tag{11.85}$$

   where $c_{a_3}\left(t + c_{a_1}\right)$ denotes the link queuing delay, and in fact is determined by the Lagrange multiplier associated with the corresponding link capacity constraint.
3. The queuing length is now determined by multiplying the queuing delay $c_{a_3}(t)$ by the exit flow rate. The maximum queuing length constraint is now defined as

   $$c_{a_3}(t)v_a(t) \le Q_a \quad \forall a, t \tag{11.86}$$

In responding to the above changes of the model formulation, the proposed solution algorithm for the DNSC problem must be modified accordingly. That is, in addition to the derivative information on **v**, the derivative information on the queuing delay $c_{a_3}\left(t + c_{a_1}\right)$ with respect to green times, **g**, is also required in performing the variational inequality sensitivity analysis in Step 6 of the solution procedure.

## 11.2 Dynamic Traffic-Responsive Signal Control System

In practice, the implementation of the DNSC system is difficult because the requirements for the computational capability are rather demanding. Therefore, a somewhat simplified signal control system may be more appropriate. In a simplified case, signal settings at intersections can be treated independently, and the choice of signal timing plans can be set in response to the approaching flow pattern which may be captured by means of traffic detectors. Such a system can be represented by the dynamic traffic-responsive signal control (DTSC) system. In the following sections, we first formulate the DTSC system and then propose the IOA algorithm for its solution. Finally, a numerical example is provided for demonstration.

### 11.2.1 Model Formulation

The DTSC system may be described as a two-player noncooperative Nash game. The *first player*, in response to the current approaching flow pattern, tries to minimize the total intersection delay by allocating appropriate green times, and thereby, determining link capacities. The *second player*, based on the fixed link capacities, and hence link travel times, searches for the shortest travel time route for use, which can be mathematically represented by the dynamic user-optimal conditions.

To formulate the *first player's* problem in a simpler manner, we make a few necessary assumptions as follows:

1. The cycle length is externally determined.
2. The queuing length and delay are not considered.
3. Each link length is long enough so that the effect of the upstream intersection signal plan on the downstream intersection signal plan can be ignored; that is, the intersections are considered to be isolated.
4. Only intersection delays are used to adjust signal timing plans.

Under the above assumptions, the *first player's* problem can be formulated as follows.

$$\min_{g} z=\sum_{t}\sum_{a}\sum_{m} v_a^m(t)\frac{9}{10}\left[\frac{C^I(t)\left(1-g^{Im}(t)/C^I(t)\right)^2}{2\left(1-v_a^m(t)/S_a^m\right)} + \frac{\left(H_a^m(t)\right)^2}{2v_a^m(t)\left(1-H_a^m(t)\right)}\right] \tag{11.87}$$

s.t.

$$H_a^m(t)=\frac{v_a^m(t)}{S_a^m}\frac{C^I(t)}{g^{Im}(t)}<1 \quad \forall I,m,a\in B(I),t \tag{11.88}$$

$$g^{Im}(t)\geq \underline{g}^{Im} \quad \forall I,m,t \tag{11.89}$$

$$\sum_{m}\left(g^{Im}(t)+l^{Im}\right)=C^I(t) \quad \forall I,t \tag{11.90}$$

Note, the *first player's* problem defined by expressions (11.87)~(11.90) is a nonlinear programming model formulated on a link-node network. The objective function is the sum of nonlinear functions corresponding to the vehicular delay related to each intersection; the linear constraints conserve cycle length at each intersection.

For the *second player's* problem, the DUO route choice model is applied.

$$\mathbf{c}^*\left[\mathbf{u}-\mathbf{u}^*\right]\geq 0 \qquad \forall \mathbf{u}\in\Omega^* \tag{11.91}$$

where $\Omega^*$ is defined by constraints (11.13) ~ (11.19).

### 11.2.2 Solution Algorithm

#### 11.2.2.1 Iterative optimization and assignment method

The DTSC system can be deemed to be a noncooperative Nash game with perfect information in which the players move in sequence. In this regard, the IOA method shown in Figure 11.1 is appropriate for obtaining solutions.

Since the *second player's* problem is formulated by the DUO route choice model, which has been extensively discussed and solved by the nested diagonalization method in Chapters 4 and 5, we restrict our focus on the solution algorithm for the *first player's* problem. In addition, since the objective function is convex (see Section 11.2.2.2 for a proof) and constraints are all linear, the *first player's* problem can thus be solved by the Frank-Wolfe (FW) method, as follows. Given a current solution to the *main problem*, all approach delays are fixed at their current values. Phase lengths of the *subproblem* are solved directly from the available optimality conditions, and subproblem green time allocations are identified by minimizing the linear portion of the objective function pertaining to intersection delays, resulting in all green times at each intersection being assigned to the least delay phase. A new main problem solution is then obtained by minimizing the objective function with respect to a step size $\lambda$, $0 \le \lambda \le 1$, where $\lambda$ is a weight for combining main and subproblem solutions.

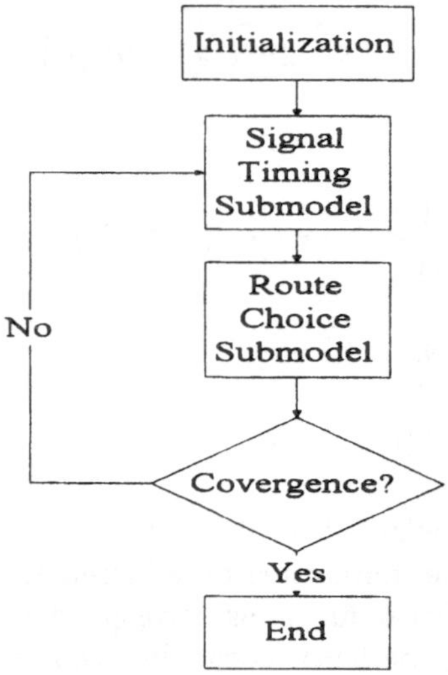


Figure 11.1: Iterative Optimization and Assignment Method

For the convenience of description, we next derive the linearized portion of the objective function as follows:

$$\nabla z(g^n)g' = \sum_I \sum_m \sum_t G^{Im}(t)^n g'^{Im}(t) \tag{11.92}$$

where

$$G^{Im}(t) = \frac{\partial z^{Im}(t)}{\partial g^{Im}(t)} \qquad \forall I, m, t \tag{11.93}$$

$$z^{Im}(t) = \sum_{a \in B(I)} v_a^m(t) d_a^m(t) \qquad \forall I, m, t \tag{11.94}$$

$$d_a^m(t) = \frac{9}{10}\left[\frac{C^I(t)\left(1 - g^{Im}(t)/C^I(t)\right)^2}{2\left(1 - v_a^m(t)/S_a^m\right)} + \frac{\left(H_a^m(t)\right)^2}{2v_a^m(t)\left(1 - H_a^m(t)\right)}\right] \qquad \forall a, m, t \tag{11.95}$$

Note that equation (11.94) is derived based on the following observation:

$$\sum_I \sum_m \sum_t z^{Im}(t) = \sum_m \sum_a \sum_t v_a^m(t) d_a^m(t) \tag{11.96}$$

With the above discussion, the linear objective function for the subproblem can be further decomposed by intersections and intervals, as follows:

$$\min_{g} \sum_{m} G^{Im}(t) g'^{Im}(t) \quad \forall I, t \tag{11.97}$$

We now present the solution algorithm for the *first player's* problem as follows:

***Algorithm***

**Step 0: Initialization.**

Find an initial solution for green times associated with phase $m$ at each intersection $I$ during interval $t$, $\{g_I^m(t)^n\}$. Set counter $n$=1.

**Step 1: Subproblem Solving.**

Perform Steps 1.1 through 1.3 for every intersection $I$ and interval $t$.

Step 1.1: Compute $\{G^{Im}(t)^n\}$ based on $\{g^{Im}(t)^n\}$.

Step 1.2: Assign the minimal green time $g_{\min}^m$ to each phase.

$$g_{\min}^m = \max\{\underline{g}^{Im}(t), \frac{v_a^m(t) C^I(t)}{S_a^m}\} \tag{11.98}$$

Step 1.3: Assign the remaining green times, $C^I(t) - \sum_m (l^{Im} + g_{\min}^m)$, to the phase with the smallest value of $\{G^{Im}(t)^n\}$. This results in green times $\{g'^{Im}(t)^n\}$ associated with each phase at an intersection.

**Step 2: Determine Move Size, $\lambda^n$.**

Solve the following mathematical problem.

$$\min_{0 \le \lambda^n \le 1} z(\mathbf{g}^n + \lambda^n (\mathbf{g}'^n - \mathbf{g}^n)) \tag{11.99}$$

**Step 3: Update the Green Times.**

$$g^{Im}(t)^{n+1} = g^{Im}(t)^n + \lambda^n (g'^{Im}(t)^n - g^{Im}(t)^n) \tag{11.100}$$

**Step 4: Convergence Test.**

If $g^{Im}(t)^{n+1} \approx g^{Im}(t)^n$, stop. Otherwise, set $n$=$n$+1 and go to Step 1.

### 11.2.2.2 Convexity of the *first player's* objective function using Webster's delay formula

The convexity of the objective function (11.87) is verified by showing that its Hessian is positive-semidefinite.

***Proof***:

Let $\mathbf{g}=\left(g^{11}(1),g^{12}(1),\cdots,g^{Im}(t),\cdots\right)=\left(g_1,g_2,\cdots,g_l,\cdots,g_n\right)$. The Hessian of $z(\mathbf{g})$ can be written in matrix form as follows:

$$\nabla^2 z(\mathbf{g})=\begin{bmatrix}\frac{\partial^2 z(\mathbf{g})}{\partial g_1^2} & \frac{\partial^2 z(\mathbf{g})}{\partial g_1 \partial g_2} & \cdots & \frac{\partial^2 z(\mathbf{g})}{\partial g_1 \partial g_n}\\ \frac{\partial^2 z(\mathbf{g})}{\partial g_2 \partial g_1} & \frac{\partial^2 z(\mathbf{g})}{\partial g_2^2} & & \\ \vdots & & & \vdots \\ \frac{\partial^2 z(\mathbf{g})}{\partial g_n \partial g_1} & \cdots & \cdots & \frac{\partial^2 z(\mathbf{g})}{\partial g_n^2}\end{bmatrix}=\begin{bmatrix}\frac{\partial^2 z(\mathbf{g})}{\partial g_1^2} & 0 & \cdots & 0\\ 0 & \frac{\partial^2 z(\mathbf{g})}{\partial g_2^2} & & 0\\ \vdots & & & \vdots\\ 0 & 0 & \cdots & \frac{\partial^2 z(\mathbf{g})}{\partial g_n^2}\end{bmatrix} \tag{11.101}$$

To prove the Hessian is positive semi-definite, we only need to show that $\frac{\partial^2 z(\mathbf{g})}{\partial g_i^2}$ is greater than or equal to zero because the Hessian is a diagonalized matrix with elements $\frac{\partial^2 z(\mathbf{g})}{\partial g_i^2}$.

$$\frac{\partial^2 z(\mathbf{g})}{\partial g_i^{\,2}}=\frac{\partial^2 z(\mathbf{g})}{\partial g^{Im}(t)^2}=\frac{\partial \frac{\partial z(\mathbf{g})}{\partial g^{Im}(t)}}{\partial g^{Im}(t)} \tag{11.102}$$

$$\frac{\partial z(\mathbf{g})}{\partial g_i}=\frac{\partial z(\mathbf{g})}{\partial g^{Im}(t)}=\frac{d\left(\sum_{a\in B(I)} v_a^m(t)d_a^m(t)\right)}{dg^{Im}(t)}$$

$$=\sum_{a\in B(I)}\frac{9v_a^m(t)}{10}\left\{\frac{\frac{g^{Im}(t)}{C^I(t)}-1}{1-\frac{v_a^m(t)}{S_a^m}}+\right.$$

$$\left.\frac{1}{2v_a^m(t)}\left[\frac{2H_a^m(t)\left(1-H_a^m(t)\right)+H_a^m(t)^2}{\left(1-H_a^m(t)\right)^2}\right]\frac{\left[-v_a^m(t)C^I(t)\right]}{S_a^m g^{Im}(t)^2}\right\}$$

$$= \sum_{a\in B(I)} \frac{9v_a^m(t)}{10}\left\{\frac{\frac{g^{Im}(t)}{C^I(t)}-1}{1-\frac{v_a^m(t)}{S_a^m}} - \frac{\left(2H_a^m(t)-H_a^m(t)^2\right)C^I(t)}{2\left(1-H_a^m(t)\right)^2 S_a^m g^{Im}(t)^2}\right\} \quad (11.103)$$

$$\frac{\partial^2 z(\mathbf{g})}{\partial g_i^2} = \frac{\partial \frac{\partial z(\mathbf{g})}{\partial g^{Im}(t)}}{\partial g^{Im}(t)}$$

$$= \sum_{a\in B(I)} \frac{9v_a^m(t)}{10}\left\{\frac{\frac{1}{C^I(t)}}{1-\frac{v_a^m(t)}{S_a^m}} - \frac{2C^I(t)\left(1-H_a^m(t)\right)\frac{-v_a^m(t)C^I(t)}{S_a^m g^{Im}(t)^2}}{\left[2\left(1-H_a^m(t)\right)^2 S_a^m g^{Im}(t)^2\right]} + \frac{\left(2H_a^m(t)-H_a^m(t)^2\right)C^I(t)\frac{d\left(2\left(1-H_a^m(t)\right)^2 S_a^m g^{Im}(t)^2\right)}{dg^{Im}(t)}}{\left[2\left(1-H_a^m(t)\right)^2 S_a^m g^{Im}(t)^2\right]^2}\right\} \quad (11.104)$$

$$\frac{d\left(2\left(1-H_a^m(t)\right)^2 S_a^m g^{Im}(t)^2\right)}{dg^{Im}(t)}$$

$$= 2S_a^m\left\{2\left(1-H_a^m(t)\right)g^{Im}(t)\left[\frac{v_a^m(t)C^I(t)}{S_a^m g^{Im}(t)^2}g^{Im}(t)+\left(1-H_a^m(t)\right)\right]\right\}$$

$$= 2S_a^m\left\{2\left(1-H_a^m(t)\right)\left[\frac{v_a^m(t)C^I(t)}{S_a^m}+\left(1-H_a^m(t)\right)g^{Im}(t)\right]\right\}$$

$$= 2S_a^m\left\{2\left(1-H_a^m(t)\right)g^{Im}(t)\right\} \qquad \left(\because H_a^m(t)g^{Im}(t) = \frac{v_a^m(t)C^I(t)}{S_a^m}\right)$$

$$= 4S_a^m\left(1-H_a^m(t)\right)g^{Im}(t) \quad (11.105)$$

$$\frac{\partial^2 z(\mathbf{g})}{\partial g_i^2} = \sum_{a\in B(I)} \frac{9v_a^m(t)}{10}\left\{\frac{\frac{1}{C^I(t)}}{1-\frac{v_a^m(t)}{S_a^m}} + \frac{2C^I(t)\left(1-H_a^m(t)\right)\frac{v_a^m(t)C^I(t)}{S_a^m g^{Im}(t)^2}}{\left[2\left(1-H_a^m(t)\right)^2 S_a^m g^{Im}(t)^2\right]} + \right.$$

$$\left. \frac{H_a^m(t)\left(2-H_a^m(t)\right)C^I(t)4S_a^m\left(1-H_a^m(t)\right)g^{Im}(t)}{\left[2\left(1-H_a^m(t)\right)^2 S_a^m g^{Im}(t)^2\right]^2}\right\} \quad (11.106)$$

Note that $H_a^m(t)<1$ , and $\frac{v_a^m(t)}{S_a^m}<\frac{g^{Im}(t)}{C^I(t)}<1$ $\left(\because H_a^m(t)=\frac{v_a^m(t)}{S_a^m}\frac{C^I(t)}{g^{Im}(t)}\right)$,

therefore $\frac{\partial^2 z(\mathbf{g})}{\partial g_i^{\,2}} \geq 0,\ \forall i$.

Since the Hessian $\nabla^2 z(\mathbf{g})$ is positive-semidefinite, the objective function $z(\mathbf{g})$ is convex. This completes the proof.

### 11.2.3 Numerical Example

#### 11.2.3.1 Input data

A simple network shown in Figure 11.2 is used for testing. The test network consists of 24 links and 9 nodes, in which nodes 2,4,6,8 represent both origins and destinations, and nodes 1,3,5,7,9 are intermediates. We further assume nodes 2,4,5,6,8 are two-phase signalized intersections on which the east-west direction is designated as the first phase and the north-south direction as the second phase. The other 4 nodes, 1,3,7,9, are dummy nodes, so no signal devices are assumed.

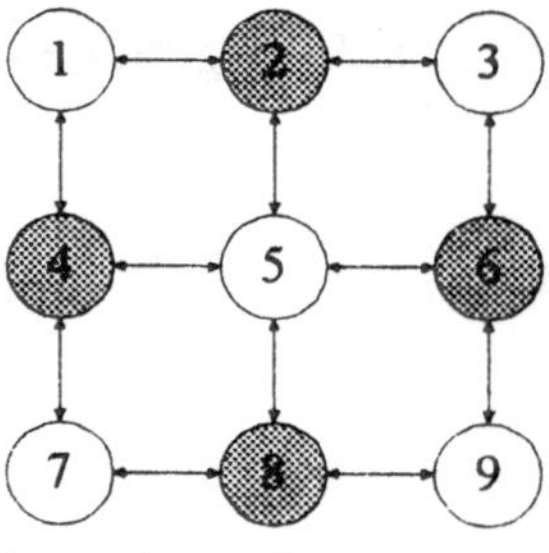


Figure 11.2: Test Network

The link travel time function is assumed as follows:

$$c_a(t) = t_{a_1}\left(1+\beta\left(\frac{u_a(t)}{Cap_a(t)}\right)^2\right) \tag{11.107}$$

The link capacity at each signalized intersection is determined by:

$$Cap_a(t) = \sum_m S_a^m \frac{g^{lm}(t)}{C^l(t)} \tag{11.108}$$

The link capacity at each unsignalized intersection is equal to a preset saturation flow rate:

$$Cap_a(t) = S_a \tag{11.109}$$

The parameters at signalized intersections are assumed in Table 11.1:

Table 11.1: Parameters for Each Signalized Intersection

| Free Travel Time $c_{a_1}$ | Coefficient $b$ | Cycle Length (sec) $C^l(t)$ | Lower Limit of Green Time (sec) $\underline{g}^{lm}(t)$ | Saturation Flow Rate $S_a^m$ | Lost Time (sec) $l^{lm}$ |
|---|---|---|---|---|---|
| 1 | 0.8 | 60 | 5 | 50 | 3 |

The time-dependent O-D demands are assumed in Table 11.2:

Table 11.2: Time-Dependent Origin-Destination Demands

| Departure Interval | O-D Pair | | | |
|---|---|---|---|---|
| | 2-8 | 8-2 | 4-6 | 6-4 |
| $k$=1 | 15 | 0 | 20 | 0 |
| $k$=2 | 0 | 15 | 0 | 20 |

#### 11.2.3.2 Test results

Two signal control systems, i.e., pretimed and traffic-responsive timing plans, are compared on the test network.

For the pretimed signal timing plan, all phases are assumed to have 27-second green times. The resulting flow pattern is summarized in Table 11.3. The route travel times are also computed and shown in Table 11.4.

Table 11.3: Results for the Pretimed Signal Timing Plan

| Objective Value = 1804.27 | | | | | | |
|---|---|---|---|---|---|---|
| Link | Entering Interval | Inflow | Exit Flow | Number of Vehicle | Link Travel Time | Exiting Interval |
| 2→5 | 1 | 15.0 | 0.0 | 0.0 | 1.36 | 2 |
| | 2 | 0.0 | 15.0 | 15.0 | 1.00 | - |
| 4→5 | 1 | 20.0 | 0.0 | 0.0 | 1.63 | 3 |
| | 2 | 0.0 | 0.0 | 20.0 | 1.00 | - |
| | 3 | 0.0 | 20.0 | 20.0 | 1.00 | - |
| 5→2 | 3 | 15.0 | 0.0 | 0.0 | 1.36 | 4 |
| | 4 | 0.0 | 15.0 | 15.0 | 1.00 | - |
| 5→4 | 4 | 20.0 | 0.0 | 0.0 | 1.63 | 6 |
| | 5 | 0.0 | 0.0 | 20.0 | 1.00 | - |
| | 6 | 0.0 | 20.0 | 20.0 | 1.00 | - |
| 5→6 | 3 | 20.0 | 0.0 | 0.0 | 1.63 | 5 |
| | 4 | 0.0 | 0.0 | 20.0 | 1.00 | - |
| | 5 | 0.0 | 20.0 | 20.0 | 1.00 | - |
| 5→8 | 2 | 15.0 | 0.0 | 0.0 | 1.36 | 3 |
| | 3 | 0.0 | 15.0 | 15.0 | 1.00 | - |
| 6→5 | 2 | 20.0 | 0.0 | 0.0 | 1.63 | 4 |
| | 3 | 0.0 | 0.0 | 20.0 | 1.00 | - |
| | 4 | 0.0 | 20.0 | 20.0 | 1.00 | - |
| 8→5 | 2 | 15.0 | 0.0 | 0.0 | 1.36 | 3 |
| | 3 | 0.0 | 15.0 | 15.0 | 1.00 | - |

Table 11.4: Route Travel Times for the Pretimed Signal Timing Plan

| Departure Time | Route | | | |
|---|---|---|---|---|
| | 2→5→8 | 8→5→2 | 4→5→6 | 6→5→4 |
| $k$=1 | 72 (15) | NA | 3.26 (20) | NA |
| $k$=2 | NA | 72 (15) | NA | 3.26 (20) |

(NA: not applicable because routes are not used; "numbers" in brackets indicate the corresponding route flow )

For the traffic-responsive signal timing plan, the computed green times associated with each phase are shown in Table 11.5. The resulting flow pattern is summarized in Table 11.6. The route travel times are also computed and shown in Table 11.7.

Table 11.5: Green Times Allocation for the DTSC Timing Plan

| Intersection Number | Time Interval | Green Times (sec)* Phase 1 | Phase 2 |
|---|---|---|---|
| 2 | 4 | 19.6 | 34.4 |
| 4 | 6 | 30.5 | 23.5 |
| 5 | 2 | 5.0 | 49.0 |
| | 3 | 32.9 | 21.1 |
| | 4 | 28.1 | 25.9 |
| | 5 | 37.9 | 16.1 |
| 6 | 5 | 49.0 | 5.0 |
| 8 | 3 | 5.0 | 49.0 |
| | 4 | 47.9 | 6.1 |

(*Other intersections are allocated equal phase green times, i.e., 27 seconds )

Table 11.6: Results for the DTSC Timing Plan

| Objective Value = 1052.73 | | | | | | |
|---|---|---|---|---|---|---|
| Link | Entering Interval | Inflow | Exit Flow | Number of Vehicle | Travel Time | Exiting Interval |
| 1→4 | 5 | 6.9 | 0.0 | 0.0 | 1.08 | 6 |
| | 6 | 0.0 | 6.9 | 6.9 | 1.00 | - |
| 2→1 | 4 | 6.9 | 0.0 | 0.0 | 1.02 | 5 |
| | 5 | 0.0 | 6.9 | 6.9 | 1.00 | - |
| 2→5 | 1 | 15.0 | 0.0 | 0.0 | 1.36 | 2 |
| | 2 | 0.0 | 15.0 | 15.0 | 1.00 | - |
| | 4 | 3.6 | 0.0 | 0.0 | 1.02 | 5 |
| | 5 | 0.0 | 3.6 | 3.6 | 1.00 | - |
| 3→2 | 3 | 10.6 | 0.0 | 0.0 | 1.18 | 4 |
| | 4 | 0.0 | 10.6 | 10.6 | 1.00 | - |
| 4→5 | 1 | 20.0 | 0.0 | 0.0 | 1.63 | 3 |
| | 2 | 0.0 | 0.0 | 20.0 | 1.00 | - |
| | 3 | 0.0 | 20.0 | 20.0 | 1.00 | - |
| 5→2 | 3 | 15.0 | 0.0 | 0.0 | 1.36 | 4 |
| | 4 | 0.0 | 15.0 | 15.0 | 1.00 | - |
| 5→4 | 5 | 10.5 | 0.0 | 0.0 | 1.17 | 6 |
| | 6 | 0.0 | 10.5 | 10.5 | 1.00 | - |

Table 11.6: Results for the DTSC Timing Plan (continued)

| Objective Value = 1052.73 | | | | | | |
|---|---|---|---|---|---|---|
| Link | Entering Interval | Inflow | Exit Flow | Number of Vehicle | Travel Time | Exiting Interval |
| 5→6 | 3 | 20.0 | 0.0 | 0.0 | 1.63 | 5 |
| | 4 | 0.0 | 0.0 | 20.0 | 1.00 | - |
| | 5 | 0.0 | 20.0 | 20.0 | 1.00 | - |
| 5→8 | 2 | 15.0 | 0.0 | 0.0 | 1.36 | 3 |
| | 3 | 0.0 | 15.0 | 15.0 | 1.00 | - |
| 6→3 | 2 | 10.6 | 0.0 | 0.0 | 1.04 | 3 |
| | 3 | 0.0 | 10.6 | 10.6 | 1.00 | - |
| 6→5 | 2 | 6.9 | 0.0 | 0.0 | 3.16 | 5 |
| | 3 | 0.0 | 0.0 | 6.9 | 1.00 | - |
| | 4 | 0.0 | 0.0 | 6.9 | 1.00 | - |
| | 5 | 0.0 | 6.9 | 6.9 | 1.00 | - |
| 6→9 | 2 | 2.6 | 0.0 | 0.0 | 1.00 | 3 |
| | 3 | 0.0 | 2.6 | 2.6 | 1.00 | - |
| 7→4 | 5 | 2.6 | 0.0 | 0.0 | 1.01 | 6 |
| | 6 | 0.0 | 2.6 | 2.6 | 1.00 | - |
| 8→5 | 2 | 15.0 | 0.0 | 0.0 | 1.11 | 3 |
| | 3 | 0.0 | 15.0 | 15.0 | 1.00 | - |
| 8→7 | 4 | 2.6 | 0.0 | 0.0 | 1.00 | 5 |
| | 5 | 0.0 | 2.6 | 2.6 | 1.00 | - |
| 9→8 | 3 | 2.6 | 0.0 | 0.0 | 1.31 | 4 |
| | 4 | 0.0 | 2.6 | 2.6 | 1.00 | - |

Table 11.7: Route Travel Times for the DTSC Timing Plan

| Departure Time | Route | | | | | | |
|---|---|---|---|---|---|---|---|
| | 2→5→8 | 4→5→6 | 6→5→4 | 6→3→2 →1→4 | 6→3→2 →5→4 | 6→9→ 8→7→4 | 8→5→2 |
| $k$=1 | 2.72 (15) | 3.26 (20) | NA | NA | NA | NA | NA |
| $k$=2 | NA | NA | 4.33 (6.9) | 4.32 (6.9) | 4.41 (3.6) | 4.32 (2.6) | 2.47 (15) |

(NA: not applicable because routes are not used; "numbers" in brackets indicate the corresponding route flow )

The objective value associated with the traffic-responsive signal timing plan (Table 11.6) is about 41.7% lower than that with the pretimed signal timing plan (Table 11.3). This is because the traffic-responsive signal timing plan assigns more green times to those approach links with higher exit flows, see Tables 5 and

6. When there is no exit flow presented at each approach link, the green times are evenly distributed, i.e., 27 seconds for each phase at the intersection.

The traffic-responsive signal timing plan (Table 11.7) does not necessarily result in lower route travel times as compared with that of the pretimed signal timing plan (Table 11.4). The reason is basically due to the nature of Webster's delay function. Webster's delay function is derived based on a computer simulation assuming the Poisson distribution for the vehicle arrival rates, which does not minimize the route travel times.

## 11.3 Notes

In this chapter we formulated the dynamic network signal control system and also proposed a solution algorithm involving a variational inequality sensitivity analysis. The proposed variational inequality sensitivity analysis is difficult to carry out because a VIP corresponding to the DUO route choice problem is embedded with constraints associated with the DNSC system.

We also formulated the DTSC system as a two player Nash game. By adopting the so-called iterative optimization and assignment (IOA) method, a numerical example is provided and analyzed. The most interesting feature of the proposed IOA procedure is that the FW method can also be employed to solve the delay minimization problem. However, due to the convexity of the objective function using Webster's delay formula, multiple solutions might occur for the *first player's* problem. Note also that if the termination criterion is set for the *first player's* problem, then the dynamic user-optimal conditions associated with the *second player's* problem may not be precisely satisfied. Caution has to be taken for the use of the comparative results, because a pretimed signal timing plan using 50/50 green splits may not necessarily reflect the real situation.

Other issues that have yet to be explored are, among others, the suitable form of dynamic travel time functions, a stability mechanism for ensuring a smooth variation in signal control, and the first-in-first-out requirement. Nevertheless, the proposed models provide a reference platform for further improvements.

# Chapter 12

# Stochastic/Dynamic User-Optimal Route Choice Model

The stochastic dynamic user-optimal route choice model assumes that the perceived travel times are incomplete and/or imprecise. To illustrate the perceived travel times, an error term is often hypothesized to accompany the actual travel times, and many probability distributions have been applied to represent the real situation. For a specified error term of the perceived travel times, we assume all drivers make their route choice decisions based on their perceptions of O-D travel times. A notable example is the Gumbel distribution which yields the logit model. The resulting traffic flows are expected to be more dispersed over parallel routes for the stochastic dynamic route choice model than for its deterministic counterpart.

The first optimization model for the static stochastic user-optimal problem was formulated by Fisk (1980). Its optimality conditions can be characterized by the principle of stochastic user optimal (SUO) (Daganzo and Sheffi, 1977), which can be stated as follows:

*In a stochastic user equilibrium network, no user believes he can improve his travel time by unilaterally changing routes.*

An equivalent notion of SUO is that no traveler can reduce his own perceived travel time by unilaterally changing routes. Applying the principle of SUO, an equivalent stochastic user-optimal route choice model was also developed by Sheffi and Powell (1982). In fact, the principle of SUO can be naturally extended to the dynamic scenario. Unfortunately, very little literature is available in this regard. Cascetta (1991) studied the variation of dynamic route choice from day to day while Vythoulkas (1990) also extended the stochastic static route choice models into the dynamic route choice framework. Based on their instantaneous (or reactive) and/or predictive (or ideal) DUO route choice models, Ran and Boyce (1994) further developed a set of stochastic DUO models. A comparison of our predictive stochastic/dynamic user-optimal route choice model with that of Ran and Boyce (1994) is made in Section 12.7.

In this chapter, the stochastic/dynamic route travel time function is analyzed in Section 12.1. The equilibrium conditions and model formulation for the stochastic/dynamic user-optimal (SDUO) route choice model are presented in Section 12.2. The nested diagonalization method embedding the method of successive averages and the stochastic dynamic method (SADA) is described in Section 12.3. A simple numerical example is presented in Section 12.4. The deterministic counterpart approximation is discussed in Section 12.5. Braess's paradox is presented in Section 12.6. Finally, concluding notes are given in Section 12.7.

## 12.1 Stochastic/Dynamic Route Travel Time Function Analysis

### 12.1.1 Stochastic/Dynamic Route Travel Time Function

The stochastic/dynamic route travel time function $\hat{c}_p(k)$ can be interpreted as the perceived route travel time for travelers between origin-destination $rs$ departing during interval $k$. It is the sum of two components: a systematic term and an error term.

$$\hat{c}_p^{rs}(k) = c_p^{rs}(k) + \varepsilon_p^{rs}(k) \qquad \forall r, s, p, k \tag{12.1}$$

Symbol $c_p^{rs}(k)$ denotes the travel time experienced by travelers between O-D pair $rs$ departing during interval $k$ over route $p$, and $\varepsilon_p^{rs}(k)$ is the random component. Depending on the probability distributions chosen for the random components, different models are obtained (Sheffi, 1985; Sheffi and Powell, 1981, 1982; Daganzo, 1982). If we assume that the random term is an independently and identically distributed (I.I.D.) Gumbel variate, the widely used multinomial logit model is obtained. Given route flow and route travel times, the stochastic/dynamic route travel time function can be expressed mathematically as follows (for details see Section 12.7):

$$\hat{c}_p^{rs}(k) = c_p^{rs}(k) + \frac{1}{\theta}\ln h_p^{rs}(k) \qquad \forall r, s, p, k \tag{12.2}$$

where $\frac{1}{\theta}\ln h_p^{rs}(k)$ is the error term associated with the travel time over route $p$, and $\theta$ is a constant reflecting the degree of uncertainty of traffic information. As $\theta$ approaches infinity, the second term vanishes. However, when $\theta$ is near zero, then the second term dominates the perceived route travel time. By expressing actual route travel time in terms of link travel time, equation (12.2) can be alternatively rewritten as:

$$\hat{c}_p^{rs}(k)=\sum_a\sum_t c_a(t)\delta_{apk}^{rs}(t)+\frac{1}{\theta}\ln h_p^{rs}(k) \qquad \forall r,s,p,k \tag{12.3}$$

where $c_a(t)$ is the actual travel time for link $a$ during interval $t$, and $\delta_{apk}^{rs}(t)$ is a zero-one indicator variable, for which $\delta_{apk}^{rs}(t)=1$ if flow departing origin $r$ during interval $k$ over route $p$ toward destination $s$ enters link $a$ during interval $t$; otherwise, $\delta_{apk}^{rs}(t)=0$. Note that equation (12.3) is essentially the dynamic counterpart of the logit-based stochastic route travel time function of Fisk (1980).

### 12.1.2 Asymmetric Link Interactions

The stochastic/dynamic route travel time function $\hat{c}_p(k)$ contains the dynamic link travel time function $c_a(t)$ as its component. As shown before, for any physical link $a$, inflows can be affected by those previously entered inflows; but, the reverse is not true. This property makes the representative believed link travel time functions asymmetric. The theorem directly follows from Green's theorem. Consequently, any SDUO route choice problem with the stochastic/dynamic link travel time function $\hat{c}_p(k)$ as shown in equation (12.2) or equation (12.3) does not have an equivalent optimization problem. The variational inequality approach emerges.

### 12.1.3 Time-Space Network

For the SDUO route choice problem, the corresponding time-space network to be constructed is essentially the same as that for the DUO route choice problem. We redraw the time-space network in Figure 12.1 for ready reference.

## 12.2 Equilibrium Conditions and Model Formulation

In this section, the stochastic/dynamic user-optimal conditions are first defined to characterize the travelers' driving behavior for using routes with the minimal perceived travel time. The perceived route travel times can be computed by adding up the perceived link travel times in consideration of the flow propagation requirements along that route. The variational inequality formulation is then presented for the SDUO route choice problem. Afterwards, the equivalence between the stochastic/dynamic user-optimal conditions and the variational inequality formulation is verified by a proof.

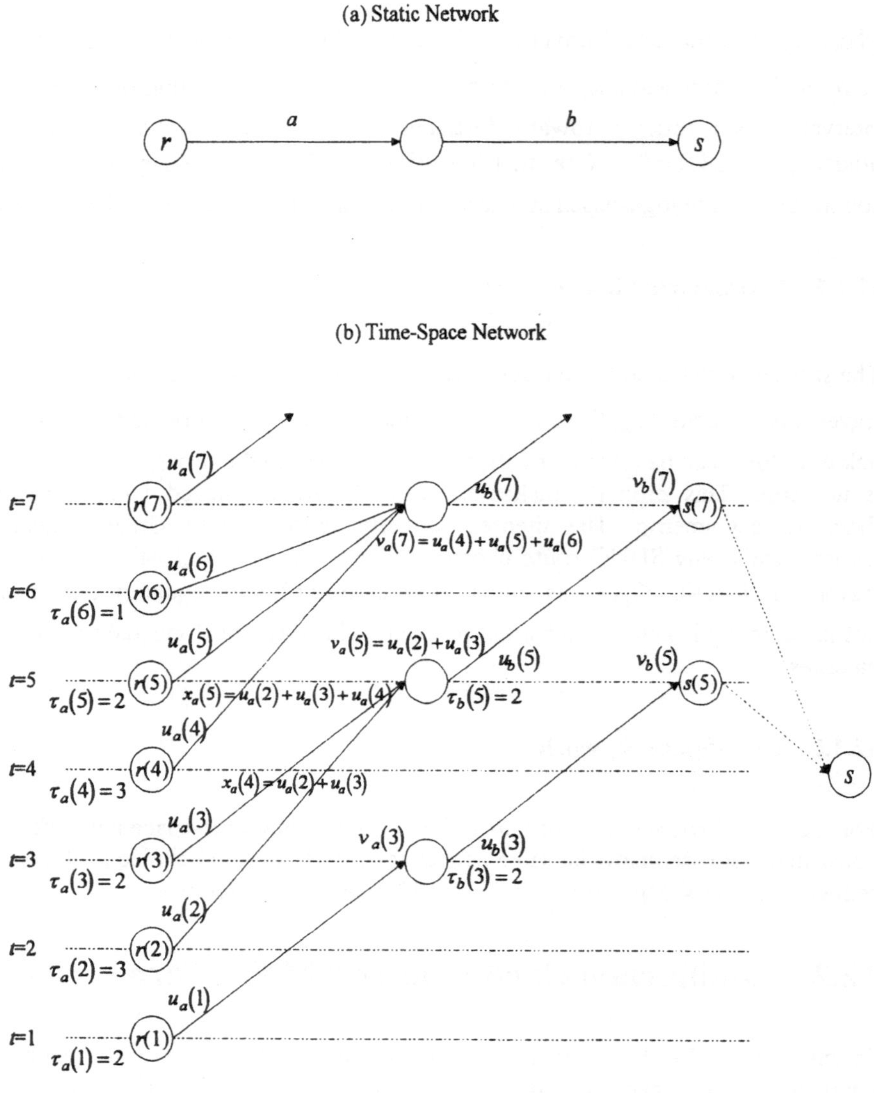


Figure 12.1: Time-Space Network

### 12.2.1 Stochastic/Dynamic User-Optimal Conditions

The stochastic/dynamic user-optimal conditions state for each O-D pair that the perceived route travel times for travelers departing during the same interval are equal to the minimal perceived route travel time; or no traveler would be better off with respect to the perceived route travel times by unilaterally changing his/her route. In other words, the perceived route travel times of any unused route for each O-D pair is greater than or equal to the minimal perceived route travel time. Therefore, for each O-D pair *rs*, if the flow over route *p* departing during interval *k* is positive, i.e., $h_p^{rs}(k) > 0$, then the corresponding perceived route travel time is minimal. However, if no flow occurs on route *p*, i.e., $h_p^{rs}(k) = 0$, then the corresponding perceived route travel time is at least as great as the minimal perceived route travel time. This equilibrium conditions can be mathematically expressed as follows:

$$\hat{c}_p^{rs*}(k)\begin{cases} = \hat{\pi}^{rs}(k) & \text{if } h_p^{rs*}(k) > 0 \\ \geq \hat{\pi}^{rs}(k) & \text{if } h_p^{rs*}(k) = 0 \end{cases} \quad \forall r,s,p,k \tag{12.4}$$

### 12.2.2 Variational Inequality Problem

The following theorem is stated for the proposed SDUO route choice model.

***Theorem 12.2***: The SDUO route choice problem is equivalent to finding a vector $\mathbf{h}^* \in \Omega$ such that the following VIP holds:

$$\hat{\mathbf{c}}^*\left[\mathbf{h} - \mathbf{h}^*\right] \geq 0 \quad \forall \mathbf{h} \in \Omega^* \tag{12.5}$$

Or, alternatively, in expanded form:

$$\sum_{rs}\sum_{p}\sum_{k}\hat{c}_p^{rs}(k)\left[h_p^{rs}(k) - h_p^{rs*}(k)\right] \geq 0 \quad \forall \mathbf{h} \in \Omega^* \tag{12.6}$$

where $\Omega^*$ is a subset of $\Omega$ with $\delta_{apk}^{rs}(t)$ being realized at equilibrium, i.e., $\left(\delta_{apk}^{rs}(t) = \delta_{apk}^{rs*}(t)\right)$, $\forall r,s,a,p,k,t$. The symbol $\Omega$ is delineated by the following constraints, including flow conservation, flow propagation, nonnegativity, and definitional constraints.

Flow conservation constraint:

$$\sum_{p} h_p^{rs}(k) = \bar{q}^{rs}(k) \quad \forall r,s,k \tag{12.7}$$

Flow propagation constraints:

$$u_{apk}^{rs}(t) = h_p^{rs}(k)\delta_{apk}^{rs}(t) \quad \forall r,s,a,p,k,t \tag{12.8}$$

$$\sum_{t}\delta_{apk}^{rs}(t) = 1 \quad \forall r,s,p,a \in p,k \tag{12.9}$$

$$\delta_{apk}^{rs}(t) = \{0,1\} \quad \forall r,s,a,p,k,t \tag{12.10}$$

Nonnegativity constraint:

$$h_p^{rs}(k) \geq 0 \quad \forall r,s,p,k \tag{12.11}$$

Definitional constraints:

$$u_a(t) = \sum_{rs}\sum_p\sum_k h_p^{rs}(k)\delta_{apk}^{rs}(t) \quad \forall a,t \tag{12.12}$$

$$c_p^{rs}(k) = \sum_a\sum_t c_a(t)\delta_{apk}^{rs}(t) \quad \forall r,s,p,k \tag{12.13}$$

$$\hat{c}_p^{rs}(k) = c_p^{rs}(k) + \frac{1}{\theta}\ln h_p^{rs}(k) \quad \forall r,s,p,k \tag{12.14}$$

Except for equation (12.14), the above constraints are the same as those for the DUO route choice model. Equation (12.14) defines the perceived route travel time in terms of actual route travel time and route flows.

### 12.2.3 Equivalence Analysis

The following theorem verifies the equivalence between the equilibrium conditions (12.4) and the proposed VIP (12.6).

***Theorem 12.3***: Under a certain flow propagation relationship $\left(\delta_{apk}^{rs}(t) = \delta_{apk}^{rs*}(t)\right)$, the SDUO route choice conditions (12.4) implies variational inequality (12.6) and vice versa.

***Proof of necessity***: We need to prove that under a certain flow propagation relationship $\left(\delta_{apk}^{rs}(t) = \delta_{apk}^{rs*}(t)\right)$, the SDUO route choice conditions (12.4) implies VIP (12.6). We first rearrange equilibrium conditions (12.4) as follows:

$$\left[\hat{c}_p^{rs*}(k) - \hat{\pi}^{rs}(k)\right]\left[h_p^{rs}(k) - h_p^{rs*}(k)\right] \geq 0 \tag{12.15}$$

By summing over $r,s,p,k,$ and then making a substitution of $\sum_p h_p^{rs}(k) = \sum_p h_p^{rs*}(k) = \bar{q}^{rs}(k)$, one obtains:

$$\sum_{rs}\sum_k\sum_p \hat{c}_p^{rs*}(k)\left[h_p^{rs}(k) - h_p^{rs*}(k)\right] \geq 0 \tag{12.16}$$

Equation (12.16) is identical to equation (12.4).

***Proof of sufficiency***: We next prove that VIP (12.6) implies the stochastic/dynamic user-optimal conditions (12.4). If we suppose the equilibrium flow pattern is characterized by $\{h_p^{rs*}(k)\}$, then for each O-D pair $rs$ with traffic demand greater than zero, we consider two situations that could arise at equilibrium:

(i) we can find any pair of two used routes $p_1^{rs}, p_2^{rs}$, with positive route flows $h_{p_1}^{rs*}(k), h_{p_2}^{rs*}(k)$, respectively. To show that two used routes have exactly the same perceived route travel time, we arbitrarily switch a small amount of flow $\Delta_1$ from route $p_1^{rs}$ to $p_2^{rs}$. The new feasible flow pattern $\{h_p^{rs}(k)\}$ would be the same as the one at equilibrium except:

$$h_{p_1}^{rs}(k) = h_{p_1}^{rs*}(k) - \Delta_1 \tag{12.17}$$

and

$$h_{p_2}^{rs}(k) = h_{p_2}^{rs*}(k) + \Delta_1 \tag{12.18}$$

where $0 < \Delta_1 \le h_{p_1}^{rs*}(k)$. By applying the new feasible solution $\{h_p^{rs}(k)\}$ into VIP (12.6), we can yield:

$$\bar{c}_{p_1}^{rs*}(k)\left[h_{p_1}^{rs}(k) - h_{p_1}^{rs*}(k)\right] + \bar{c}_{p_2}^{rs*}(k)\left[h_{p_2}^{rs}(k) - h_{p_2}^{rs*}(k)\right] \ge 0 \tag{12.19}$$

By using equation (12.17) and equation (12.18), one obtains:

$$\bar{c}_{p_2}^{rs*}(k) \ge \bar{c}_{p_1}^{rs*}(k) \tag{12.20}$$

Similarly, by switching a small amount of flow $\Delta_2$ with $0 < \Delta_2 \le h_{p_2}^{rs*}(k)$ from route $p_2^{rs}$ to $p_1^{rs}$, we have:

$$\bar{c}_{p_1}^{rs*}(k) \ge \bar{c}_{p_2}^{rs*}(k) \tag{12.21}$$

Since equation (12.20) and equation (12.21) must hold simultaneously, it implies:

$$\bar{c}_{p_2}^{rs*}(k) = \bar{c}_{p_1}^{rs*}(k) \tag{12.22}$$

We can repeat this procedure to verify that, for each O-D pair, all used routes with positive flow will have the same perceived route travel time.

(ii) One route flow is positive and the other route flow is nil. We arbitrarily assume, without loss of generality, $h_{p_1}^{rs*}(k) > 0$, and $h_{p_2}^{rs*}(k) = 0$. We switch a small amount of flow $\Delta_1$ from route $p_1^{rs}$ to $p_2^{rs}$ with $0 < \Delta_1 \le h_{p_1}^{rs*}(k)$. By the same argument shown in (i), we have $\bar{c}_{p_2}^{rs*}(k) \ge \bar{c}_{p_1}^{rs*}(k)$. We repeat this procedure to verify that, for each O-D pair, all unused routes with zero flow will have the perceived route travel time no lower than the minimal perceived route travel time.

Since both (i) and (ii) must hold, it follows that VIP (12.6) implies equilibrium conditions (12.4). This completes the proof.

## 12.3 Nested Diagonalization Method

### 12.3.1 Solution Algorithm

A nested diagonalization method is proposed to solve the SDUO route choice model. The nested diagonalization method is defined as an algorithm that consists of the method in its solution procedure. This diagonalization method in turn embeds the method of successive averages (MSA). The nested diagonalization method is extremely useful for problems with two types of link flow interactions. Because one type of link flow interactions can be relaxed to yield a subproblem that can be solved by the diagonalization method, within the diagonalization solution procedure, the second type of link flow interactions can be further relaxed to result in a second level diagonalized subproblem that can be solved by the method of successive averages.

For the SDUO route choice model, the source of link flow interactions is twofold. One is from the actual link travel time $\tau_a(t)$, which is not known in advance, and hence, needs to be estimated, and the other interaction is from the interference among inflows under the estimated actual link travel times. Once these two types of link flow interaction are temporarily fixed, a diagonalized subproblem results. This diagonalized subproblem can be reformulated as a convex optimization problem and solved by the MSA method. The MSA method in turn consists of a linearized subproblem to be solved by a heuristic method, called SADA (stochastic dynamic method). The nested diagonalization method is formally stated as follows.

***The Nested Diagonalization Algorithm***

**Step 0: Initialization.**

Step 0.1: Let $m$=0. Set $\tau_a^0(t) = NINT[c_{a_0}(t)], \forall a,t$.

Step 0.2: Let $n$=1. Find an initial feasible solution $\{u_a^1(t)\}$. Compute the associated link travel times $\{c_a^1(t)\}$.

**Step 1: *First Loop* Operation.**

Let $m$=$m$+1. Update the estimated actual link travel times by

$$\tau_a^m(t) = NINT[(1-\gamma)\tau_a^{m-1}(t) + \gamma c_a^n(t)] \quad \forall a,t \tag{12.23}$$

where $0 < \gamma \leq 1$.

Construct the corresponding feasible time-space network based on the estimated actual link travel times.

**Step 2: *Second Loop* Operation.**

Step 2.1: Let $n$=1. Compute and reset the initial feasible solution $\{u_a^n(t)\}$, based on the time-space network, constructed by the estimated actual link travel times $\left\{\tau_a^m(t)\right\}$.

Step 2.2: Fix the inflows for all time-space links other than that on the subject time-space link at the current level, yielding the following optimization problem.

$$\min z(\mathbf{u},\mathbf{h}) = \sum_a \sum_t \int_0^{u_a^{n+1}(t)} c_a\left(u_a^n(1), u_a^n(2), \cdots, u_a^n(t-1), \omega\right) d\omega + \frac{1}{\theta} \sum_{rs} \sum_p \sum_k \left[h_p^{rs}(k) \ln\left(h_p^{rs}(k)\right) - h_p^{rs}(k)\right] \tag{12.24}$$

Flow conservation constraint:

$$\sum_p h_p^{rs}(k) = \bar{q}^{rs}(k) \quad \forall r,s,k \tag{12.25}$$

Nonnegativity constraint:

$$h_p^{rs}(k) \geq 0 \quad \forall r,s,p,k \tag{12.26}$$

Definitional constraints:

$$u_{apk}^{rs}(t) = h_p^{rs}(k) \bar{\delta}_{apk}^{rs}(t) \quad \forall r,s,a,p,k,t \tag{12.27}$$

$$\bar{\delta}_{apk}^{rs}(t) = \{0,1\} \quad \forall r,s,a,p,k,t \tag{12.28}$$

$$u_a(t) = \sum_{rs} \sum_p \sum_k h_p^{rs}(k) \bar{\delta}_{apk}^{rs}(t) \quad \forall a,t \tag{12.29}$$

$$c_p^{rs}(k) = \sum_a \sum_t c_a(t) \bar{\delta}_{apk}^{rs}(t) \quad \forall r,s,p,k \tag{12.30}$$

$$\hat{c}_p^{rs}(k) = c_p^{rs}(k) + \frac{1}{\theta} \ln h_p^{rs}(k) \quad \forall r,s,p,k \tag{12.31}$$

**Step 3: *Third Loop* Operation.**

Solve for the solution $\left\{u_a^{n+1}(t)\right\}$ in the optimization problem (12.24)~(12.31) using the method of successive averages (MSA). Compute the resulting link travel times $\left\{c_a^{n+1}(t)\right\}$.

**Step 4: Convergence Check for the *Second Loop* Operation.**

If $u_a^{n+1}(t) \approx u_a^n(t), \forall a,t$, go to Step 5; otherwise, set $n=n+1$, and go to Step 2.2.

**Step 5: Convergence Check for the *First Loop* Operation.**

If $\tau_a^m(t) \approx c_a^{n+1}(t), \forall a,t$, stop; the current solution is optimal. Otherwise, set $n=n+1$, and go to Step 1.

### 12.3.2 Method of Successive Averages

The diagonalized subproblem of the SDUO route choice model is essentially equivalent to a logit model. The optimality conditions are derived as follows.

#### 12.3.2.1 Optimality conditions

The optimality conditions of the diagonalized subproblem to be solved by the MSA method are also equivalent to a logit model. This result can be derived from the first order conditions for the diagonalized subproblem. We can construct the Lagrange function from equation (12.24) by relaxing the flow conservation constraints as follows.

$$L=\sum_a\sum_t\int_0^{u_a^{n+1}(t)} c_a\left(u_a^n(1),u_a^n(2),\cdots,u_a^n(t-1),\varpi\right)d\omega$$
$$+\frac{1}{\theta}\sum_{rs}\sum_p\sum_k\left[h_p^{rs}(k)\ln\left(h_p^{rs}(k)\right)-h_p^{rs}(k)\right] \tag{12.32}$$
$$+\sum_{rs}\sum_k\bar{\pi}^{rs}(k)\left[\bar{q}^{rs}(k)-\sum_p h_p^{rs}(k)\right]$$

The first order conditions with respect to route flow is as follows:

$$\frac{\partial L}{\partial h_p^{rs}(k)}=c_p^{rs}(k)+\frac{1}{\theta}\ln h_p^{rs}(k)-\bar{\pi}^{rs}(k)\geq 0 \tag{12.33}$$

$$h_p^{rs}(k)\frac{\partial L}{\partial h_p^{rs}(k)}=0 \tag{12.34}$$

When $h_p^{rs}(k)>0$, the second term in the complementary slackness condition (12.34) must be zero, i.e., $\frac{\partial L}{\partial h_p^{rs}(k)}=0$.

$$\left[c_p^{rs}(k)+\frac{1}{\theta}\ln h_p^{rs}(k)\right]-\bar{\pi}^{rs}(k)=0 \quad \forall r,s,p,k \tag{12.35}$$

By manipulation:

$$h_p^{rs}(k)=e^{-\theta\left(c_p^{rs}(k)-\bar{\pi}^{rs}(k)\right)} \tag{12.36}$$

Summing over $p$, one obtains:

$$\sum_p h_p^{rs}(k)=\bar{q}^{rs}(k) \quad \forall r,s,k \tag{12.37}$$

Dividing (12.36) by (12.37) yields:

$$\frac{h_p^{rs}(k)}{\bar{q}^{rs}(k)} = \frac{e^{-\theta\left(c_p^{rs}(k)-\bar{\pi}^{rs}(k)\right)}}{\sum_{p'} e^{-\theta\left(c_{p'}^{rs}(k)-\bar{\pi}^{rs}(k)\right)}} = \frac{e^{-\theta c_p^{rs}(k)}}{\sum_{p'} e^{-\theta c_{p'}^{rs}(k)}} \tag{12.38}$$

Since the first order condition for the diagonalized subproblem is equivalent to a logit model, the main problem, at equilibrium, must have the same solution as that of a diagonalized subproblem; implying the solution for the entire problem must be equivalent to a logit model. Note also that if $h_p^{rs}(k)=0$ in equation (12.36), then $c_p^{rs}(k)$ must be equal to $\infty$; implying that no one would choose route $p$ during interval $k$ (Boyce, 1984) .

#### 12.3.2.2 Solution algorithm

Note that objective function (12.24) cannot be converted into a function with link flow variables. Thus, the associated diagonalized problem cannot be solved by the well known Frank-Wolfe method, because route enumeration is required in this model to determine the move size. To cope with the computational complexity implied by the requirement of route enumeration, the method of successive averages (MSA) is commonly applied (Sheffi, 1985). The MSA method, in the dynamic sense, can be stated as follows:

***MSA Algorithm***

**Step 1:** Update link travel time based on the previous feasible solution $\{u_a^n(t)\}$:

$$c_a(t) = c_a\left(u_a^n(1), u_a^n(2), \cdots, u_a^n(t)\right) \quad \forall a,t \tag{12.39}$$

**Step 2:** Perform the SADA method (to be described below) to solve for the following linearized mathematical problem, yielding the inflows $\{p_a^n(t)\}$.

$$\min z(\mathbf{w}) = \sum_{rs}\sum_{p}\sum_{k}\left[c_p^{rs}(k)^n + \frac{1}{\theta}\ln\left(h_p^{rs}(k)^n\right)\right]w_p^{rs}(k) \tag{12.40}$$

Subject to:

Flow conservation constraint:

$$\sum_p w_p^{rs}(k) = \bar{q}^{rs}(k) \quad \forall r,s,k \tag{12.41}$$

Nonnegativity constraint:

$$w_p^{rs}(k) \geq 0 \quad \forall r,s,p,k \tag{12.42}$$

Definitional constraints:

$$p_{apk}^{rs}(t) = w_p^{rs}(k)\bar{\delta}_{apk}^{rs}(t) \quad \forall r,s,a,p,k,t \tag{12.43}$$

$$\bar{\delta}_{apk}^{rs}(t)=\{0,1\} \quad \forall r,s,a,p,k,t \tag{12.44}$$

$$p_a(t)=\sum_{rs}\sum_{p}\sum_{k} w_p^{rs}(k)\bar{\delta}_{apk}^{rs}(t) \quad \forall a,t \tag{12.45}$$

$$c_p^{rs}(k)=\sum_{a}\sum_{t} c_a(t)\bar{\delta}_{apk}^{rs}(t) \quad \forall r,s,p,k \tag{12.46}$$

$$\hat{c}_p^{rs}(k)=c_p^{rs}(k)+\frac{1}{\theta}\ln\left(w_p^{rs}(k)\right) \quad \forall r,s,p,k \tag{12.47}$$

**Step 3:** Update the flow pattern:

$$u_a^{n+1}(t)=u_a^n(t)+\lambda^n\left(p_a^n(t)-u_a^n(t)\right) \quad \forall a,t \tag{12.48}$$

$$\lambda^n=\frac{l_1}{l_2+n} \tag{12.49}$$

where $l_1$ is a positive constant and $l_2$ a nonnegative number.

**Step 4:** If convergence criterion is met, stop. Otherwise, set $n$=$n$+1, and go to step 1.

### 12.3.3 Stochastic Dynamic Method (SADA)

The optimization problem (12.40)~(12.47) involves route variables and can be theoretically solved by route enumeration. This approach is prohibitive for large networks, so we employ a link-based heuristic procedure called SADA to cope with this problem. The SADA algorithm is performed in Step 2 of the MSA method. Given the updated link travel time $c_a(t)=c_a\left(u_a^n(1),u_a^n(2),\cdots,u_a^n(t)\right), \forall a,t$, and suppose link $a$ in Step 1 of the MSA method is alternatively represented by a tail node $i(t)$ and a head node $j$, i.e., $a=\left(i(t),j\right)$, then the steps of the SADA algorithm for one origin-destination pair $rs$ and departure interval $k$ are outlined below. These steps should be repeated for each origin-destination pair in the time-space network.

***SADA Algorithm:***

**Step 0 : Preliminaries**

(a) Compute the minimum perceived travel time $\pi_{ri(t)}^{rs}(k)$ from node $r$ to all other nodes $i$.

(b) Define $m\in\beta_i(t)$ as the set of upstream nodes of all links arriving at node $i$ during interval $t$.

(c) Define $n\in\alpha_i(t)$ as the set of downstream nodes of all links leaving node $i$ during interval $t$.

(d) For each link $ij$, compute the *link likelihood*, $L^{rs}_{ij,k}(t)$:

$$L^{rs}_{ij,k}(t)\begin{cases} = e^{\theta\left[\pi^{rs}_{rj}(k)-\pi^{rs}_{ri(t)}(k)-c_{ij}(t)\right]} & \text{if } \pi^{rs}_{ri(t)}(k) < \pi^{rs}_{rj}(k) \text{ and} \\ & \quad \pi^{rs}_{is}(t) > \pi^{rs}_{js}\left(t + c_{ij}(t)\right) \\ = 0 & \text{otherwise} \end{cases} \tag{12.50}$$

where $t = k + \tau^{rs}_{ri(t)}(k)$

Figure 12.2: Graphical Illustration for Equation (12.50)

**Step 1 : Forward Pass**

Consider nodes in ascending order of $\pi^{rs}_{ri(t)}(k)$, starting from the origin $r$. For each node $i$, calculate the *link weight*, $w^{rs}_{ij,k}(t)$, for each $j$ (i.e., for each link emanating from $i$) as follows:

$$w^{rs}_{ij,k}(t)\begin{cases} = L^{rs}_{ij,k}(t) & \text{if } i = r \\ = L^{rs}_{ij,k}(t) \sum_{m\in\beta_i(t)} w^{rs}_{mi,k}(t') & \text{otherwise} \end{cases} \tag{12.51}$$

where $t = t' + \tau^{rs}_{mi}(t')$

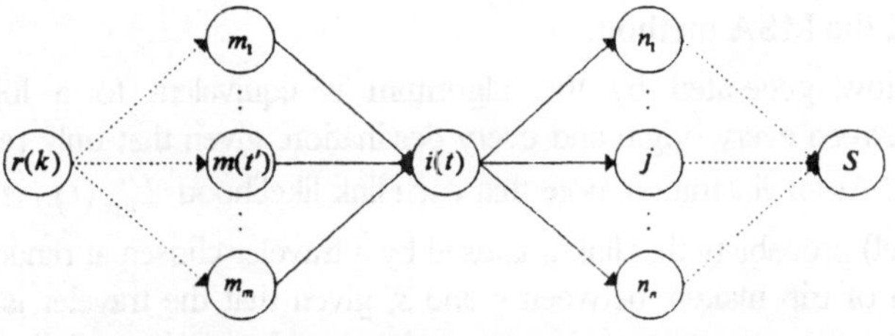


Figure 12.3: Graphical Illustration for Equation (12.51)

When the destination node $s$ is reached, this step is completed.

**Step 2 : Backward Pass**

Consider nodes in ascending values of $\pi^{rs}_{js}(t)$, starting from the destination $s$. When each node, $j$, is considered, compute the link flow

$u^{rs}_{ij,k}(t)$ for each $i$ (i.e., for each link entering $j$), by following the assignment:

$$u^{rs}_{ij,k}(t)\begin{cases} = \bar{q}^{rs}(k)\dfrac{w^{rs}_{ij,k}(t)}{\sum_{m\in\beta_j(t'')} w^{rs}_{mj,k}(t)} & \text{for } j = s \\ = \left[\sum_{n\in\alpha_j(t'')} u^{rs}_{jn,k}(t'')\right]\dfrac{w^{rs}_{ij,k}(t)}{\sum_{m\in\beta_j(t'')} w^{rs}_{mj,k}(t)} & \text{for all other links} \end{cases} \tag{12.52}$$

where $t'' = t + \tau^{rs}_{ij}(t)$

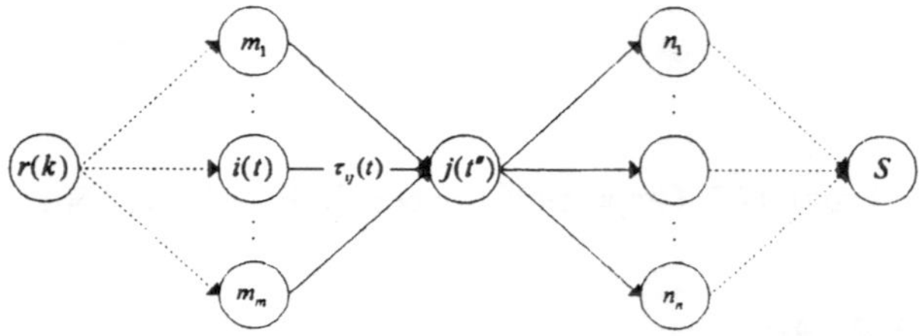


Figure 12.4: Graphical Illustration for Equation (12.52)

This step is applied iteratively until the origin $r$ is reached.

**Step 3:** Compute the inflow rate entering link $a$ during interval $t$ by the following formula:

$$u_a(t) = \sum_{rs}\sum_{k} u^{rs}_{a,k}(t) \tag{12.53}$$

The output obtained from equation (12.53) is then renamed as $p^n_a(t)$ and inserted into Step 2 of the MSA method.

The flow generated by this algorithm is equivalent to a logit-based route assignment between every origin and every destination, given that only reasonable routes are considered. As an illustration, note that each link likelihood $L^{rs}_{ij,k}(t)$, is proportional to the (logit model) probability that link $ij$ is used by a traveler chosen at random from among the population of trip makers between $r$ and $s$, given that the traveler is at node $i$. The probability that a given route is used is proportional to the product of all of the likelihoods of the links comprising this route. The probability of using route $k$ between $r$ and $s$ departing the origin during interval $k$, $P^{rs}_p(k) = (r,1,2,\cdots,i-1,i,\cdots,n,s)$, is then:

$$P^{rs}_p(k) = G^{rs}(k)\prod_{ij}\left\{L^{rs}_{ij,k}(t)\right\}^{\delta^{rs}_{ij,p}} \tag{12.54}$$

where $G^{rs}(k)$ is a proportionality constant, and the product is taken over all links in the time-space network. The incidence variable, $\delta^{rs}_{ij,p}(t)$, ensures that $P^{rs}_p(k)$

includes only those links in the $p$th route between $r$ and $s$ departing the origin during interval $k$. If we further assume that $c_p^{rs}(k)$ is the travel time over route $p$ between O-D $rs$ departing the origin during interval $k$, then $\tau_{ri,p}^{rs}(k)$ is the actual route travel time rounded off as an integer between origin $r$ and node $i$ departing the origin during interval $k$. For any interval $t = k + \tau_{ri,p}^{rs}(k)$, substituting the expression for the likelihood equation (12.50) into equation (12.54), the choice probability of choosing a particular (reasonable) route becomes:

$$\begin{aligned} P_p^{rs}(k) &= G^{rs}(k)\prod_{ij}\left\{L_{ij,k}^{rs}(t)\right\}^{\delta_{ij,p}^{rs}} \\ &= G^{rs}(k)\prod_{ij} e^{\theta\left[\pi_{rj}^{rs}(k)-\pi_{ri}^{rs}(k)-c_{ij}(t)\right]\delta_{ij,p}^{rs}} \\ &= G^{rs}(k)e^{\theta\sum_{ij}\left[\pi_{rj}^{rs}(k)-\pi_{ri}^{rs}(k)-c_{ij}(t)\right]\delta_{ij,p}^{rs}} \\ &= G^{rs}(k)e^{\theta\left[\pi^{rs}(k)-c_{ij}(t)\right]} \end{aligned} \tag{12.55}$$

where

$$\begin{aligned} &\sum_{ij}\left[\pi_{rj}^{rs}(k)-\pi_{ri(t)}^{rs}(k)-c_{ij}(t)\right]\delta_{ij,p}^{rs} \\ &\quad= \pi_{r1}^{rs}(k)-\pi_{rr}^{rs}(k)-c_{r1}(k) \\ &\qquad+\pi_{r2}^{rs}(k)-\pi_{r1}^{rs}(k)-c_{12}\left(k+\tau_{r1,p}^{rs}(k)\right) \\ &\qquad+\ \vdots \qquad\qquad \vdots \\ &\qquad+\pi_{rs}^{rs}(k)-\pi_{rn}^{rs}(k)-c_{ns}\left(k+\tau_{rn,p}^{rs}(k)\right) \\ &\quad= \pi^{rs}(k)-c_p(k) \end{aligned} \tag{12.56}$$

In order for equation (12.55) to be a proper probability statement, the proportionality constant has to be set such that:

$$\sum_p P_p^{rs}(k) = 1.0 \tag{12.57}$$

This means that:

$$G(k) = \frac{1}{\sum_p e^{-\theta\left[c_p^{rs}(k)+\pi^{rs}(k)\right]}} \tag{12.58}$$

whence

$$P_p^{rs}(k) = \frac{e^{-\theta\left[c_p^{rs}(k)+\pi^{rs}(k)\right]}}{\sum_{p'} e^{-\theta\left[c_{p'}^{rs}(k)+\pi^{rs}(k)\right]}} = \frac{e^{-\theta c_p^{rs}(k)}}{\sum_{p'} e^{-\theta c_{p'}^{rs}(k)}} \tag{12.59}$$

Equation (12.59) depicts a logit model of route choice among the reasonable routes connecting O-D pair *rs*.

## 12.4 Numerical Example

### 12.4.1 Input Data

A simple network shown in Figure 12.5 is used for testing. The test network consists of 7 links and 6 nodes, in which node 1 is the origin, node 6 is the destination and all other nodes are intermediates.

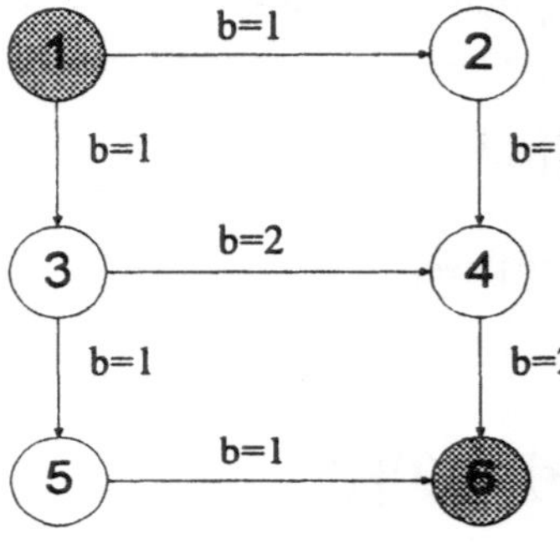


Figure 12.5: Test Network 1

The adopted dynamic travel time function is arbitrarily constructed as follows:

$$c_a(t) = b + 0.01(u_a(t))^2 + 0.01(x_a(t))^2 \tag{12.60}$$

The assumed origin-destination (O-D) demand is shown in Table 12.1:

Table 12.1: Time-Dependent O-D Demand ($k$=1)

| O-D Pair | Departure Rate |
|---|---|
| 1-6 | 30 |

### 12.4.2 Test Results

A computer program coded with Borland $C^{++}$ is used for solving the SDUO route choice model, when given the input data. The obtained results are summarized as shown in Table 12.2.

Table 12.2: Results for Test Network 1 ($\theta = 1$)

| Link | Entering Time Interval | Inflow | Exit Flow | Number of Vehicles | Link Travel Time | Exiting Time Interval |
|---|---|---|---|---|---|---|
| 1→2 | 1 | 12.5 | 0 | 0 | 2.56 | 3 |
| | 2~3 | 0 | 0 | 12.5 | 2.56 | - |
| | 4 | 0 | 12.5 | 12.5 | 2.56 | - |
| 1→3 | 1 | 17.5 | 0 | 0 | 4.06 | 5 |
| | 2~4 | 0 | 0 | 17.5 | 4.06 | - |
| | 5 | 0 | 17.5 | 17.5 | 4.06 | - |
| 2→4 | 4 | 12.5 | 0 | 0 | 2.56 | 7 |
| | 5~6 | 0 | 0 | 12.5 | 2.56 | - |
| | 7 | 0 | 12.5 | 12.5 | 2.56 | - |
| 3→4 | 5 | 3.9 | 0 | 0 | 2.15 | 7 |
| | 6 | 0 | 0 | 3.9 | 2.15 | - |
| | 7 | 0 | 3.9 | 3.9 | 2.15 | - |
| 3→5 | 5 | 13.6 | 0 | 0 | 2.84 | 8 |
| | 6~7 | 0 | 0 | 13.6 | 2.84 | - |
| | 8 | 0 | 13.6 | 13.6 | 2.84 | - |
| 4→6 | 7 | 16.4 | 0 | 0 | 4.70 | 12 |
| | 8~11 | 0 | 0 | 16.4 | 4.70 | - |
| | 12 | 0 | 16.4 | 16.4 | 4.70 | - |
| 5→6 | 8 | 13.6 | 0 | 0 | 2.84 | 11 |
| | 9~10 | 0 | 0 | 13.6 | 2.84 | - |
| | 11 | 0 | 13.6 | 13.6 | 2.84 | - |

The rationale of the proposed model and associated solution algorithm can be verified by checking if the resulting perceived route travel times satisfy the SDUO conditions. If we consider route 1→2→4→6 to be departing origin 1 during interval 1, then the corresponding perceived route travel time can be obtained by summing up the actual link travel time on link 1→2 during interval 1, and the actual link travel time on link 2→4 during interval $1 + c_{1\to2}(1)$, and the actual link travel time on link 4→6 during interval $\left(1 + c_{1\to2}(1) + c_{2\to4}\left(1 + c_{1\to2}(1)\right)\right)$, and the error term $\ln\left(h_{1\to2\to4\to6}\right)$ as follows:

$$\begin{aligned}
&\hat{c}_{1\to2\to4\to6}(1)\\
&= c_{1\to2}(1) + c_{2\to4}\left(1 + c_{1\to2}(1)\right) + c_{4\to6}\left(1 + c_{1\to2}(1) + c_{2\to4}\left(1 + c_{1\to2}(1)\right)\right) + \ln\left(h_{1\to2\to4\to6}\right)\\
&= 2.56 + c_{2\to4}(3.56) + c_{4\to6}\left(3.56 + c_{2\to4}(3.56)\right) + \ln(12.50)\\
&\approx 2.56 + c_{2\to4}(4) + c_{4\to6}\left(4 + c_{2\to4}(4)\right) + \ln(12.50) \qquad \text{(By integration)}\\
&= 2.56 + 2.56 + c_{4\to6}(4 + 2.56) + \ln(12.50)\\
&\approx 5.12 + c_{4\to6}(7) + \ln(12.50) \qquad \text{(By integration)}\\
&= 5.12 + 4.70 + 2.53 = 9.82 + 2.53 = 12.35
\end{aligned} \tag{12.61}$$

The remaining used perceived route travel times are also computed and summarized in Table 12.3. Note that trips departing from the same origin during the same interval have approximately the same perceived route travel time. For the sake of comparison, an exact logit-based local solution, based on the same actual travel times $\tau_a(t), \forall a, t$ obtained from the computed solution, is also included; it shows the difference between the computed and exact solutions is negligible. The nil difference is basically due to round-off errors accruing from arithmetic operations.

Table 12.3: Stochastic Route Travel Times for Test Network 1 ($\theta = 1$)

| Route | Computed Solution | | Exact Local Solution* | |
|---|---|---|---|---|
| | Flow | Perceived Travel Time | Flow | Perceived Travel Time |
| 1→2→4→6 | 12.50 | 12.35 | 12.46 | 12.34 |
| 1→3→4→6 | 3.90 | 12.27 | 4.01 | 12.34 |
| 1→3→5→6 | 13.60 | 12.35 | 13.53 | 12.34 |
| Total | 30.00 | | 30.00 | |

* Based on the same actual travel times $\{\tau_a(t)\}$ obtained from the computed solution.

## 12.5 Deterministic Counterpart Approximation

A special case of the SDUO route choice model can be reduced to the deterministic DUO route choice model. This result can be obtained by choosing appropriate values for the dispersion parameter $\theta$. The larger the value of $\theta$, the more certain the traffic information will be, whereas, the smaller the value of $\theta$, the more uncertain the traffic information will be. To show how to derive the deterministic DUO solution from our SDUO model, we set the parameter

$\theta = 10000$ (very large), and summarize the results in Table 12.4:

Table 12.4: Deterministic Counterpart Approximation for Test Network 1

| Route | Computed Solution ($\theta = 10000$) | | Exact Local Solution* | |
|---|---|---|---|---|
| | Route Flow | Perceived Travel Time | Route Flow | Perceived Travel Time |
| 1→2→4→6 | 13.50 | 10.02 | 13.62 | 10.04 |
| 1→3→4→6 | 1.80 | 10.11 | 1.63 | 10.04 |
| 1→3→5→6 | 14.60 | 10.02 | 14.75 | 10.04 |
| Total | 29.90 | -- | 30.00 | -- |

*Based on the same actual travel times $\{\tau_a(t)\}$ obtained from the computed solution.

From Table 12.4, it can be seen that the obtained perceived route travel times are almost identical to the exact logit-based solutions. The slight difference is due to the round-off error accruing from the arithmetic operations. The static counterpart approximation can be obtained by mathematically setting the dispersion parameter $\theta = \infty$ and actual link travel times $\tau_a = 0$.

## 12.6 Braess's Paradox

Braess's paradox in general illustrates that additions (deletions) of links to a network can make the congestion and delays worse (better off). In the deterministic scenario, this counter-intuitive result can happen only when the travel time is a function of link flows. In other words, the increase (or decrease) in travel time is rooted in the essence of the user equilibrium, where each motorist minimizes his or her own travel time. However, in the stochastic scenario, there is an additional counter-intuitive phenomenon which may arise even when link travel times are invariant. This phenomenon is not rooted in the dependence of the links' travel time on the flow, but in the stochastic nature of the network assignment model, that is, in the randomness of the perceived route travel times. If the variance of the actual travel time is zero, then the stochastic network assignment model is reduced to be its deterministic counterpart, and the Braess-like paradox discussed here does not occur. In the following section, we show by numerical examples the two different cases of Braess's paradox.

### 12.6.1 First Case of Braess's Paradox Caused by Variable Link Travel Times

The first case of Braess's paradox is basically due to the user equilibrium realized by the perceived route travel times. To show this, the following network is used for testing. Figure 12.6(a) denotes the network before improvement, and 12.6(b) after improvement (with an additional link 4→5 ).

The adopted dynamic travel time function is arbitrarily constructed as follows:

$$c_a(t) = b + 0.01(u_a(t))^2 + 0.01(x_a(t))^2 \quad \forall a,t \tag{12.62}$$

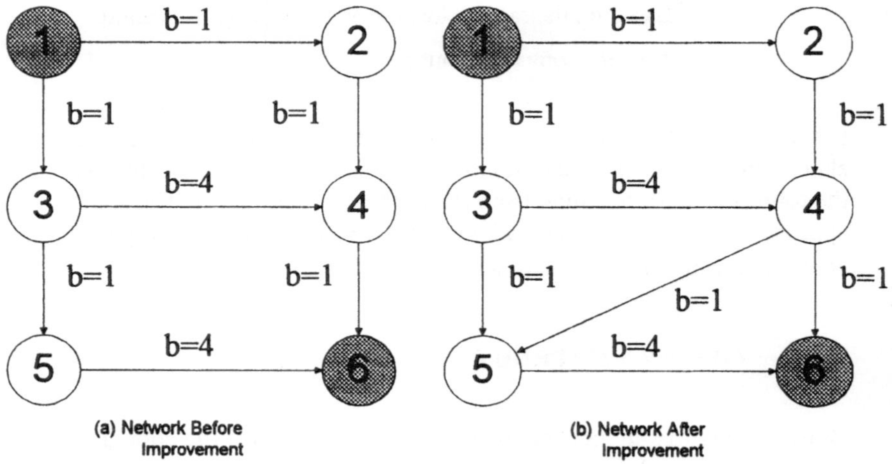


Figure 12.6: Test Network 2

The assumed origin-destination (O-D) demand is given in Table 12.5:

Table 12.5: Time-dependent O-D Demand ($k$=1)

| O-D Pair | Departure Rate |
|---|---|
| 1-6 | 20 |

The corresponding route travel times are computed and summarized in Table 12.6.

Table 12.6: Route Travel Times for Test Network 2 (First Case of Braess's Paradox)

| Route | Solution for Figure 12.6(a)* | | Solution for Figure 12.6(b)* | |
|---|---|---|---|---|
| | Flow | Perceived Travel Time | Flow | Perceived Travel Time |
| 1→2→4→6 | 7.7 | 9.88 (7.84) | 4.9 | 9.59 (8.00)** |
| 1→2→4→5→6 | NA | NA | 4.4 | 9.59 (8.11) |
| 1→3→4→6 | 0.3 | 9.95 (11.15) | 0.4 | 9.52 (10.44) |
| 1→3→4→5→6 | NA | NA | 0.4 | 9.63 (10.55) |
| 1→3→5→6 | 12.0 | 9.87 (7.39) | 9.9 | 9.58 (7.29) |
| Total Actual Travel Time | 152.39 | | 155.45 | |

* Based on the same actual travel times $\{\tau_a(t)\}$ obtained from the computed solution.

** "Numbers" in brackets denote the actual route travel time.

By comparing the total actual travel times for Figures 12.6(a) and 12.6(b), it is observed that an addition of a link results in higher total travel time; implying a Braess's paradox.

### 12.6.2 Second Case of Braess's Paradox Caused by Constant Link Travel Times

The second case of Braess's paradox is basically due to randomness of perceived route travel times. To show this, the following network is used for testing. Figure 12.7(a) denotes the network before improvement, and 12.7(b) after improvement (with travel time one unit less on link 3→5 ).

The adopted dynamic travel time function is arbitrarily constructed as follows:

$$c_a(t) = b \quad \forall a,t \tag{12.63}$$

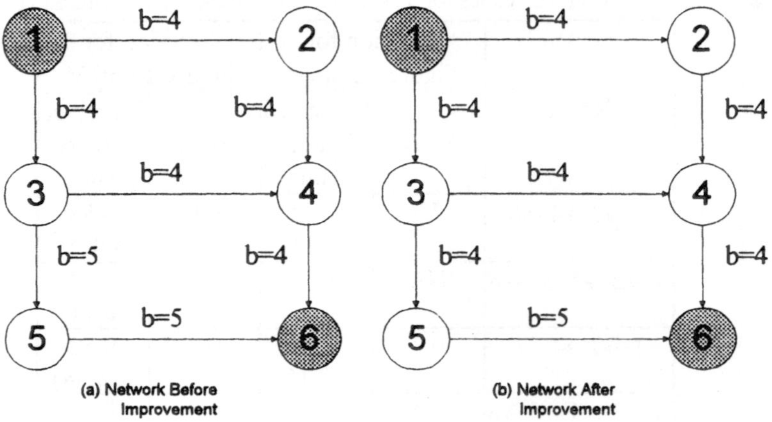


Figure 12.7: Test Network 3

The origin-destination (O-D) demand is assumed in Table 12.7.

Table 12.7: Time-dependent O-D Demand ($k$=1)

| O-D Pair | Departure Rate |
|---|---|
| 1-6 | 30 |

The corresponding route travel times are computed and summarized in Table 12.8.

Table 12.8: Route Travel Times for Test Network 3 (Second Case of Braess's Paradox)

| Route | Solution for Figure 12.7(a)* | | Solution for Figure 12.7(b)* | |
|---|---|---|---|---|
| | Flow | Perceived Travel Time | Flow | Perceived Travel Time |
| 1→2→4→6 | 14.4 | 14.64(12.0)** | 12.7 | 14.54 (12.0)* |
| 1→3→4→6 | 14.0 | 14.64 (12.0) | 12.7 | 14.54 (12.0) |
| 1→3→5→6 | 1.9 | 14.64 (12.0) | 4.7 | 14.55 (13.0) |
| Total Actual Travel Time | 362.6 | | 365.9 | |

* Based on the same actual travel times $\{\tau_a(t)\}$ obtained from the computed solution.

** "Numbers" in brackets denote the actual route travel time.

By comparing the total actual travel times for Figures 12.7(a) and 12.7(b), it observed that an improvement of a link results in higher total travel time, implying Braess's paradox.

## 12.7 Notes

The stochastic/dynamic route travel time may be defined as the derivative of the "objective function" of the logit-type stochastic/dynamic user-optimal model (a dynamic counterpart of Fisk (1980)) with respect to a route flow. The "objective function" contains two components: the deterministic use-optimal "objective function" and an entropy term, as follows:

$$z(\mathbf{u})=\sum_a\sum_t\int_0^{u_a(t)} c_a(\varpi)d\omega+\frac{1}{\theta}\left[\sum_{rs}\sum_p\sum_k h_p^{rs}(k)\ln h_p^{rs}(k)-1\right] \quad (12.64)$$

Suppose the link travel time function involves both temporal and topological interactions, then the stochastic/dynamic route travel time can be derived as follows:

$$\begin{aligned}
\frac{\partial z(\mathbf{u})}{\partial h_p^{rs}(k)} &= \frac{\partial}{\partial h_p^{rs}(k)}\left[\sum_a\sum_t\int^{u_a(t)} c_a(\varpi)d\varpi+\frac{1}{\theta}\left(\sum_{rs}\sum_p\sum_k h_p^{rs}(k)\ln h_p^{rs}(k)-h_p^{rs}(k)\right)\right] \\
&=\sum_b\sum_{t'}\left(\frac{\partial\sum_a\sum_t\int^{u_a(t)} c_a(\varpi)d\varpi}{\partial u_b(t')}\times\frac{\partial u_b(t')}{\partial h_p^{rs}(k)}\right) \\
&\quad+\frac{\partial}{\partial h_p^{rs}(k)}\left[\frac{1}{\theta}\left(\sum_{rs}\sum_p\sum_k h_p^{rs}(k)\ln h_p^{rs}(k)-h_p^{rs}(k)\right)\right] \\
&=\sum_b\sum_{t'} c_b(t')\delta_{bpk}^{rs}(t')+\frac{1}{\theta}\left[h_p^{rs}(k)\frac{\partial \ln h_p^{rs}(k)}{\partial h_p^{rs}(k)}+\ln h_p^{rs}(k)\frac{\partial h_p^{rs}(k)}{\partial h_p^{rs}(k)}-\frac{\partial h_p^{rs}(k)}{\partial h_p^{rs}(k)}\right] \\
&=c_p^{rs}(k)+\frac{1}{\theta}\left[1+\ln h_p^{rs}(k)-1\right] \\
&=c_p^{rs}(k)+\frac{1}{\theta}\ln h_p^{rs}(k) \qquad \forall r,s,p,k
\end{aligned} \quad (12.65)$$

Note that equation (10.1) had been used for the derivation of the above equation.

In Section 12.3.3, we used the double-pass method within the SADA algorithm to define a reasonable route, including only links $ij$ such that $\pi_{ri(t)}^{rs}(k)<\pi_{rj}^{rs}(k)$ and $\pi_{is}^{rs}(t)>\pi_{js}^{rs}(t+c_{ij}(t))$. In fact, a single-pass method can also be accommodated within the SADA algorithm to define a reasonable route (Sheffi, 1985). We studied the former approach because it makes more sense in terms of the drivers' behavior; however, the latter approach is more efficient in terms of computational time, as fewer shortest route searching operations are needed.

In Section 12.6.1, we discussed the first case of Braess's paradox due to the user equilibrium. The cause of the second case of Braess's paradox shown in Section 12.6.2 may also be embedded in this numerical example as these two different factors cannot be separated easily.

Most stochastic route choice models are static in nature (Daganzo and Sheffi, 1977; de Palma et al, 1983; Damberg et al, 1996, Maher and Hughes, 1997). The only predictive (or ideal) stochastic/dynamic user-optimal route choice model in the literature is presented by Ran and Boyce (1994), which is of interest for comparison with our model. The major difference between these two models is summarized in the following table.

Table 12.9: Comparison between Two Stochastic/Dynamic Route Choice Models

| Item | Ran and Boyce | Chen and Tu |
|---|---|---|
| Model Formulation | Optimal control theory approach with optimization form | Variational inequality approach |
| Active Variables | Inflow, exit flow and number of vehicles on a link; cumulative number of vehicles arriving at destination $s$ from origin $r$ at the beginning of interval $k$. | Link inflow |
| Constraint Categories | 1. Relationship between state and control variables.<br>2. Flow conservation.<br>3. Logit route flow.<br>4. Constraints for mean actual route travel time.<br>5. Flow propagation.<br>6. Definitional.<br>7. Nonegativity.<br>8. Boundary conditions. | 1. Flow conservation.<br>2. Flow propagation.<br>3. Nonegativity.<br>4. Definitional. |
| Indicator Variable | $\delta_{ap}^{rs}$ : realized as 1 if link $a$ on route $p$ between O-D pair $rs$. | $\delta_{apk}^{rs}(t)$ : realized as 1 if inflow rate on link $a$ during time interval $t$ departs from origin $r$ over route $p$ toward destination $s$ during time interval $k$. |
| | DYNASTOCH2 (a dynamic variation of STOCH)<br>1. Backward pass proceeds forward pass.<br>2. Calculation of link likelihood needs to know | SADA (a dynamic variation of STOCH)<br>1. Forward pass proceeds backward pass.<br>2. Calculation of link likelihood needs to know |

| | | |
|---|---|---|
| Solution Algorithm | the actual link travel time for link *a*, i.e., $\tau_a(t)$.<br>3. Calculation of link weight needs to know the actual link travel time for link *a*, i.e., $\tau_a(t)$.<br>4. The relationship between their proposed model and the corresponding algorithm is unclear, especially in terms of the number of decision variables. | the actual link travel time for link *a*, i.e., $c_a(t)$.<br>3. Calculation of link weight needs to know the entering time interval and actual link travel time for link *a*, i.e., $t' = t - \tau_a(t')$ and $\tau_a(t)$.<br>4. The relationship between their proposed model and the corresponding algorithm is clearly identified. |
| Convergence Criterion | $\tau_a^{(m)}(k) \approx \tau_a^{(m+1)}(k)$ | $\tau_a^{(m)}(t) \approx c_a^{(m+1)}(t)$ |

With the above comparison, it is clear that our stochastic/dynamic user-optimal model is fundamentally *different* from that of Ran and Boyce (1994).

# Chapter 13

# Fuzzy/Dynamic User-Optimal Route Choice Model

Fuzzy set theory was first developed in order to solve imprecise or vague problems in the field of artificial intelligence, especially for imprecise reasoning and modeling linguistic terms. Stemming from research by Tanaka et al (1974), a number of fuzzy linear programming models have been developed. Lai and Hwang (1992) have classified linear programming models with imprecise information into two main classes: fuzzy linear programming, and possibilistic programming. As an extension, nonlinear programming models with imprecise information should be transferred into an equivalent crisp nonlinear programming problem, and then solved by conventional solution techniques or packages of nonlinear programming.

The possibilistic nonlinear programming (PNLP) and fuzzy nonlinear programming (FNLP) problems are different in that the former is based on the degree of the occurrence of an event to obtain possibility distributions, whereas the latter is based on the preferred concept to establish membership functions. In other words, the PNLP is similar to probability methods and the FNLP is analogous to utility approaches.

Since Zadeh (1978), there has been much research on the possibility theory, and possibilistic decision making models have provided an important aspect for handling practical decision making problems. Unlike stochastic linear programming problems, PNLP models provide computational efficiency and flexible doctrines. The possibility measure of an event might be interpreted as the degree of possibility of its occurrence under a possibility distribution that is an analogous to a probability distribution.

The fuzzy/dynamic user-optimal route choice model assumes that the link travel times are imprecise. Various possibility distributions that have been exploited in different fields may be accommodated to represent the real situations, among which the most popular possibility distributions are triangular $\tilde{c}_a(t)=\left(\tilde{c}_a^L(t),\tilde{c}_a^M(t),\tilde{c}_a^U(t)\right)$ and trapezoidal (L-R type), $\tilde{c}_a(t)=\left(\tilde{c}_a^L(t),\tilde{c}_a^{M_1}(t),\tilde{c}_a^{M_2}(t),\tilde{c}_a^U(t)\right)$. Under a specific assumption on the possibility distributions of the link travel times, it is reasonable to assume that all drivers make their route choice decisions based on their representative believed route travel times. The resulting traffic flows are expected to be more dispersed over parallel routes for the fuzzy/dynamic route choice model than for its deterministic counterpart.

In this chapter, the fuzzy/dynamic link travel time function is analyzed in Section 13.1. The equilibrium conditions and model formulation for the fuzzy/dynamic user-optimal (FDUO) route choice problem are presented in Section 13.2. The nested diagonalization method is described in Section 13.3. Numerical examples corresponding to three possibility distributions are demonstrated in Section 13.4. A deterministic counterpart approximation is discussed in Section 13.5. Finally, concluding notes are given in Section 13.6.

## 13.1 Fuzzy/Dynamic Link Travel Time Function Analysis

### 13.1.1 Fuzzy/Dynamic Link Travel Time Function

We assume that the fuzziness of subjective link travel times $\tilde{c}_a(t)$ can be described by the possibility theory with distribution denoted by $\Pi_{\tilde{c}_a(t)}$. A realistic assumption on link travel times is that their range variations are bounded; an example would be to limit the link travel times to be no lower than their corresponding free flow link travel time and no greater than some predetermined upper limit (by human's judgment). If the possibility distribution is triangular in shape, then the grade function can be described by the following formula, also graphed in Figure 13.1.

$$\Pi_{c_a(t)}(\chi)=\begin{cases}0, & \chi\le\tilde{c}_a^L(t)\\[2ex] \dfrac{\chi-\tilde{c}_a^L(t)}{\tilde{c}_a^M(t)-\tilde{c}_a^L(t)}, & \tilde{c}_a^L(t)<\chi\le\tilde{c}_a^M(t)\\[2ex] \dfrac{\tilde{c}_a^U(t)-\chi}{\tilde{c}_a^U(t)-\tilde{c}_a^M(t)}, & \tilde{c}_a^M(t)<\chi\le\tilde{c}_a^U(t)\\[2ex] 0, & \tilde{c}_a^U(t)<\chi\end{cases}\qquad \forall a,t \qquad (13.1)$$

For simplicity, we denote this possibility distribution of the fuzzy/dynamic link travel time function, $\tilde{c}_a(t)$, as $\left(\tilde{c}_a^L(t),\tilde{c}_a^M(t),\tilde{c}_a^U(t)\right)$. If the minimal acceptable possibility is set at $\alpha$-cut level ($\ge\alpha$), the corresponding possibility distribution of the fuzzy/dynamic link travel time function $\tilde{c}_{a,\alpha}(t)$ is denoted as:

$$\begin{aligned}&\left(\tilde{c}^{L}_{a,\alpha}(t),\tilde{c}^{M}_{a}(t),\tilde{c}^{U}_{a,\alpha}(t)\right)\\&\quad=\left\{\left[\chi,\Pi_{c_a(t)}(\chi)\right]\Pi_{c_a(t)}(\chi)\geq\alpha\right\},\alpha\in[0,1]\qquad\forall a,t\\&\text{and}\quad\tilde{c}^{L}_{a,\alpha}(t)=\min\{\chi\},\quad\Pi_{c_a(t)}(\chi)=\alpha\\&\qquad\quad\tilde{c}^{U}_{a,\alpha}(t)=\max\{\chi\},\quad\Pi_{c_a(t)}(\chi)=\alpha\end{aligned}\tag{13.2}$$

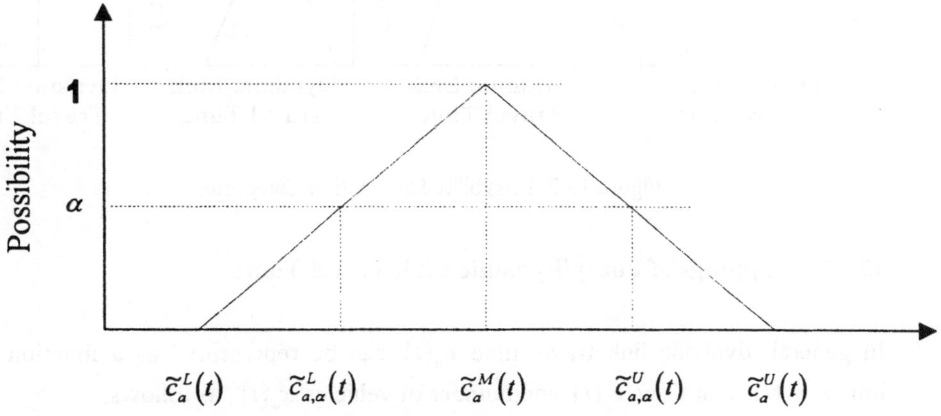


Figure 13.1: Triangular Possibility Distribution

In general, possibility distribution shape can be classified into the following four categories, in which each distribution represents different types of subjective traffic information:

1. Right-skewed possibility distribution: this type of possibility distribution has a wider right spread than left spread. This assumption is applicable to road scenarios where recurrent traffic congestion prevails. Therefore, drivers expect to take a longer time (as compared with actual commuter experience) to traverse a route.
2. Left-skewed possibility distribution: this type of possibility distribution has a wider left spread than right spread. This assumption may be appropriate for those road segments where traffic congestion hardly appears. Therefore, drivers expect to take a shorter time (as compared with actual commuter experience) to traverse a route.
3. Symmetric possibility distribution: this type of possibility distribution assumes that its right spread equals its left spread. Road segments with few opportunities of congestion occurrence, fall into this category.
4. Crisp possibility distribution: this type of possibility distribution assumes that

believed traffic information is perfect and thus both its right and left spreads equal zero. In a well-equipped closed system, such as a freeway where traffic information is likely to be precise, the travel time variation distribution can sometimes be treated as crisp distribution.

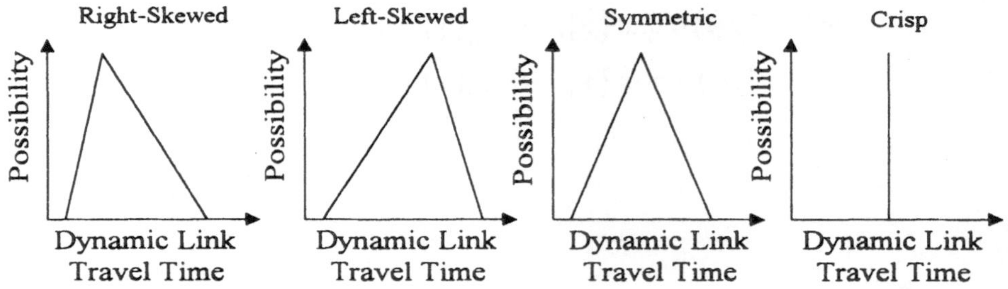


Figure 13.2: Possibility Distribution Categories

### 13.1.2 Mappings of Fuzzy/Dynamic Link Travel Times

In general, dynamic link travel time $c_a(t)$ can be represented as a function of inflow $u_a(t)$, exit flow $v_a(t)$ and number of vehicles $x_a(t)$, as follows:

$$c_a(t) = c_a\big(u_a(t), v_a(t), x_a(t)\big) \quad \forall a, t \tag{13.3}$$

However, in consideration of flow propagation, both exit flow $v_a(t)$ and number of vehicles $x_a(t)$ can be substituted by the previously entered inflows. Therefore, the above function can be rewritten, without loss of generality, as follows (Chen and Hsueh, 1996):

$$c_a(t) = c_a\big(u_a(1), u_a(2), \cdots, u_a(t)\big) \quad \forall a, t \tag{13.4}$$

This dynamic travel function can be extended to the fuzzy/dynamic travel function $\tilde{c}_a(t)$ as follows:

$$\tilde{c}_a(t) = \tilde{c}_a\big(u_a(1), u_a(2), \cdots, u_a(t)\big) \quad \forall a, t \tag{13.5}$$

Accordingly, the fuzzy/dynamic travel time function at $\alpha$-cut level can be denoted as:

$$\tilde{c}_{a,\alpha}(t) = \tilde{c}_{a,\alpha}\big(u_a(1), u_a(2), \cdots, u_a(t)\big) \quad \forall a, t, \alpha \tag{13.6}$$

Under the assumption of triangular possibility distribution, the fuzzy/dynamic link travel time function at $\alpha$-cut level $\tilde{c}_{a,\alpha}(t) = \tilde{c}_{a,\alpha}\big(u_a(1), \cdots, u_a(t)\big)$ embeds three important parameters, i.e., lower bound $\tilde{c}^L_{a,\alpha}(t)$, main value $\tilde{c}^M_a(t)$ and upper bound $\tilde{c}^U_{a,\alpha}(t)$. Accordingly, the most likely value $\overline{\tilde{c}}_{a,\alpha}(t)$ of

$\tilde{c}_{a,\alpha}(t)=\tilde{c}_{a,\alpha}\left(u_a(1),\cdots,u_a(t)\right)$ (Lai and Hwang , 1992) can be written as:

$$\bar{\tilde{c}}_{a,\alpha}(t)=\frac{\tilde{c}^{L}_{a,\alpha}\left(u_a(1),\cdots,u_a(t)\right)+4\tilde{c}^{M}_{a}\left(u_a(1),\cdots,u_a(t)\right)+\tilde{c}^{U}_{a,\alpha}\left(u_a(1),\cdots,u_a(t)\right)}{6} \tag{13.7}$$

Throughout this chapter, we use the representative believed value instead of the most likely value for simplicity; it is denoted as follows:

$$\bar{\tilde{c}}_{a,\alpha}(t)=Rep\left[\tilde{c}_{a,\alpha}\left(u_a(1),\cdots,u_a(t)\right)\right] \tag{13.8}$$

where

$$\bar{\tilde{c}}_{a,\alpha}(t)\in\tilde{c}_{a,\alpha}\left(u_a(1),\cdots,u_a(t)\right) \tag{13.9}$$

At $\alpha$ -cut level, the representative link travel time is also a function of inflows on link $a$ at different intervals.

### 13.1.3 Asymmetric Link Interactions

The representative believed link travel time function $\bar{\tilde{c}}_{a,\alpha}(t)$ contains the dynamic link travel time function $c_a(t)$ as its component. As shown before, for any physical link $a$, inflows can be affected by those previously entered inflows; but, the reverse is not true. This property makes the representative believed link travel time functions asymmetric. The theorem directly follows from Green's theorem. Consequently, any FDUO route choice problem with the fuzzy/dynamic link travel time function represented by a representative believed link travel time $\bar{\tilde{c}}_{a,\alpha}(t)$ as shown in equation (13.1) does not have an equivalent optimization problem. The variational inequality approach emerges.

### 13.1.4 Time-Space Network

For the FDUO route choice problem, the corresponding time-space network is essentially the same as that for the DUO route choice problem. We show the time-space network in Figure 13.3 for ready reference.

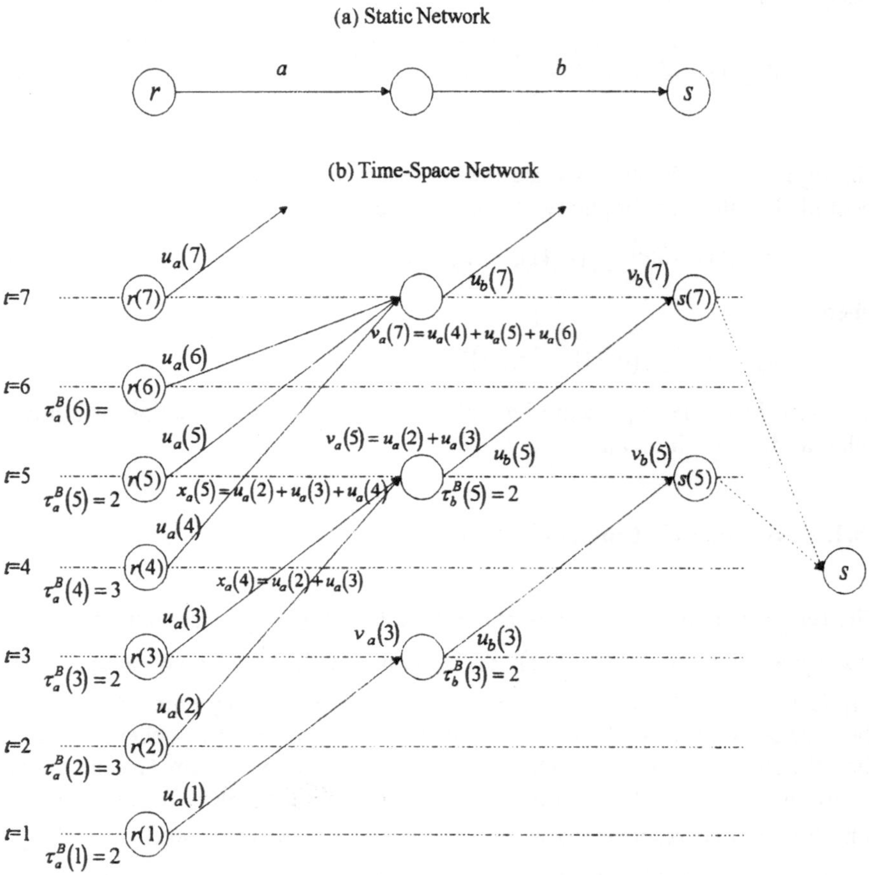


Figure 13.3: Time-Space Network

## 13.2 Equilibrium Conditions and Model Formulation

In this section, the fuzzy/dynamic user-optimal conditions are first defined to characterize the travelers' driving behavior for using routes with the minimal representative believed travel time. The representative believed route travel times can be computed by adding up the representative believed link travel times in consideration of flow propagation requirements along that route. The variational inequality formulation is then presented for the FDUO route choice problem. After that, the equivalence between the fuzzy/dynamic user-optimal conditions and the variational inequality formulation is verified by a proof.

### 13.2.1 Fuzzy/Dynamic User-Optimal Conditions

Analogous to the stochastic/dynamic user-optimal route choice problem (Chen et al, 1996), the optimality condition associated with the fuzzy/dynamic user-optimal route choice problem can be stated as follows:

*In a fuzzy/dynamic user equilibrium network, no user believes he can improve his travel time by unilaterally changing routes.*

The fuzzy/dynamic user-optimal conditions state for each O-D pair during each interval that, at a certain $\alpha$-cut level, the representative believed route travel times for travelers departing during the same interval are equal and minimal; or no traveler would be better off in terms of representative believed route travel times by unilaterally changing his/her route. In other words, at a certain $\alpha$-cut level, the representative believed route travel time of any unused route for each O-D pair is greater than or equal to the minimal representative believed route travel time. Therefore, at a certain $\alpha$-cut level, for each O-D pair ($rs$), if the flow over route $p$ departing during interval $k$ is positive, i.e., $h_p^{rs}(k) > 0$, then the corresponding representative believed route travel time is minimal. However, if no flow occurs on route $p$, i.e., $h_p^{rs}(k) = 0$, then the corresponding representative believed route travel time is at least as great as the minimal representative believed route travel time. This equilibrium conditions can be mathematically expressed as follows:

$$\bar{\bar{c}}_{p,\alpha}^{rs*}(k) \begin{cases} = \bar{\bar{\pi}}_\alpha^{rs}(k) & \text{if } h_p^{rs*}(k) > 0 \\ \geq \bar{\bar{\pi}}_\alpha^{rs}(k) & \text{if } h_p^{rs*}(k) = 0 \end{cases} \quad \forall r,s,p,k \tag{13.10}$$

### 13.2.2 Variational Inequality Problem

The following theorem is stated for the proposed FDUO route choice model.

***Theorem 13.1***: The FDUO route choice problem is equivalent to finding a vector $\mathbf{u}^* \in \Omega$ such that the following VIP holds:

$$\bar{\bar{\mathbf{c}}}_\alpha^*[\mathbf{u} - \mathbf{u}^*] \geq 0 \quad \forall \mathbf{u} \in \Omega^* \tag{13.11}$$

Or, alternatively, in an expanded form:

$$\sum_a \sum_t \bar{\bar{c}}_{a,\alpha}^*(t)[u_a(t) - u_a^*(t)] \geq 0 \quad \forall \mathbf{u} \in \Omega^* \tag{13.12}$$

where $\Omega^*$ is a subset of $\Omega$ with $\delta_{apk}^{rs,B}(t)$ being realized at equilibrium, i.e., $\left(\delta_{apk}^{rs,B}(t) = \delta_{apk}^{rs,B*}(t)\right)$, $\forall r,s,a,p,k,t$. The symbol $\Omega$ is delineated by the following flow conservation, flow propagation, nonnegativity, and definitional constraints.

Flow conservation constraint:

$$\sum_p h_p^{rs}(k) = \overline{q}^{rs}(k) \qquad \forall r,s,k \tag{13.13}$$

Flow propagation constraints:

$$u_{apk}^{rs}(t) = h_p^{rs}(k)\delta_{apk}^{rs,B}(t) \qquad \forall r,s,a,p,k,t \tag{13.14}$$

$$\sum_t \delta_{apk}^{rs,B}(t) = 1 \qquad \forall r,s,p,a \in p,k \tag{13.15}$$

$$\delta_{apk}^{rs,B}(t) = \{0,1\} \qquad \forall r,s,a,p,k,t \tag{13.16}$$

Nonnegativity constraint:

$$h_p^{rs}(k) \geq 0 \qquad \forall r,s,p,k \tag{13.17}$$

Definitional Constraints:

$$u_a(t) = \sum_{rs}\sum_p\sum_k h_p^{rs}(k)\delta_{apk}^{rs,B}(t) \qquad \forall a,t \tag{13.18}$$

$$\overline{\overline{c}}_{p,\alpha}^{rs}(k) = \sum_a\sum_t \overline{\overline{c}}_{a,\alpha}(t)\delta_{apk}^{rs,B}(t) \qquad \forall r,s,p,k \tag{13.19}$$

Note that the FDUO route choice model is essentially the same as the DUO route choice model but with the link travel time replaced by the representative believed link travel time, which is also used in flow propagation requirement

### 13.2.3 Equivalence Analysis

The following theorem verifies the equivalence between the equilibrium conditions (13.10) and the proposed VIP (13.12).

***Theorem 13.2***: Under a certain flow propagation relationship $\left(\delta_{apk}^{rs,B}(t) = \delta_{apk}^{rs,B*}(t)\right)$, the FDUO route choice conditions (13.10) imply variational inequality (13.12) and vice versa.

***Proof of necessity***: We need to prove that under a certain flow propagation relationship $\left(\delta_{apk}^{rs,B}(t) = \delta_{apk}^{rs,B*}(t)\right)$, the FDUO route choice conditions (13.10) imply VIP (13.12). We first rearrange the equilibrium conditions (13.12) as follows:

$$\left[\overline{\overline{c}}_{p,\alpha}^{rs*}(k) - \overline{\overline{\pi}}_{\alpha}^{rs}(k)\right]\left[h_p^{rs}(k) - h_p^{rs*}(k)\right] \geq 0 \tag{13.20}$$

By summing over $r,s,p,k,$ and then making a substitution of $\sum_p h_p^{rs}(k) = \sum_p h_p^{rs*}(k) = \overline{q}^{rs}(k)$, one obtains:

$$\sum_{rs}\sum_{k}\sum_{p}\overline{\overline{c}}_{p,\alpha}^{rs*}(k)\left[h_p^{rs}(k)-h_p^{rs*}(k)\right]\geq 0 \tag{13.21}$$

Equation (13.21), accompanied with the relevant constraints, is in fact a route-based FDUO route choice model. By applying equation (13.19) with $\left(\delta_{apk}^{rs,B}(t)=\delta_{apk}^{rs,B*}(t)\right)$ into equation (13.21), one obtains:

$$\sum_{rs}\sum_{k}\sum_{p}\left[\sum_{a}\sum_{t}\overline{\overline{c}}_{a,\alpha}^{*}(t)\delta_{apk}^{rs,B*}(t)\right]\left[h_p^{rs}(k)-h_p^{rs*}(k)\right]\geq 0 \tag{13.22}$$

By changing the order of summation, it follows:

$$\sum_{a}\sum_{t}\overline{\overline{c}}_{a,\alpha}^{*}(t)\sum_{rs}\sum_{k}\sum_{p}\delta_{apk}^{rs,B*}(t)\left[h_p^{rs}(k)-h_p^{rs*}(k)\right]\geq 0 \tag{13.23}$$

By using equation (13.18), we have:

$$\sum_{a}\sum_{t}\overline{\overline{c}}_{a,\alpha}^{*}(t)\left[u_a(t)-u_a^{*}(t)\right]\geq 0 \tag{13.24}$$

Equation (13.24) is identical to VIP (13.12).

***Proof of sufficiency***: We next prove that VIP (13.12) implies the fuzzy/dynamic user-optimal conditions (13.10). If, we suppose that the equilibrium flow pattern is characterized by $\left\{h_p^{rs*}(k)\right\}$, then for each O-D pair *rs*, with traffic demand greater than zero, we consider two situations that could arise at equilibrium:

(i) we can find any pair of two used routes $p_1^{rs}, p_2^{rs}$, with positive route flows $h_{p_1}^{rs*}(k)$, $h_{p_2}^{rs*}(k)$, respectively. To show that two used routes have exactly the same representative believed route travel time, we arbitrarily switch a small amount of flow $\Delta_1$ from route $p_1^{rs}$ to $p_2^{rs}$. The new feasible flow pattern $\left\{h_p^{rs}(k)\right\}$ would be the same as the one at equilibrium except:

$$h_{p_1}^{rs}(k)=h_{p_1}^{rs*}(k)-\Delta_1 \tag{13.25}$$

and

$$h_{p_2}^{rs}(k)=h_{p_2}^{rs*}(k)+\Delta_1 \tag{13.26}$$

where $0<\Delta_1\leq h_{p_1}^{rs*}(k)$. Applying the new feasible solution $\left\{h_p^{rs}(k)\right\}$ into VIP (13.12) will yield:

$$\overline{\overline{c}}_{p_1,\alpha}^{rs*}(k)\left[h_{p_1}^{rs}(k)-h_{p_1}^{rs*}(k)\right]+\overline{\overline{c}}_{p_2,\alpha}^{rs*}(k)\left[h_{p_2}^{rs}(k)-h_{p_2}^{rs*}(k)\right]\geq 0 \tag{13.27}$$

By using equation (13.25) and equation (13.26), one obtains:

$$\overline{\overline{c}}_{p_2,\alpha}^{rs*}(k)\geq\overline{\overline{c}}_{p_1,\alpha}^{rs*}(k) \tag{13.28}$$

Similarly, by switching a small amount of flow $\Delta_2$ with $0<\Delta_2\leq h_{p_2}^{rs*}(k)$ from route $p_2^{rs}$ to $p_1^{rs}$, we have:

$$\bar{\tilde{c}}_{p_1,\alpha}^{rs*}(k) \geq \bar{\tilde{c}}_{p_2,\alpha}^{rs*}(k) \tag{13.29}$$

Since equation (13.28) and equation (13.29) must hold simultaneously, it implies:

$$\bar{\tilde{c}}_{p_2,\alpha}^{rs*}(k) = \bar{\tilde{c}}_{p_1,\alpha}^{rs*}(k) \tag{13.30}$$

We can repeat this procedure to verify that, at a certain $\alpha$-cut level, for each O-D pair, all used routes with positive flow will have the same representative believed route travel time.

(ii) One route flow is positive and the other route flow is nil. We arbitrarily assume, without loss of generality, $h_{p_1}^{rs*}(k) > 0$, and $h_{p_2}^{rs*}(k) = 0$. We switch a small amount of flow $\Delta_1$ from route $p_1^{rs}$ to $p_2^{rs}$ with $0 < \Delta_1 \leq h_{p_1}^{rs*}(k)$. By the same argument shown in (i), we have $\bar{\tilde{c}}_{p_2,\alpha}^{rs*}(k) \geq \bar{\tilde{c}}_{p_1,\alpha}^{rs*}(k)$. We repeat this procedure to verify that, at a certain $\alpha$-cut level, for each O-D pair, all unused routes with zero flow will have the representative believed route travel time no lower than the minimal representative believed route travel time.

Since both (i) and (ii) must hold, it follows that VIP (13.12) implies equilibrium conditions (13.10). This completes the proof.

## 13.3 Nested Diagonalization Method

A nested diagonalization method is proposed to solve the FDUO route choice model. The nested diagonalization method is defined as an algorithm that consists of the diagonalization method in its solution procedure, and is extremely useful for problems with two types of link flow interactions. Since one type of link flow interaction can be relaxed to yield a subproblem that can be solved by the diagonalization method. Then, within the diagonalization solution procedure, the second type of link flow interaction can be further relaxed to result in a second level diagonalized subproblem, which can be solved by the FW method.

For the FDUO route choice model, the source of link flow interactions is twofold. One is from the believed actual link travel time $\tau_a^B(t)$, which is not known in advance, and hence, needs to be estimated and the other interaction is from the interference among inflows under the estimated believed actual link travel times. Once these two types of link flow interactions are temporarily fixed, a diagonalized subproblem is yielded, which can be reformulated as a convex optimization problem and solved by the FW method. The nested diagonalization method is formally stated as follows.

***The Nested Diagonalization Algorithm***

**Step 0: Initialization.**

Step 0.1: Set the minimum acceptance level, $\alpha$-cut.

Step 0.2: Let $m$=0. Set $\left(\tau_a^B\right)^0(t) = NINT\left[\overline{\tilde{c}}_{a_0}(t)\right], \forall a,t$.

Step 0.3: Let $n$=1. Find an initial feasible solution $\left\{u_a^1(t)\right\}$. Compute the associated link travel times $\left\{\overline{\tilde{c}}_{a,\alpha}^1(t)\right\}$.

**Step 1:** ***First Loop*** **Operation.**

Let $m$=$m$+1. Update the estimated believed actual link travel times by

$$\left(\tau_a^B\right)^m(t) = NINT\left[(1-\gamma)\left(\tau_a^B\right)^{m-1}(t) + \gamma\overline{\tilde{c}}_{a,\alpha}^n(t)\right] \quad \forall a,t. \tag{13.31}$$

where $0 < \gamma \le 1$.

Construct the corresponding feasible time-space network based on the estimated believed actual link travel times.

**Step 2:** ***Second Loop*** **Operation.**

Step 2.1: Let $n$=1. Compute and reset the initial feasible solution $\left\{u_a^n(t)\right\}$, based on the time-space network constructed by the estimated believed actual link travel times $\left\{\left(\tau_a^B\right)^m(t)\right\}$.

Step 2.2: Fix the inflows for all time-space links other than that on the subject time-space link at the current level, yielding the following optimization problem.

$$\min\ z(\mathbf{u}) = \sum_a \sum_t \int_0^{u_a^{n+1}(t)} \overline{\tilde{c}}_{a,\alpha}\left(u_a^n(1), u_a^n(2), \cdots, u_a^n(t-1), \omega\right) d\omega \tag{13.32}$$

Flow conservation constraint:

$$\sum_p h_p^{rs}(k) = \overline{q}^{rs}(k) \quad \forall r,s,k \tag{13.33}$$

Nonnegativity constraint:

$$h_p^{rs}(k) \ge 0 \quad \forall r,s,p,k \tag{13.34}$$

Definitional constraints:

$$u_{apk}^{rs}(t) = h_p^{rs}(k)\overline{\delta}_{apk}^{rs,B}(t) \quad \forall r,s,a,p,k,t \tag{13.35}$$

$$\overline{\delta}_{apk}^{rs,B}(t) = \{0,1\} \quad \forall r,s,a,p,k,t \tag{13.36}$$

$$u_a(t) = \sum_{rs}\sum_p\sum_k h_p^{rs}(k)\overline{\delta}_{apk}^{rs,B}(t) \quad \forall a,t \tag{13.37}$$

$$\overline{\tilde{c}}_{p,\alpha}^{rs}(k) = \sum_a\sum_t \overline{\tilde{c}}_{a,\alpha}(t)\overline{\delta}_{apk}^{rs,B}(t) \quad \forall r,s,p,k \tag{13.38}$$

**Step 3:** ***Third Loop*** **Operation.**

Solve for the solution, $u_a^{n+1}(t), \forall a,t$, in the optimization problem (13.32)~(13.38) using the FW method. Compute the resulting link travel times $\overline{\tilde{c}}_{a,\alpha}^{n+1}(t), \forall a,t$.

**Step 4: Convergence Check for the *Second Loop* Operation.**

If $u_a^{n+1}(t) \approx u_a^n(t), \forall a,t$, go to Step 5; otherwise, set $n=n+1$, and go to Step 2.2.

**Step 5: Convergence Check for the *First Loop* Operation.**

If $\left(\tau_a^B\right)^m(t) \approx \overline{\tilde{c}}_{a,\alpha}^{n+1}(t), \forall a,t$, stop; the current solution is optimal. Otherwise, set $n=n+1$, and go to Step 1.

## 13.4 Numerical Example

### 13.4.1 Input Data

A simple network shown in Figure 13.4 is used for testing. The test network consists of 6 links and 5 nodes, in which nodes 1 and 3 are origins, node 5 is a destination, and nodes 2 and 4 are intermediate nodes.

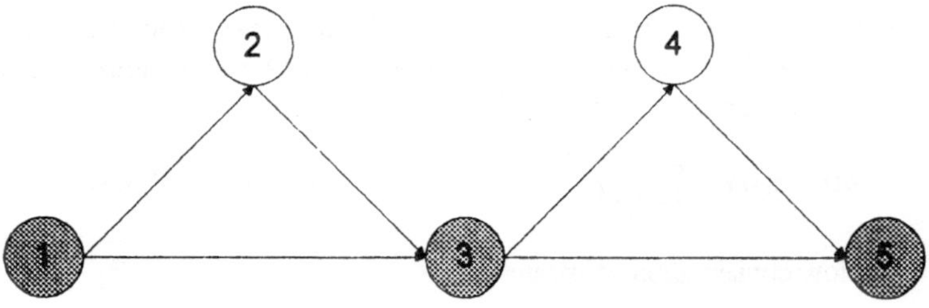


Figure 13.4: Test Network

The assumed origin-destination (O-D) demands are shown in Table 13.1:

Table 13.1: Time-Dependent O-D Demands

| O-D Pair | Time Interval $k$=1 | $k$=2 | $k$=3 | $k$=4 |
|---|---|---|---|---|
| 1-5 | 15 | 20 | 0 | 0 |
| 3-5 | 0 | 0 | 15 | 20 |

The adopted dynamic travel time function is arbitrarily constructed as follows:

$$\tilde{c}_a(t) = \tilde{1} \;\tilde{\oplus}\; 0.\tilde{0}1 \tilde{\otimes} \left(u_a(t)\right)^2 \;\tilde{\oplus}\; 0.\tilde{0}1 \tilde{\otimes} \left(x_a(t)\right)^2 \quad \forall a,t \tag{13.39}$$

To explore how the traffic information influences the flow pattern, we analyze the following three cases. The supports and spreads of coefficients used in the corresponding fuzzy/dynamic link travel time functions are summarized in Table

13.2:

1. Case I: right-skewed possibility distribution is assumed for links 1→2, 2→3, 3→4 and 4→5, and symmetric possibility distribution for links 1→3 and 3→5.
2. Case II: left-skewed possibility distribution is assumed for links 1→2, 2→3, 3→4 and 4→5, and symmetric possibility distribution for links 1→3 and 3→5.
3. Case III: all links are assumed to be symmetrically distributed.

Table 13.2: Possibility Distribution of Link Travel Time Coefficients

| Case | Link | Type of Imperfect Information | Possibility Distribution of Link Travel Time Coefficients | |
|---|---|---|---|---|
| | | | $\tilde{1}$ | $\widetilde{0.01}$ |
| I | 1→2<br>2→3<br>3→4<br>4→5 | Right-Skewed | (0.5,1.0,2.0) | (0.005,0.01,0.02) |
| | 1→3<br>3→5 | Symmetric | (0.5,1.0,1.5) | (0.005,0.01,0.015) |
| II | 1→2<br>2→3<br>3→4<br>4→5 | Left-Skewed | (0.2,1.0,1.5) | (0.002,0.01,0.015) |
| | 1→3<br>3→5 | Symmetric | (0.5,1.0,1.5) | (0.005,0.01,0.015) |
| III | All | Symmetric | (0.5,1.0,1.5) | (0.005,0.01,0.015) |

For the numerical examples demonstrated below, we chose the minimal acceptable level $\alpha$=0. A computer program coded in Borland $C^{++}$ was used to solve the FDUO route choice model for the input data.

### 13.4.2 Test Results for Case I

The obtained results are summarized in Table 13.3.

Table 13.3: Results for Case I

| Link | Entering Time Interval | Inflow | Exit Flow | Number of Vehicles | Rep. Believed Link Travel Time | Exiting Time Interval |
|---|---|---|---|---|---|---|
| 1→2 | 1 | 3.20 | 0.00 | 0.00 | 1.20 | 2 |
| | 2 | 8.21 | 3.20 | 3.20 | 1.95 | 4 |
| | 3 | 0.00 | 0.00 | 8.21 | 1.84 | - |
| | 4 | 0.00 | 8.21 | 8.21 | 1.84 | 6 |
| 1→3 | 1 | 11.80 | 0.00 | 0.00 | 2.39 | 3 |
| | 2 | 11.79 | 0.00 | 11.80 | 3.78 | 6 |
| | 3 | 0.00 | 11.80 | 23.59 | 6.57 | - |
| | 4~5 | 0.00 | 0.00 | 11.79 | 2.39 | - |
| | 6 | 0.00 | 11.79 | 11.79 | 2.39 | - |
| 2→3 | 2 | 3.20 | 0.00 | 0.00 | 1.20 | 3 |
| | 3 | 0.00 | 3.20 | 3.20 | 1.20 | - |
| | 4 | 8.21 | 0.00 | 0.00 | 1.84 | 6 |
| | 5 | 0.00 | 0.00 | 8.21 | 1.84 | - |
| | 6 | 0.00 | 8.21 | 8.21 | 1.84 | - |
| 3→4 | 3 | 10.70 | 0.00 | 0.00 | 2.36 | 5 |
| | 4 | 8.03 | 0.00 | 10.70 | 3.08 | 7 |
| | 5 | 0.00 | 10.70 | 18.73 | 5.00 | - |
| | 6 | 18.77 | 0.00 | 8.03 | 5.74 | 12 |
| | 7 | 0.00 | 8.03 | 26.80 | 9.10 | 16 |
| | 8~11 | 0.00 | 0.00 | 18.77 | 5.02 | - |
| | 12 | 0.00 | 18.77 | 18.77 | 5.02 | 17 |
| 3→5 | 3 | 19.30 | 0.00 | 0.00 | 4.72 | 8 |
| | 4 | 11.97 | 0.00 | 19.30 | 6.16 | 10 |
| | 5 | 0.00 | 0.00 | 31.27 | 10.78 | - |
| | 6 | 1.23 | 0.00 | 31.27 | 10.79 | 17 |
| | 7 | 0.00 | 0.00 | 32.50 | 11.56 | - |
| | 8 | 0.00 | 19.30 | 32.50 | 11.56 | - |
| | 9 | 0.00 | 0.00 | 13.20 | 2.74 | - |
| | 10 | 0.00 | 11.97 | 13.20 | 2.74 | - |
| | 11~16 | 0.00 | 0.00 | 1.23 | 1.02 | - |
| | 17 | 0.00 | 1.23 | 1.23 | 1.02 | - |
| 4→5 | 5 | 10.70 | 0.00 | 0.00 | 2.36 | 7 |
| | 6 | 0.00 | 0.00 | 10.70 | 2.36 | - |
| | 7 | 8.03 | 10.70 | 10.70 | 3.08 | 10 |
| | 8~9 | 0.00 | 0.00 | 8.03 | 1.80 | - |
| | 10 | 0.00 | 8.03 | 8.03 | 1.80 | - |
| | 12 | 18.77 | 0.00 | 0.00 | 5.02 | 17 |
| | 13~16 | 0.00 | 0.00 | 18.77 | 5.02 | - |
| | 17 | 0.00 | 18.77 | 18.77 | 5.02 | - |

The rationale of the proposed model and associated solution algorithm can be verified by checking if the resulting representative believed route travel times satisfy the FDUO equilibrium conditions. If we consider route 1→3→5 departing origin 1 during interval 1, then the corresponding representative believed route travel time can be obtained by summing up the representative believed link travel time on link 1→3 during interval 1 with the representative believed link travel time on link 3→5 during interval $1+\bar{\bar{c}}_{1\to3}(1)$ as follows:

$$\begin{aligned}\bar{\bar{c}}_{1\to3\to5}(1) &= \bar{\bar{c}}_{1\to3}(1) + \bar{\bar{c}}_{3\to5}\left(1+\bar{\bar{c}}_{1\to3}(1)\right) \\ &= 2.39 + \bar{\bar{c}}_{3\to5}(3.39) \approx 2.39 + \bar{\bar{c}}_{3\to5}(3) = 7.12\end{aligned} \qquad (13.40)$$

The remaining used representative believed route travel times are also computed and summarized in Table 13.4. It is observed that the trips departing the same origin during the same interval have approximately the same representative believed route travel time.

Table 13.4: Representative Believed Route Travel Times for Case I

| Time Interval | Route* | | | | | |
|---|---|---|---|---|---|---|
| | 3→5 | 3→4→5 | 1→3→5 | 1→2→3 →5 | 1→3→4 →5 | 1→2→3 →4→5 |
| $k$=1 | NA | NA | 7.12 | 7.12 | 7.12 | 7.12 |
| $k$=2 | NA | NA | 14.58 | 14.58 | 14.52 | 14.52 |
| $k$=3 | 4.72 | 4.73 | NA | NA | NA | NA |
| $k$=4 | 6.16 | 6.16 | NA | NA | NA | NA |

NA: means the route not being used, therefore the representative believed link travel time is not available.

*Based on the same believed actual travel times, $\{\tau_a^B(t)\}$, obtained from the computed solution.

### 13.4.3 Test Results for Case II

The test results for case II are summarized in Tables 13.5 and 13.6.

Table 13.5: Results for Case II

| Link | Entering Time Interval | Inflow | Exit Flow | Number of Vehicles | Rep. Believed Link Travel Time | Exiting Time Interval |
|---|---|---|---|---|---|---|
| 1→2 | 1 | 4.37 | 0.00 | 0.00 | 1.13 | 2 |
| | 2 | 8.74 | 4.37 | 4.37 | 1.86 | 4 |
| | 3 | 0.00 | 0.00 | 8.74 | 1.67 | - |
| | 4 | 0.00 | 8.74 | 8.74 | 1.67 | - |
| 1→3 | 1 | 10.63 | 0.00 | 0.00 | 2.33 | 3 |
| | 2 | 11.27 | 0.00 | 10.63 | 3.60 | 6 |
| | 3 | 0.00 | 10.63 | 21.89 | 5.99 | - |
| | 4~5 | 0.00 | 0.00 | 11.27 | 2.47 | - |
| | 6 | 0.00 | 11.27 | 11.27 | 2.47 | 8 |
| 2→3 | 2 | 4.37 | 0.00 | 0.00 | 1.20 | 3 |
| | 3 | 0.00 | 4.37 | 4.37 | 1.20 | - |
| | 4 | 8.74 | 0.00 | 0.00 | 1.74 | 6 |
| | 5 | 0.00 | 0.00 | 8.74 | 1.74 | - |
| | 6 | 0.00 | 8.74 | 8.74 | 1.74 | - |
| 3→4 | 3 | 11.52 | 0.00 | 0.00 | 2.21 | 5 |
| | 4 | 8.41 | 0.00 | 11.52 | 2.88 | 7 |
| | 5 | 0.00 | 11.52 | 19.93 | 4.72 | - |
| | 6 | 19.54 | 0.00 | 8.41 | 5.25 | 11 |
| | 7 | 0.00 | 8.41 | 27.95 | 8.37 | - |
| | 8~10 | 0.00 | 0.00 | 19.54 | 4.58 | - |
| | 11 | 0.00 | 19.54 | 19.54 | 4.58 | - |
| 3→5 | 3 | 18.48 | 0.00 | 0.00 | 4.42 | 7 |
| | 4 | 11.59 | 0.00 | 18.48 | 5.76 | 10 |
| | 5 | 0.00 | 0.00 | 30.07 | 10.04 | - |
| | 6 | 0.46 | 0.00 | 30.07 | 10.04 | 16 |
| | 7 | 0.00 | 18.48 | 30.53 | 10.32 | - |
| | 8~9 | 0.00 | 0.00 | 12.05 | 2.45 | - |
| | 10 | 0.00 | 11.59 | 12.05 | 2.45 | - |
| | 11 | 0.00 | 0.00 | 0.46 | 1.00 | 12 |
| | 12~15 | 0.00 | 0.00 | 0.46 | 1.00 | - |
| | 16 | 0.00 | 0.46 | 0.46 | 1.00 | - |
| 4→5 | 5 | 11.52 | 0.00 | 0.00 | 2.21 | 7 |
| | 6 | 0.00 | 0.00 | 11.52 | 2.21 | - |
| | 7 | 8.41 | 11.52 | 11.52 | 2.88 | 10 |
| | 8~9 | 0.00 | 0.00 | 8.41 | 1.62 | - |
| | 10 | 0.00 | 8.41 | 8.41 | 1.62 | - |
| | 11 | 19.54 | 0.00 | 0.00 | 4.58 | 16 |
| | 12~15 | 0.00 | 0.00 | 19.54 | 4.58 | - |
| | 16 | 0.00 | 19.54 | 19.54 | 4.58 | - |

Table 13.6: Representative Believed Route Travel Times for Case II

| Time Interval | Route* | | | | | |
|---|---|---|---|---|---|---|
| | 3→5 | 3→4→5 | 1→3→5 | 1→2→3 →5 | 1→3→4 →5 | 1→2→3 →4→5 |
| $k$=1 | NA | NA | 6.75 | 6.75 | 6.75 | 6.75 |
| $k$=2 | NA | NA | 13.58 | 13.58 | 13.37 | 13.37 |
| $k$=3 | 4.42 | 4.42 | NA | NA | NA | NA |
| $k$=4 | 5.76 | 5.77 | NA | NA | NA | NA |

NA: means the route not being used, therefore the representative believed link travel time is not available.

*Based on the same believed actual travel times, $\{\tau_a^B(t)\}$, obtained from the computed solution.

### 13.4.4 Test Results for Case III

The test results for case II are summarized in Tables 13.7 and 13.8.

Table 13.7: Results for Case III

| Link | Entering Time Interval | Inflow | Exit Flow | Number of Vehicles | Rep. Believed Link Travel Time | Exiting Time Interval |
|---|---|---|---|---|---|---|
| 1→2 | 1 | 3.71 | 0.00 | 0.00 | 1.14 | 2 |
| | 2 | 8.52 | 3.71 | 3.71 | 1.86 | 4 |
| | 3 | 0.00 | 0.00 | 8.52 | 1.73 | - |
| | 4 | 0.00 | 8.52 | 8.52 | 1.73 | - |
| 1→3 | 1 | 11.29 | 0.00 | 0.00 | 2.27 | 3 |
| | 2 | 11.48 | 0.00 | 11.29 | 3.59 | 6 |
| | 3 | 0.00 | 11.29 | 22.76 | 6.18 | - |
| | 4~5 | 0.00 | 0.00 | 11.48 | 2.32 | - |
| | 6 | 0.00 | 11.48 | 11.48 | 2.32 | - |
| 2→3 | 2 | 3.71 | 0.00 | 0.00 | 1.14 | 3 |
| | 3 | 0.00 | 3.71 | 3.71 | 1.14 | - |
| | 4 | 8.52 | 0.00 | 0.00 | 1.73 | 6 |
| | 5 | 0.00 | 0.00 | 8.52 | 1.73 | - |
| | 6 | 0.00 | 8.52 | 8.52 | 1.73 | - |

Table 13.7: Results for Case III (continued)

| Link | Entering Time Interval | Inflow | Exit Flow | Number of Vehicles | Rep. Believed Link Travel Time | Exiting Time Interval |
|---|---|---|---|---|---|---|
| 3→4 | 3 | 11.23 | 0.00 | 0.00 | 2.26 | 5 |
| | 4 | 8.28 | 0.00 | 11.23 | 2.95 | 7 |
| | 5 | 0.00 | 11.23 | 19.51 | 4.81 | - |
| | 6 | 19.51 | 0.00 | 8.28 | 5.49 | 11 |
| | 7 | 0.00 | 8.28 | 27.79 | 8.72 | - |
| | 8~10 | 0.00 | 0.00 | 19.51 | 4.80 | - |
| | 11 | 0.00 | 19.51 | 19.51 | 4.80 | - |
| 3→5 | 3 | 18.77 | 0.00 | 0.00 | 4.52 | 8 |
| | 4 | 11.72 | 0.00 | 18.77 | 5.90 | 10 |
| | 5 | 0.00 | 0.00 | 30.49 | 10.29 | - |
| | 6 | 0.49 | 0.00 | 30.49 | 10.30 | 16 |
| | 7 | 0.00 | 0.00 | 30.98 | 10.60 | - |
| | 8 | 0.00 | 18.77 | 30.98 | 10.60 | - |
| | 9 | 0.00 | 0.00 | 12.22 | 2.49 | - |
| | 10 | 0.00 | 11.72 | 12.22 | 2.49 | - |
| | 11~15 | 0.00 | 0.00 | 0.49 | 1.00 | - |
| | 16 | 0.00 | 0.49 | 0.49 | 1.00 | - |
| 4→5 | 5 | 11.23 | 0.00 | 0.00 | 2.26 | 7 |
| | 6 | 0.00 | 0.00 | 11.23 | 2.26 | - |
| | 7 | 8.28 | 11.23 | 11.23 | 2.95 | 10 |
| | 8~9 | 0.00 | 0.00 | 8.28 | 1.69 | - |
| | 10 | 0.00 | 8.28 | 8.28 | 1.69 | - |
| | 11 | 19.51 | 0.00 | 0.00 | 4.80 | 16 |
| | 12~15 | 0.00 | 0.00 | 19.51 | 4.80 | - |
| | 16 | 0.00 | 19.51 | 19.51 | 4.80 | - |

Table 13.8: Representative Believed Route Travel Times for Case III

| Time Interval | Route* | | | | | |
|---|---|---|---|---|---|---|
| | 3→5 | 3→4→5 | 1→3→5 | 1→2→3 →5 | 1→3→4 →5 | 1→2→3 →4→5 |
| $k$=1 | NA | NA | 6.80 | 6.80 | 6.80 | 6.80 |
| $k$=2 | NA | NA | 13.89 | 13.89 | 13.89 | 13.89 |
| $k$=3 | 4.52 | 4.52 | NA | NA | NA | NA |
| $k$=4 | 5.89 | 5.90 | NA | NA | NA | NA |

NA: means the route not being used, therefore the representative believed link travel time is not available.

* Based on the same believed actual travel times, $\{\tau_a^B(t)\}$, obtained from the computed solution.

By comparing all three cases, the links characterized by a symmetric possibility distribution attract more traffic flow when compared with the right-skewed possibility distribution. As shown in column 3 of Tables 13.3 and 13.7, the former has lower flows on links 1→2, 2→3, 3→4 and 4→5, and higher flows on links 1→3 and 3→5. Similarly, the links characterized by a left-skewed possibility distribution are intended to attract more traffic flows when compared with the symmetric possibility distribution. As shown in column 3 of Tables 13.5 and 13.7, the former has lower flows on links 1→2, 2→3, 3→4 and 4→5, and higher flows on links 1→3 and 3→5.

## 13.5 Fuzzy/Static Counterpart Approximation

Fuzzy/static counterpart approximation can be obtained by removing the flow propagation constraint and thus, mathematically setting the main link travel times equal to zero. By using the input data given in Section 13.4 and exercising the assertion, we can obtain the static/fuzzy approximation as shown in Table 13.9.

Table 13.9: Fuzzy/Static Approximation

| Link | Entering Time Interval | Inflow | Exit Flow | Number of Vehicles | Link Travel Time | Exiting Time Interval |
|---|---|---|---|---|---|---|
| 1→2 | 1 | 13.02 | 13.02 | 0.00 | 2.92 | - |
| 1→3 | 1 | 21.98 | 21.98 | 0.00 | 5.83 | - |
| 2→3 | 1 | 13.02 | 13.02 | 0.00 | 2.92 | - |
| 3→4 | 1 | 27.75 | 27.75 | 0.00 | 9.42 | - |
| 3→5 | 1 | 42.25 | 42.25 | 0.00 | 18.85 | - |
| 4→5 | 1 | 27.75 | 27.75 | 0.00 | 9.42 | - |

The corresponding representative route travel time can be computed as follows:

Table 13.10: Representative Believed Route Travel Times for the Fuzzy/Static Model

| Time Interval | Route | | | | | |
|---|---|---|---|---|---|---|
| | 3→5 | 3→4→5 | 1→3→5 | 1→2→3 5 | 1→3→4 5 | 1→2→3 4→5 |
| $k$=1 | 18.85 | 18.85 | 24.68 | 24.69 | 24.68 | 24.69 |

## 13.6 Notes

Special cases of the FDUO model can be reduced to be static and/or deterministic user-optimal models. The FDUO model employs subjective possibility distribution;

at each $\alpha$-cut level, representative believed link travel times are used in the traffic assignment. This treatment of the representative believed link travel time function, though not general in the fuzzy sense, is extremely convenient and simple in computation. It is also worth noting that at each $\alpha$-cut level (the events with the possibility $\geq \alpha$), the skewness of the possibility distributions are still kept; therefore, different solutions may result at different $\alpha$-cut levels of the possibility distribution.

A permissive alternative approach to formulating the FDUO route choice problem is the fuzzy variational inequality, which has been applied to the fuzzy/static user-optimal problem (Wang and Liao, 1994). However, a conversion of the fuzzy variational inequality into the equivalent multiple objective decision problems is required for attaining the optimal solution, which is complicate. Moreover, interval analysis is inevitably involved in the solution procedure, which makes the actual applications of this approach difficult. It is also noted that other than the aforementioned analytical models, the simulation-based models using reactive traffic information may be worth to explore, as route switching decisions can be made at each intermediate node according to the prevailing traffic condition.

As a final remark, an alternative shape of possibility distribution for the fuzzy/dynamic link travel time function $\tilde{c}_a(t)$ is trapezoidal. The trapezoidal possibility distribution (see Figure 13.5) is usually symbolized as $\tilde{c}_a(t) = \left(\tilde{c}_a^L(t), \tilde{c}_a^{M_1}(t), \tilde{c}_a^{M_2}(t), \tilde{c}_a^U(t)\right)$, where $\tilde{c}_a^L(t)$ donates the lower bound, $\tilde{c}_a^U(t)$ donates the upper bound, $\tilde{c}_a^{M_1}(t)$ donates the left main value, and $\tilde{c}_a^{M_2}(t)$ donates the right main value. The left spread can be calculated by $\tilde{c}_a^{M_1}(t) - \tilde{c}_a^L(t)$ and the right spread by $\tilde{c}_a^U(t) - \tilde{c}_a^{M_2}(t)$.

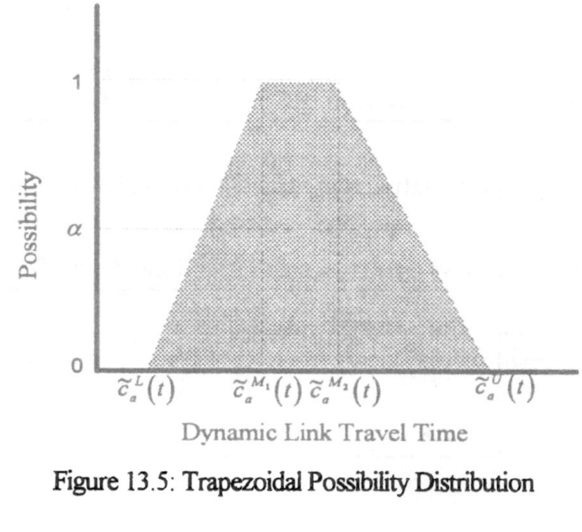


Figure 13.5: Trapezoidal Possibility Distribution

The most likely solution is found by solving an auxiliary nonlinear programming problem by using a weighted average of lower, main and upper values. In detail, at the level of $\alpha$-cut (a minimal acceptable possibility), the most likely link travel time can be represented by:

$$\bar{\tilde{c}}_{a,\alpha}(t)=\left(\frac{\tilde{c}_{a,\alpha}^{L}(t)+2\tilde{c}_{a}^{M_1}(t)+2\tilde{c}_{a}^{M_2}(t)+\tilde{c}_{a,\alpha}^{U}(t)}{6}\right) \tag{13.41}$$

Thus, with the above representative believed link travel time, the model and solution algorithm presented in this chapter is equally applicable.

# Chapter 14

# Future Research and Applications

In this book, we have presented a fresh look on modeling discrete-time dynamic transportation networks mainly using the variational inequality approach. However, in addition to the most difficult accounting problems (non-convergence or oscillation behavior due to discretization of the temporal dimension), and the nonconvex property of the feasible solution, many other modeling issues still need more clarification including: dynamic link travel time functions, link capacity constraints, first-in-first-out requirements, computational efficiency, multiple user classes, nonconvex feasible regions, integration of analytical- and simulation-based dynamic models and others. Nevertheless, in their present form, the proposed dynamic travel choice models are still applicable to various engineering and economic problems, such as transportation planning, route guidance, economic networks and logistics. In this chapter, we address the remaining research issues in Section 14.1, and then discuss the possible applications in Section 14.2 before adding some final notes in Section 14.3.

## 14.1 Research Directions

### 14.1.1 Dynamic Link Travel Time Functions

Intuitively, the dynamic link travel time may be denoted as a function of three time-dependent link-related variables, i.e., inflow, number of vehicles and exit flow. However, by using the estimated actual link travel times, the dynamic link travel time function can be simplified to function of time-dependent inflows. This observation was first presented by Chen et al. (1996) for the dynamic user-optimal route choice problem and then extended to many other dynamic travel choice models (Chen et al., 1997a, 1997b).

For the dynamic link travel time function to be practical, the function form must be first determined (probably in terms of generalized cost or utility), and then the associated coefficients calibrated from the time-dependent field data, or by using some sophisticated simulation model. The work involved should not be underestimated, and extreme caution needs to be taken in applying the best experimental design.

In some circumstances, one may prefer to embed a link capacity limitation implicitly in the dynamic link travel time function. To this end, a possible direction is to apply the entropy formulation (Chabini and Florian, 1995) as follows:

$$c_a(t) = b_0 + \cdots + b_n\left(CAP_a(t) - u_a(t)\right)\ln\left(\frac{CAP_a(t) - u_a(t)}{CAP_a(t) - u_a^0(t)}\right) \qquad (14.1)$$

It is observed that when the amount of link inflow reaches the link capacity, the corresponding link travel time increases abruptly, which is impossible in many mathematical problems. Hence, the link capacity constraint can be automatically satisfied.

### 14.1.2 Link Capacity Constraints

The link capacity may also be imposed as an explicit constraint in dynamic travel choice models.

$$u_a(t) \le CAP_a(t) \quad \forall a, t \qquad (14.2)$$

This additional side constraint causes a nice property of the original network structure to no longer hold. Moreover, the corresponding equilibrium conditions must be changed accordingly. We take the DUO route choice model as an example to illustrate the resulting dynamic user-optimal conditions and the corresponding nested diagonalization methods hereafter.

When O-D demands are fixed and time-dependent, the dynamic user-optimal conditions state for each O-D pair that the *capacitated* route travel times (including route travel times and queuing delays) experienced by travelers departing during the same interval are equal and minimal; or no traveler would be better off by unilaterally changing his/her route. In contrast, the *capacitated* route travel time of any unused route for each O-D pair is greater than or equal to the minimal *capacitated* route travel time. In other words, at equilibrium, if the flow departing from origin $r$ during interval $k$ over route $p$ toward destination $s$ is positive, i.e., $h_p^{rs*}(k) > 0$, then the corresponding *capacitated* route travel time is minimal. To the contrary, if no flow occurs on route $p$, i.e., $h_p^{rs*}(k) = 0$, then the corresponding *capacitated* route travel time is at least as great as the minimal *capacitated* route travel time. This equilibrium conditions can be mathematically expressed as follows.

$$\breve{c}_p^{rs*}(k)\begin{cases} = \breve{\pi}^{rs}(k) & \text{if } h_p^{rs*}(k) > 0 \\ \geq \breve{\pi}^{rs}(k) & \text{if } h_p^{rs*}(k) = 0 \end{cases} \tag{14.3}$$

where

$$\breve{c}_p^{rs*}(k) = \sum_a \sum_t \left(c_a^*(t) + \beta_a^*(t)\right)\delta_{apk}^{rs*}(t) = \sum_a \sum_t \breve{c}_a^*(t)\delta_{apk}^{rs*}(t) \tag{14.4}$$

$\beta_a^*(t)$ denotes the optimal value of the Lagrangian multiplier for the link capacity constraint (14.2). If we interpret this Lagrangian multiplier as the link queuing delay (or link toll), then the capacitated link travel time is comprised of the link travel time and the link queuing delay.

The corresponding nested diagonalization method is essentially the same as before, except a modification is required for the *third loop* as follows.

***The Nested Diagonalization Method***

**Step 0: Initialization.**

Step 0.1: Let $m$=0. Set $\tau_a^0(t) = NINT\left[c_{a_0}(t)\right], \forall a, t$.

Step 0.2: Let $n$=1. Find an initial feasible solution $\{u_a^1(t)\}$. Compute the associated link travel times $\{c_a^1(t)\}$.

**Step 1: *First Loop* Operation.**

Let $m$=$m$+1. Update the estimated actual link travel times by

$$\tau_a^m(t) = NINT\left((1-\gamma)\tau_a^{m-1}(t) + \gamma c_a^n(t)\right) \quad \forall a, t. \tag{14.5}$$

where $0 < \gamma \leq 1$.

Construct the corresponding feasible time-space network based on the estimated actual link travel times.

**Step 2: *Second Loop* Operation.**

Step 2.1: Let $n$=1. Compute and reset the initial feasible solution $\{u_a^n(t)\}$, based on the time-space network constructed by the estimated actual link travel times $\{\tau_a^m(t)\}$.

Step 2.2: Fix the inflows for each physical link other than on the subject time-space link at the current level, yielding the following optimization problem.

$$\min\ z(\mathbf{u}) = \sum_a \sum_t \int_0^{u_a^{n+1}(t)} c_a\left(u_a^n(1), u_a^n(2), \cdots, u_a^n(t-1), \omega\right) d\omega \tag{14.6}$$

Flow conservation constraint:

$$\sum_{p} h_p^{rs}(k) = \bar{q}^{rs}(k) \qquad \forall r,s,k \tag{14.7}$$

Nonnegativity constraint:

$$h_p^{rs}(k) \geq 0 \qquad \forall r,s,p,k \tag{14.8}$$

Side constraint:

$$u_a(t) \leq CAP_a(t) \qquad \forall a,t \tag{14.9}$$

Definitional constraints:

$$u_{apk}^{rs}(t) = h_p^{rs}(k)\bar{\delta}_{apk}^{rs}(t) \qquad \forall r,s,a,p,k,t \tag{14.10}$$

$$\bar{\delta}_{apk}^{rs}(t) = \{0,1\} \qquad \forall r,s,a,p,k,t \tag{14.11}$$

$$u_a(t) = \sum_{rs}\sum_{p}\sum_{k} h_p^{rs}(k)\bar{\delta}_{apk}^{rs}(t) \qquad \forall a,t \tag{14.12}$$

$$c_p^{rs}(k) = \sum_{a}\sum_{t} c_a(t)\bar{\delta}_{apk}^{rs}(t) \qquad \forall r,s,p,k \tag{14.13}$$

**Step 3: *Third Loop* Operation.**

Solve for the solution, $\{u_a^{n+1}(t)\}$, in the optimization problem (14.6)~(14.13) by an appropriate method. Compute the resulting link travel times $\{c_a^{n+1}(t)\}$.

**Step 4: Convergence Check for the *Second Loop* Operation.**

If $u_a^{n+1}(t) \approx u_a^n(t), \forall a,t$, go to Step 5; otherwise, set $n=n+1$, and go to Step 2.2.

**Step 5: Convergence Check for the *First Loop* Operation.**

If $\tau_a^m(t) \approx c_a^{n+1}(t), \forall a,t$, stop; the current solution is optimal. Otherwise, set $n=n+1$, and go to Step 1.

Note that the optimization problem (14.6)~(14.13) cannot be solved by the FW method as before because the inclusion of the link capacity constraint (14.9) makes the nice property of network structure vanish. In detail, the linearized subproblem of problem (14.6)~(14.13) becomes a linear multi-commodity flow problem, rather than a shortest route problem, which is prohibitively expensive to solve repeatedly. Based on the recognition of this fact, two categories of solution methods emerge (Larsson et al., 1995). In the first category, attempts are made to use the shortest route subproblems to generate search directions, while in the second approach, the capacitated problem is converted into a sequence of incapacitated problems through a penalization/dualization of the link capacity constraint (14.9). An application of the augmented Lagrangian dual method which embeds the disaggregate simplicial decomposition (DSD) method to the DUO route choice problem was presented by Chen and Wang (1998).

### 14.1.3 First-In-First-Out Conditions

The first-in-first-out (FIFO) conditions require that travelers cannot arrive earlier by leaving later. To preserve the FIFO conditions, two approaches may be worth exploring. The first approach is to develop a suitable dynamic travel time function that implicitly satisfies the FIFO requirement. Toward this end, Ran and Boyce (1994) used six link-based variables to represent the dynamics of traffic on a link better, and Daganzo (1993) applied a cell transmission model to describe the sequential movement behavior of traffic. However, there is insufficient evidence that their results can be readily applied to dynamic travel choice models.

The second approach is to impose the FIFO as a constraint on the DUO travel choice models. Again, the inclusion of the side constraint of the FIFO into any DUO travel choice model destroys the nice property of the network structure. In addition, the estimation of actual travel times using the link travel time function within the nested diagonalization method lose their applicability since the FIFO conditions are hard to justify within the solution procedure. To the best of the author's knowledge, no efficient solution algorithm is available to solve such nonconvex mathematical problems. Even if one existed, we are not sure whether a local solution exists. Hence, this difficulty indeed requires further research.

### 14.1.4 Computational Efficiency

A deep concern raised by transportation practitioners for the implementation of the dynamic travel choice models is their computational efficiency in large urban networks. Finding the shortest route is often the most computationally-intensive component of most user-optimal solution procedures. The other components (loading the shortest routes, performing the line search, and checking for the convergence) do require smaller proportions of the total CPU time (Sheffi, 1985). Consequently, the computational effort associated with the application of the nested diagonalization method is proportional to the product of the number of the *first loop* iterations $m$, the number of the *second loop* iterations $n$, the number of the FW iterations $l$, the number of time intervals $K$, the number of origins $R$ (which determines the number of minimum-route trees to be calculated at each iteration), the number of links $A$ and the number of nodes $O$ (which determines the effort needed to calculate each tree). In other words,

$$\text{(Arithmetic Operations)} \approx m \times n \times l \times K \times R \times A \times O \qquad (14.14)$$

To reduce the total number of searches in finding the minimum shortest routes within the nested diagonalization method proposed in this book, and hence improve the computational efficiency, two strategies may be appropriate:

1. Relaxing the convergence criteria of the *second loop* and *third loop* operations that are associated with the early iterations of the *first loop* operation. In an extreme case, only one iteration is performed for the *second loop* and *third loop* operations, yielding a *streamlined* variation of the nested diagonalization method.
2. Employing a good initial solution, an effective search direction, a suitable move step and/or a combination of the above three strategies. For the initial solution, the incremental assignment technique may perform better than the all-or-nothing assignment. To find a effective search direction, Fukushima (1984) used the previous search directions to generate a weighted new search direction to improve the speed of the convergence for the FW method. LeBlanc et al (1985) developed a paralell tangent (PARTAN) technique which applied the current solution and the one before last to form a new search direction so as to speed up the convergence for the FW method. Arezki (1990) further exploited PARTAN's benefit, and Harker (1988) presented an acceleration step for diagonalization and projection algorithms for finite-dimensional variational inequalities, which is reminiscent of a PARTAN step in nonlinear programming problems. For the move size, Weintraub et al (1985) increased the step size so as to alleviate the zigzagging problem for the FW method near the optimal solution.

Certainly, alternative solution algorithms to the FW method can be accommodated in the nested diagonalization method. Examples include some route-based solution algorithms, such as the gradient projection (GP) method, aggregated simplicial decomposition (ASD) method and disaggregated simplicial decomposition (DSD) methods. A comparison of the three route-based algorithms mentioned above was made by Chen and Chang (1998).

In consideration of both software and hardware advancement in the computer industry, employing a parallel processing technique in shortening the turn-around execution times is also permissible. The usefulness of parallel machines, including both execution times and memory requirements, for static large-scale location and travel choice models defined on the transportation network have been demonstrated by Chen and Boyce (1988). This technique can naturally be extended to the time-space network of dynamic travel choice models.

### 14.1.5 Multiple Traveler Classes

Multiple traveler classes are commonly seen in models applied to real world problems. There are several approaches to stratifying travelers. Chen et al (1994) classified travelers into three classes based on the preciseness of traffic information available to them, i.e., static, instantaneous and actual. Ran and Boyce (1996) stratified travelers into 18 combinations, according to route diversion willingness (three combinations), income and age (nine combinations), driving behavior (three combinations) and route choice behavior (three combinations). Ran et al (1997) modeled three classes of travelers, i.e., fixed route, stochastic/dynamic user-optimal and dynamic user-optimal, in an analytical dynamic traffic assignment model using the variational inequality approach and further proposed a heuristic method for solutions.

To illustrate the concept of the multi-class DUO route choice model, we consider the equilibrium conditions for two traveler classes, denoted as $d$ and $\hat{d}$, according to either dynamic or stochastic/dynamic traffic information, as follows.

$$c_{dp}^{rs*}(k)\begin{cases} = \pi_{d}^{rs}(k) & \text{if } h_{dp}^{rs*}(k) > 0 \\ \geq \pi_{d}^{rs}(k) & \text{if } h_{dp}^{rs*}(k) = 0 \end{cases} \quad \forall r,s,p,k \tag{14.15}$$

$$\hat{c}_{\hat{d}p}^{rs*}(k)\begin{cases} = \hat{\pi}_{\hat{d}}^{rs}(k) & \text{if } h_{\hat{d}p}^{rs*}(k) > 0 \\ \geq \hat{\pi}_{\hat{d}}^{rs}(k) & \text{if } h_{\hat{d}p}^{rs*}(k) = 0 \end{cases} \quad \forall r,s,p,k \tag{14.16}$$

The corresponding multi-class DUO route choice problem may be described by the following variational inequality formulation.

$$\begin{aligned} &\sum_{rs}\sum_{p}\sum_{k} c_{dp}^{rs*}(k)\left[h_{dp}^{rs}(k) - h_{dp}^{rs*}(k)\right] \\ &+\sum_{rs}\sum_{p}\sum_{k} \hat{c}_{\hat{d}p}^{rs*}(k)\left[h_{\hat{d}p}^{rs}(k) - h_{\hat{d}p}^{rs*}(k)\right] \geq 0 \quad \forall \mathbf{h} \in \Omega^* \end{aligned} \tag{14.17}$$

where $\Omega^*$ is a subset of $\Omega$ with $\delta_{apk}^{rs}(t)$ being realized at equilibrium, i.e., $\left(\delta_{apk}^{rs}(t) = \delta_{apk}^{rs*}(t)\right), \forall r,s,a,p,k,t$. The symbol $\Omega$ is delineated by the following constraints, including flow conservation, flow propagation, nonnegativity, and definitional constraints.

Flow conservation constraint:

$$\sum_{p} h_{d'p}^{rs}(k) = \bar{q}_{d'}^{rs}(k) \qquad \forall r,s,d' \in \left(d,\hat{d}\right),k \tag{14.18}$$

Flow propagation constraints:

$$u_{apk}^{rs}(t) = h_{p}^{rs}(k)\delta_{apk}^{rs}(t) \qquad \forall r,s,a,p,k,t \tag{14.19}$$

$$\sum_{t} \delta_{apk}^{rs}(t) = 1 \qquad \forall r,s,p,a \in p,k \tag{14.20}$$

$$\delta_{apk}^{rs}(t) = \{0,1\} \qquad \forall r,s,a,p,k,t \tag{14.21}$$

Nonnegativity constraint:

$$h_{d'p}^{rs}(k) \geq 0 \qquad \forall r,s,d' \in \left(d,\hat{d}\right),p,k \tag{14.22}$$

Definitional constraints:

$$\sum_{d'} h_{d'p}^{rs}(k) = h_p^{rs}(k) \quad \forall r,s,p,k \tag{14.23}$$

$$u_a(t) = \sum_{rs}\sum_{p}\sum_{d'}\sum_{k} h_{d'p}^{rs}(k)\delta_{ad'pk}^{rs}(t) \quad \forall a,t \tag{14.24}$$

$$c_{dp}^{rs}(k) = \sum_{a}\sum_{t} c_a(t)\delta_{adpt}^{rs}(t) \quad \forall r,s,p,k \tag{14.25}$$

$$\hat{c}_{dp}^{rs}(k) = c_{dp}^{rs}(k) + \frac{1}{\theta}\ln h_{dp}^{rs}(k) \quad \forall r,s,p,k \tag{14.26}$$

Unfortunately, the equivalence analysis between the equilibrium conditions (14.15)~(14.16) and the variational inequality model (14.17)~(14.26) are not proven. Furthermore, the nested diagonalization method cannot be directly applied for solutions. These unresolved issues are also left for future research.

### 14.1.6 Nonconvex Feasible Region

The feasible regions for dynamic travel choice models are essentially nonconvex, implying multiple local solutions can exist. According to our experience, using different initial solutions and/or adopting different algorithms are apt to result in different local solutions. Perhaps, meta-heuristic methods, such as a Tabu search, a genetic algorithm and a simulated annealing method have a better chance of obtaining the global solution.

### 14.1.7 Integration of Analytical- and Simulation-Based Dynamic Models

In Chapter 11, we presented two dynamic signal control schemes, i.e., network-wide and traffic-responsive. The advantages of these two analytical-based dynamic models are that various traffic control strategies can be properly evaluated, and the travelers' behavior can be easily emulated. However, in many instances, various measures of effectiveness are also needed, and random traffic events, such as non-recurrent incidents, are appropriately represented. In consideration of the sophistication of the state-of-the-art traffic simulation models, the integration of analytical- and simulation-based dynamic models becomes a natural choice for future exploration. Unless the interface can be properly set up, the integration of analytical- and simulation-based dynamic models does not necessarily end up with a convergent result.

### 14.1.8 Dynamic Origin-Destination Estimation

Given time-dependent link flows (such as 5, 10, 15 or 30 minutes intervals) for the entire analysis period, the corresponding time-space network representation can be constructed for the dedicated transportation network. Practically, this time-space network can then be deemed a *static* transportation network, and employ conventional techniques, either regression-based models or mathematical programming-based methodologies (Nguyen, 1984), to estimate origin-destination demands.

### 14.1.9 Reactive Dynamic Travel Choice Models

In this book, we have presented a set of dynamic travel choice models using *predictive* traffic information. Theoretically, another category of dynamic travel choice models using the *reactive* (or *instantaneous*) traffic information can also be formulated following the same approach. *Reactive* traffic information is especially useful for shorter trips and unpredictable future conditions. However, we found that if we accommodate reactive dynamic travel choice problems into an analytical-based model, some difficulties may be encountered with diagonalization-type solution algorithms (Chen et al, 1996):

1. The equivalence between the predictive dynamic travel choice models and the corresponding equilibrium conditions is hard to prove.
2. The shortest route searching procedure in the linearized subproblem must be performed through the corresponding time-space network, which inevitably involves flow propagation movements over several time intervals; not a coincidence with the definition of *reactive* traffic information.
3. The solution for the linearized subproblem is obtained by assigning all departure flow rates onto the time-dependent shortest routes for each O-D pair. With this assignment, travelers are not allowed to change their subroute at each intermediate decision node toward the destination, even if a better subroute than the current one is generated due to the ever-changing traffic condtions. This also contradicts the *reactive* assumption that travelers can change their subroute at each intermediate decision node.

To avoid any difficulties caused by the analytical-based model formulation, Chen and Chiu (1995, 1997) employed the simulation-based approach to deal with the predictive dynamic route choice problem; in addition, the results were compared with those obtained from an analytical model using optimal control theory. Kuwahara and Akamatsu (1997) discussed the formulation and solution algorithm of the reactive dynamic traffic assignment with the link travel time by explicitly taking into account the effects of queues under the point queue concept. To ensure that the route choice of the vehicles is clearly dependent only upon the instantaneous link travel times at the present interval $t$, but independent of the future link travel times, the assignment is decomposed with respect to the present time $t$. The assignment is then sequentially performed from the beginning of the study time period. This solution procedure is essentially the simulation-based approach. To develop an analytical-based solution algorithm, more effort is still needed.

## 14.2 Applications

### 14.2.1 Transportation Planning

Analogous to static travel choice models, dynamic travel choice models can be readily applied in transportation planning. In responding to the prevailing stochastic and incomplete/uncertain traffic information, stochastic/dynamic and fuzzy/dynamic travel choice models may provide more solid results. In fact, the application of the dynamic travel choice models in this regard is not restricted for the sake of planning only, but is also applicable for operational purposes, as shorter period traffic variations can be better captured and/or represented.

### 14.2.2 Route Guidance/Disaster Evacuation Plan

Route guidance is a fundamental function of dynamic travel choice models in intelligent transportation systems. By guiding travelers (drivers) through in-vehicle equipment to appropriate alternative routes, both the system (in terms of network utilization and MOEs) and individuals (in terms of departure time and *en route* travel cost and time) can benefit, and more benefit can be expected when real-time traffic control is operating. This function can be easily extended to a disaster evacuation plan without any difficulty. To implement route guidance functions, shortest route information has to be stored; therefore, the FW algorithm embedded in the nested diagonalization method must be modified accordingly or simply replaced by some route-based algorithm.

### 14.2.3 Economic Networks

The equilibrium principle plays an important role in economics. Since our dynamic travel choice models and associated algorithms are rooted in the equilibrated travel behavior, all equilibrium-relevant economic problems, such as spatial price, migration, Walrasian price, financial and constrained matrix problems (Nagurney, 1993), can readily adopt our dynamic model formulations and methodologies for use.

### 14.2.4 Logistics

The field of Logistics is very broad. It includes minimum cost network flow, traveling salesman, Chinese postman, covering and matching, location, vehicle routing problems, etc. In the past, temporal dynamics did not particularly address these problems; however, in today's society, without considering route travel time variation over time (time-dependent traffic congestion), or the store's operating hours (time window), the full usefulness of Logistics models will not come into play. By adopting the techniques of temporal treatment developed in this book, the importance of Logistics modeling may be enhanced to a large extent.

## 14.3 Notes

To justify the usefulness of our dynamic travel choice models, full scale implementation on a real transportation network is necessary. Toward this end, time-dependent traffic data, such as link travel time functions and O-D traffic demands, need to be prepared, which may be expensive. In addition, the traditional relationships among three link variables, i.e., flow, speed, and density, also need to be re-examined so as to comply with this new scenario.

# References


Arezki Y. and van Vlient D. 1990. A Full Analytical Implementation of PARTAN/Frank-Wolfe Algorithm for Equilibrium Assignment. *Transportation Science*, **24**, 58-62.

Akiyama T., Shao C. F. and Sasaki T. 1992. Traffic Flow on Urban Networks under Fuzzy Information. *Journal of Civil Engineering Association*, No. **449**/IV-17, 145-164. (in Japanese)

Ben-Akiva M., de Palma A. and Kanaroglu P. 1986. Dynamic Model of Peak Period Traffic Congestion with Elastic Arrival Rates. *Transportation Science*, **20**, 164-181.

Berka S., Raj J. and Tarko A. 1993. Turning Specific Link Travel Time Functions for Network Modeling. *ADVANCE Working Paper Series*, TRF-BD-03, Urban Transportation Center, University of Illinois at Chicago.

Boyce D. E., 1984. Urban Transportation Network-Equilibrium and Design Models: Recent Achievement and Future Prospects, *Environment Planning*, **A16**, 1445-1474.

Boyce D. E., Ran B. and LeBlanc L. J. 1991. Dynamic User-optimal Traffic Assignment: A New Model and Solution Techniques. *Presented at First Triennial Symposium on Transportation Analysis*, Montreal, Canada.

Boyce D. E., Ran B. and LeBlanc L. J. 1995. Solving an Instantaneous Dynamic User-Optimal Traffic Assignment Model. *Transportation Science*, **29**, 128-142.

Boyce D. E., Lee D. H. and Janson B. N. 1996. A Variational Inequality Model of an Ideal Dynamic User-Optimal Route Choice Problem. Paper will be published as a book chapter by the 4th meeting of the EURO working group on Transportation, Newcastle upon Tyne, UK.

Caprani O., Madsen, K. and Rall L. B. 1981. Integration of Interval Functions. *Society for Industrial Applied Mathematics*, **12**(3), 321-341.

Carey M. 1992. Nonconvexity of the Dynamic Traffic Assignment Problem. *Transportation Research*, **26B**, 127-133.

Carey M. 1986. A Constraint Qualification for a Dynamic Traffic Assignment Model. *Transportation Science*, **20**, 55-58.

Carey M. 1987. Optimal Time-Varying Flows on Congested Network. *Operations Research*, **35**, 58-69.

Cascetta E. 1991. A Day-to-Day and Within-Day Dynamic Stochastic Assignment Model. *Transportation Research*, **25A**, 277-291.

Chabini I., Drissi-Kaitouni O. and Florian M. 1994. Parallel Implementations of Primal and Dual Algorithms for Matrix Balancing. *Computational Techniques for Econometrics and Economic Analysis* (Edited A. A. Belslev). Kluwer Academic Publishers. Netherlands, 173-185.

Chabini I. and Florian M. 1995. *An Entropy Based Primal-Dual Algorithm for Convex and Linear Cost Transportation Problems*. Report **CRT-95-17**, University of Montreal.

Chang C. J. and Chen H. K. 1996. Autonomous Vehicles Separation Control with Fuzzy Logic Multimode Control (FLMC). *Journal of the Chinese Fuzzy Systems Association*, **2**(2), 89-106.

Chang G. L., Junchaya T. and Santiago A. J. 1993. A Real-Time Network Traffic Simulation Model for ATMS Applications. *Paper Presented at the 72nd Annual Meeting of Transportation Research Board*, Washington DC, USA.

Chang M. S. and Chen H. K. 1997. A Study on Fuzzy Algorithm of Traffic Assignment Model. *Journal of the Chinese Institute of Engineers*, **20**(2), 139-150.

Chang S. and Zhu Y. 1989. On Variational Inequalities for Fuzzy Mappings. *Fuzzy Sets and Systems*, **32**, 359-367.

Chen H. K. and Boyce D. E. 1988. Code Optimization for a Nonlinear Transportation Network Model: A Case Study. *Paper presented at the 75th North American Meetings of the Regional Science Association.*

Chen H. K., Lee D. H. and Fu C. T. 1994. The Application of A Route Diversion Strategy to a Vehicle Route Guidance System Using Multiple Driver Classes. *The Journal of Transportation Planning and Technology*, **18**, 81-105.

Chen H. K. and Wang C. Y. 1994. Methods for Asymmetric Traffic Assignment Model with Variable Demand. *Proceedings of the NSC-Part A: Physical Science and Engineering*, **18**(6). 551-560.

Chen H. K., Chiu Y. L. and Ran B. 1995. Solution Algorithms for an Instantaneous Dynamic User-optimal Route Choice Model. *Journal of Chinese Civil and Hydraulic Engineering*, **9**(1), 159-170. (in Chinese)

Chen H. K., Wang C. Y. and Boyce D. E. 1995. New Solution Algorithms for the Asymmetric Traffic Assignment Model. *Journal of the Chinese Institute of Engineers*, **18**(3), 411-426.

Chen H. K. and Tu M. L. 1996. A Note on Link Travel Time Functions under Dynamic Loads by Daganzo. Paper Presented at the 5th World Congress of the RSAI Conference, Tokyo, Japan.

Chen H. K. and Hsueh C. F. 1996. A Dynamic User-Optimal Route Choice Problem Using A Link-Based Variational, Paper Presented at the 5th World Congress of the RSAI, Tokyo, May 2-6.

Chen H. K., Hsu W. K. and Chiang W. L. 1997. A Comparison of Vertex Method with JHE Method. *Fuzzy Sets and Systems*, **86**, 155-168.

Chen H. K., Chiang W. L., Chang M. S. and Tseng H. C. 1997. A Study On Automatic Incident Detection – Application of Fuzzy Sets and Neural Networks. *Journal of the Chinese Fuzzy Systems Association*, **3**(1), 91-108.

Chen H. K. and Hsueh C. F. 1997. Combining Signal Timing Plan and Dynamic Traffic Assignment. *Presented at the 76th Annual Meeting of the Transportation Research Board*, Washington DC.

Chen, H.K. and Hsueh, C.F. 1997. A Model and an Algorithm for the Dynamic User-Optimal Route Choice Problem. *Transportation Research*, 32**B**(3), 219-234.

Chen, H.K., Chang C.W., Chang M. S. and Wang C. Y. 1998. A Comparison of Link-Based versus Route-Based Algorithms with the Dynamic User-Optimal Route Choice Problem, *77th Transportation Research Board Annual Meeting*, Washington D. C. (also Published in *Transportation Research Record*)

Chen H. K. and Hsueh C. F. 1998.03. A Discrete-Time Dynamic User-Optimal Departure Time/Route Choice Problem Route Choice Model. *Transportation Engineering*, ASCE , **124**(3), 246-254.

Chen Y. 1993. *Bilevel Programming Problems: Analysis, Algorithms and Applications.* Ph.D Thesis, University of Montreal.

Codina E. and J. Barcelo. 1995. Dynamic Traffic Assignment: Considerations on Some Deterministic Modeling Approaches. *Annals of Operations Research* **60**, 1-58.

Dafermos S. and F. T. Sparrow. 1969. The Traffic Assignment Problem for a General Network. *Journal of Research of the National Bureau of Standards-B. Mathematical Sciences*, **73B**(2), 191-217.

Dafermos S. 1980. Traffic Equilibrium and Variational Inequalities. *Transportation Science*, **14**, 24-54.

Dafermos S. 1982. Relaxation Algorithms for the General Asymmetric Traffic Equilibrium Problem, *Transportation Science*, **16**(2), 231-240.

Dafermos S. 1982. The General Multimodal Network Equilibrium Problem with Elastic Demand. *Networks*, **12**, 57-72.

Dafermos S. 1983. An Iterative Scheme for Variational Inequalities. *Mathematical Programming*, **26**, 40-47.

Dafermos S. and Nagurney A. 1984. Sensitivity Analysis for the Asymmetric Network Equilibrium Problem. *Mathematical Programming*, **28**, 174-184.

Daganzo C. F. and Sheffi Y. 1977. On Stochastic Models of Traffic Assignment. *Transportation Science*, **11**(3), 253-274.

Daganzo C. F. 1993. The Cell Transmission Model. Part I: A Simple Dynamic Representation of Highway Traffic. *California PATH Program*, **UCB-ITS-PRR-93-7**, Institute of Transportation Studies, University of California, Berkeley.

Daganzo C. F. 1995. Properties of Link Travel Time Functions Under Dynamic Load. *Transportation Research*, **29B**(2), 95-98.

Damberg O., Lundgren J. T. and Patriksson M. 1996. An Algorithm for the Stochastic User Equilibrium Problem. *Transportation Research*, **30B**(2), 115-131.

Debreu G. 1959. *Theory of Value – An Axiomatic Analysis of Economic Equilibrium*, John Wiley & Sons.

de Palma A., Ben-Akiva M., Lefevre C. and Litinas N. 1983. Stochastic Equilibrium Model of Peak Period Traffic Congestion. *Transportation Science*, **17**, 430-453.

Dial R. B. 1971. A Probabilistic Multipath Traffic Assignment Algorithm which Obviates

Path Enumeration. *Transportation Research*, **5**(2), 83-111.

Dong W. M. and Shah H. C. 1987. Vertex Methods for Computing Functions of Fuzzy Variables. *Fuzzy Sets and Systems*, **24**, 65-78.

Eaves B. C. and Saigal R. 1972. Homotopies for Computation of Fixed Points on Unbounded Regions, *Mathematical Programming*, **3**, 225-237.

Evans S. P. 1976. Derivation and Analysis of Some Models for Combining Trip Distribution and Assignment. *Transportation Research*, **10**, 37-57.

Fang S. C. 1979. *Generalized Variational Inequality, Complementarity and Fixed Point Problems: Theory and Application*, Ph.D. Dissertation, Northwestern University.

Fang S. C. and Peterson E L. 1982a. Generalized Variational Inequalities. *Journal of Optimization Theory and Applications*, **38**(3), 363-383.

Fang S. C. and Peterson E. L. 1982b. General Network Equilibrium Analysis. *International Journal of System Science*, **14**(11), 1249-1257.

Federal Highway Administration. 1992. *FRESIM User Guide*, Beta Version 3.1, McLean, VA.

Ferrari P., Treglia P., Cascetta E., Nuzzolo A. and Olivotto P. 1982. A New Method for Measuring The Quality of Circulation on Motorways. *Transportation Research*, **16B**(5), 399-418.

Ferrari P. 1988. The Reliability of the Motorway Transportation System. *Transportation Research*, **22B**(4), 291 -310.

Ferrari P. 1991. The Control of Motorway Reliability. *Transportation Research*, **25A**(6), 419-427.

Fisk C. and Boyce D. E. 1983. Alternative Variational Inequality Formulations of Network Equilibrium Travel Choice Problem. *Transportation Science*, **17**, 454-463.

Fisk C. and Nguyen A. 1982. Solution Algorithms for Network Equilibrium Models with Asymmetric User Costs. *Transportation Science*, **16**, 361-381.

Fisk C. 1980. Some Developments in Equilibrium Traffic Assignment. *Transportation Science*, **14B**, 243-255.

Florian M. and Spiess H. 1983. On Binary Mode Choice/Assignment Models. *Transportation Science*, **17**(1), 32-47.

Florian M. (Ed) 1984. *Transportation Planning Models*, Elsevier Science Publishers BV, The Netherlands.

Frank M. and Wolfe P. 1956. An Algorithm for Quadratic Programming. *Naval Research Logistics Quarterly*, **3**, 95-110.

Friesz T. L. Luque F. J. Tobin R. L. and Wie B. W. 1989. Dynamic Network Traffic Assignment Considered as a Continuous Time Optimal Control Problem. *Operations Research*, **37**, 893-901.

Friesz T. L., Tobin R. L., Cho H. J. and Mehta N. J. 1990. Sensitivity Analysis Based Heuristic Algorithms for Mathematical Programs with Variational Inequality Constraints. *Mathematical Programming*, **48**, 265-284.

Freisz T. L., Bernstein D., Smith T. E., Tobin R. L. and Wei B. W. 1993. A Variational Inequality Formulation of the Dynamic Network User Equilibrium Problem. *Operations Research*, **41**, 179-191.

Fukushima M. 1984. A Modified Frank-Wolfe Algorithm for Solving the Traffic Assignment Problem. *Transportation Research*, **18B**, 169-177

Fukushima M. 1986. A Relaxed Projection Method for Variational Inequalities.

*Mathematical Programming*, **35**, 58-70.

Gartner N.H. 1980a. Optimal Traffic Assignment with Elastic Demands: A Review. Part I: Analysis Framework. *Transportation Science*, 14, 174-191.

Gartner N.H. 1980b Optimal Traffic Assignment with Elastic Demands: A Review. Part II: Algorithmic Approaches. *Transportation Science*, 14, 192-208.

Ghali M. and Smith M. J. 1993. Traffic Assignment, Traffic Control and Road Pricing. *Proceedings of 12th International Symposium on Transportation and Traffic Theory*, Elsevier Science, Amsterda, 147-170.

Hansen E. 1992. *Global Optimization Using Interval Analysis*, Marcel Dekker.

Harker P. T. 1988. Accelerating the Convergence of the Diagonalization and Projection Algorithms for Finite-Dimensional Variational Inequalities. *Mathematical Programming*, **41**, 29-59.

Harker P. T. and Pang J. S. 1990. Finite-Dimensional Variational Inequality and Nonlinear Complementarity Problems: A Survey of Theory, Algorithms and Applications. *Mathematical Programming*, **48**, 161-220.

Ho J. K. 1980. A Successive Linear Optimization Approach to the Dynamic Traffic Assignment Problem. *Transportation Science*, **14**, 295-305.

Hurdle V. F. 1992. A Theory of Traffic Flow for Congested Conditions on Urban Arterial Streets II: Four Illustrative Examples. *Transportation Research*, **26B**(5), 381-396.

Janson B. N. 1991. Dynamic Traffic Assignment for Urban Road Networks, *Transportation Research*, **25B**, 143-161.

Janson B. N. 1993. Dynamic Traffic Assignment with Arrival Time Costs. *Proceedings of 12th International Symposium on Transportation and Traffic Theory*, Elsevier Science, Amsterdam, 127-146.

Janson B. N. and Robles J. 1995. A Quasi-Continuous Dynamic Traffic Assignment Model. *Transportation Research Record*, **1493**, 199-206.

Kinderlehrer D. and Stampacchia G. 1980. *An Introduction to Variational Inequalities and Their Applications*. Academic Press, New York.

Kluge R. and Telschow G. 1977. *Theory of Nonlinear Operators*, Edited by R. Kluge and A. Muller, Berlin: Akademe-Verlag, 135-163.

Korpelevich G. M. 1977. The Extragradient Method for Finding Saddle Points and Other Problem. *Matekon*, **13**, 35-49. (Cited from Nagurney, 1993)

Kurt G. 1979. *Approximation of Fixed Points and Functional Differential Inequalities*, Edited by H. O. Peitgen, Springer, 126-135.

Kuwahara M. and Akamatsu T. 1997. Decomposition of the Reactive Dynamic Assignments with Queues for a Many-to-Many Origin-Destination Pattern. *Transportation Research*, **31B**(1), 1-10.

Kvanli A. H., Guynes C. S. and Pavur R. J. 1989. *Introduction to Business Statistics : A Computer Integrated Approach*, Second Edition, West Publishing Company, United States of America.

Lai Y. J. and Hwang C. L. 1992. *Fuzzy Mathematical Programming-- Methods and Applications*, Springer-Verlag, Berlin, Germany.

Larsson T. and Patriksson M. 1995. An Augmented Lagrangean Dual Algorithm for Link Capacity Side Constrained Traffic Assignment Problems. *Transportation Research*, **29B**(6), 433-455.

Larsson T. and Patriksson M. 1992. Simplicial Decomposition with Disaggregated

Representation for the Traffic Assignment Problem. *Transportation Science*, **26**(1), 4-17.

Lee B. S. Lee G. M. Cho S. J. and Kim D. S. 1993. A Variational Inequality for Fuzzy Mappings. *Proceedings of the 5th International Fuzzy Systems Association*, World Congress, Seoul, 326-329.

Lee G. M. Kim D. S. and Lee B. S. 1996. Strongly Quasivariational Inequalities for Fuzzy Mappings. *Fuzzy Sets and Systems*, **78**, 381-386.

LeBlanc L. J., Helgason R. V. and Boyce D. E. 1985. Efficient Algorithms for Solving Elastic Demand Traffic Assignment Problems and Mode Split-Assignment Problems. *Transportation Science*, **15**, 306-317.

Leonard D. R., Tough J. B. and Baguley P.C. 1978. *CONTRAM: A Traffic Assignment Model for Predicting Flows and Queues During Peak Periods*. TRRL Report **LR841**, Transport and Road Research Laboratory, Crowthorne, Berkshire, England.

Leonard D. R. Gower P. and Taylor N. B. 1989. *CONTRAM: Structure of the Model*. TRRL Report **RR178**, Transport and Road Research Laboratory, Crowthorne, Berkshire, England.

Liao H. L. and Wang H. F. 1995. User Equilibrium in Traffic Assignment – An Application of Variational Inequality with Fuzzy Functions. *Proceedings of the International Joint Conference of CFSA/IFIS/SOFT on Fuzzy Theory and Applications*, Taipei, 356-361.

Lieberman E. B. 1972. Simulation of Corridor Traffic-The SCOT Model. *Highway Research Record*, **409**, HRB, National Research Council, Washington DC, 34-45.

Lieberman E. B. and Wicks D. A. 1980. TRAFLO: A New Tool to Evaluate Transportation System Management Strategies. *Transportation Research Record*, **772**, TRB, National Research Council, Washington DC, 9-15.

Lieberman E. B. 1993. An Advanced Approach to Meeting Saturated Flow Requirements. *Presented at the 72nd Annual Transportation Research Board Meeting*, Washington, DC.

Luck J. C. 1972. Mathematical Models for Landform Evaluation. *Journal of Geophysical Research*, 77, 2460-2464.

Luque F. J. and T. L. Friesz. 1980. Dynamic Traffic Assignment Considered as a Continuous Time Optimal Control Problem. Presented at the *TIMS / ORSA Joint National Meeting*, Washington, D. C.

Maher M. J. and Hughes P. C. 1997. A Probit-Based Stochastic User Equilibrium Assignment Model, *Transportation Research*, **31B**, 341-355.

Mahmassani H. S., Hu T. and R. Jayakrishnan. 1992. Dynamic Traffic Assignment and Simulation for Advanced Network Informatics (DYNASMART). *Presented at the 2nd International Seminar on Urban Traffic Networks*, Capri, Italy.

Makigami k., Newell G. F. and Rothery R. 1972. Three-dimensional Representations of Traffic Flow. *Transportation Science*, **5**, 302-313.

Matsui H. 1987. A Model of Dynamic Traffic Assignment. *Text of Infrastructure Planning Lectures*, JSCE, **18**, 84-96.

May A. D. 1981. Models for Freeway Corridor Analysis. *Special Report **194**: The Application of Traffic Simulation Models*, TRB, National Research Council, Washington DC.

May A. D. 1990. *Traffic Flow Fundamentals.* Prentice Hall, Englewood Cliffs, NJ.

Meneguzzer C. 1990. *Implementation and Evaluation of an Asymmetric Equilibrium Route Choice Model Incorporating Intersection-Related Travel Times*, Ph.D. Dissertation, Department of Civil Engineering, University of Illinois at Urbana-Champaign, Urbana, Illinois.

Merchant D. K. and Nemhauser G. L. 1978a. A Model and an Algorithm for the Dynamic Traffic Assignment Problems. *Transportation Science*, **12**, 183-199.

Merchant D. K. and Nemhauser G. L. 1978b. Optimality Conditions for a Dynamic Traffic Assignment Model. *Transportation Science*, **12**, 200-207.

Merrill O. H. 1972. *Applications and Extensions of an Algorithm that Computes Fixed Points of Certain Upper Semicontinuous Point-to-Set Mappings*, University of Michigan, Ph.D. Thesis.

Moore R. E. 1985. *Computational Functional Analysis*, Ellis Horwood Limited.

Nagurney A. 1984. Comparative Tests of Multimodal Traffic Equilibrium Methods. *Transportation Research*, **18B**(6), 469-485.

Nagurney A. 1986. Computational Comparisons of Algorithms for General Asymmetric Traffic Equilibrium Problems with Fixed and Elastic Demands. *Transportation Research*, **20B**(1), 78-84.

Nagurney A. 1993. *Network Economics : A Variational Inequality Approach*, Kluwer Academic Publishers, Massachusetts.

Newell G. F. 1988. Traffic Flow in the Morning Commute. *Transportation Science*, **22**, 47-58.

Newell G. F. 1993. A Simplified Theory of Kinematic Waves in Highway Traffic: I. General Theory; II. Queuing at Freeway Bottlenecks; III. Multi-Destination Flows. *Transportation Research*, **27B**(4), 281-303.

Noor M. A. 1993. Variational Inequalities for Fuzzy Mappings (I), *Fuzzy Sets and Systems*, **55**, 309-312.

Nguyen S. and Dupuis C. 1984. An Efficient Method for Computing Traffic Equilibria in Networks with Asymmetric Transportation Costs. *Transportation Science*, **18**, 185-202.

Orda A. and R. Raphael. 1991. Minimum Weight Paths in Time-Dependent Networks. *Networks 21*, 295-319.

Ortega J. M. and Rheinboldt W. C. 1970. *Iterative Solution of Nonlinear Equations in Several Variables*, Academic Press, New York, NY.

Patriksson M. 1994. *The Traffic Assignment Problem - Models and Methods*, VSP BV, The Netherlands.

Payne H. J. and Thompsom W. A. 1975. Traffic Assignment on Transportation Networks with Capacity Constraints and Queueing. *Paper presented at the 47th National ORSA Meeting/TIMS 1975 North-American Meeting*. Chicago, IL.

Ran B. and T. Shimazaki. 1989a. A General Model and Algorithm for the Dynamic Traffic Assignment Problem. Proceedings of the *Fifth World Conference on Transport Research*, Yokohama, Japan.

Ran B. and Shimazaki T. 1989b. Dynamic User Equilibrium Traffic Assignment for Congested Transportation Networks. Presented at the *Fifth World Conference on Transport Research*, Yokohama, Japan.

Ran B., Boyce D. E. and LeBlanc L. J. 1992. Dynamic User-Optimal Route Choice Models Based on Stochastic Route Travel Times. Presented at The Second International Capri Seminar on Urban Traffic Nerworks, Capri, Italy.

Ran B. Rouphail N. M. Tarko A. and Boyce D. E. 1994. Toward a Class of Link Travel Time Function for Dynamic Network Model. Submitted to *Transportation Research.*

Ran B. and Boyce D. E.. 1994. *Dynamic Urban Transportation Network Models : Theory and Implications for Intelligent Vehicle Highway Systems*, Lecture Notes in Economics and Mathematical Systems **417**, Springer-Verlag, New York.

Ran B. and D. E. Boyce. 1996. *Modeling Dynamic Transportation Networks : An Intelligent Transportation System Oriented Approach*, Second Revised Edition, Springer, New York.

Ran B., Li Y. I. and Soetopo B. D. 1997, An Analytical Path-Based Multi_Class Dynamic Traffic Assignment Model. *Paper Presented at the 76th Annual Meeting of Transportation Research Board*, Washington DC, USA.

Richards P. I. 1965. Shock Wave on the Highway. *Operations Research*, **4**, 42-51.

Roberson J. A. and Crowe C. T. 1988. *Engineering Fluid Mechanics.* Third Edition, Houghton Mifflin Company.

Rockafellar R. T. 1968. Convex Functions, Monotone operators, and Variational Inequalities. *Theory and Applications of Monotone Operators*, Edited by Ghizzetti A., Oderisi E., Gubbio, Italy. Also appeared in *Proceedings of NATO Advanced Study Institute*, 35-65.

Rockafellar R. T. 1970. *Convex Analysis*, Princeton University Press.

Rouphail N. M. and Sisiopiku V. 1993. Travel Time and Loop Detector Output Analysis on Dundee Road Closed-Loop Signal System. *ADVANCE Working Paper Series*, TRF-DF-02, Urban Transportation Center, University of Illinois at Chicago.

Sasaki, T., Fukuyama, M. and Namikawa Y. 1984. An Approximative Analysis of the Hydrodynamic Theory on Traffic Flow and a Formulation of a Traffic Simulation Model. *Proceedings of 9th International Symposium on Transportation and Traffic Theory*, Edited by Vollmuller, J. and Hamerslag, R., VNU Science Press, Utretch, 1-20.

Sen, A. Thakuriah P. V. and Liu N. 1993. Design of the Travel Time Forecasting Procedure. *ADVANCE Working Paper Series*, TRF-TT-11, Urban Transportation Center, University of Illinois at Chicago.

Sheffi, Y. 1985. *Urban Transportation Networks : Equilibrium Analysis with Mathematical Programming Methods*, Prentice-Hall, Englewood Cliffs, NJ.

Sheffi, Y. and Powell W. B. 1982. An Algorithm for the Equilibrium Assignment Problem with Random Link Times. *Networks* **12**, 191-207.

Sheffi, Y. and Powell W. B. 1981. A Comparison of Stochastic and Deterministic Traffic Assignment over Congested Networks. *Transportation Research*, **15B**, 53-64.

Sisiopiku, V. P. and Rouphail N. M. 1993. Exploratory Analysis of the Correlation between Arterial Travel Times and Detector Data from Simulation and Field Studies. *ADVANCE Working Paper Series*, TRF-DF-02, Urban Transportation Center, University of Illinois at Chicago.

Sisiopiku V. P. Rouphail N. M. and Tarko A. 1994. Estimating Travel Times on Freeway Segments. *ADVANCE Working Paper Series*, TRF-DF-101, Urban Transportation Center, University of Illinois at Chicago.

Smith M. J. 1979. The Existence, Uniqueness and Stability of Traffic Equilibria. *Transportation Research*, **13B**, 259-304.

Smith M. J. 1987. Traffic Control and Traffic Assignment in A Signal-Controlled Network with Queueing. *Proceedings of 11th International Symposium on Transportation and Traffic Theory*, Elsevier Science, Boston, MA.

Smith M. J. and Ghali M. 1990. The Dynamics of Route Choice and Traffic Control: A Theoretical Study. *Transportation Research*, **24B**, 409-422.

Smith M. J. 1993. A New Dynamic Traffic Model and the Existence and Calculation of Dynamic User Equilibria on Congested Capacity-Constrained Road Networks. *Transportation Research*, **27B**, 49-63.

Taylor N. B. 1990. *CONTRAM 5: An Enhanced Traffic Assignment Model*. TRRL Report RR249, Transport and Road Research Laboratory, Crowthorne, Berkshire, England.

Tanaka H., Okuda T. and Asai K, 1974. On Fuzzy Mathematical Programming. *Journal of Cybernetics* **3**, 37-46.

Thomas, N. 1994. *A Traffic Incident Detection Model for Arterial Streets Equipped with Fixed and Mobile Detection Systems*, Thesis for the Degree of Doctor of Philosophy in Civil Engineering, University of Illinois at Chicago.

Tobin R. L. 1986. Sensitivity Analysis for Variational Inequalities. *Journal of Optimization Theory and Applications*, **48**(1). 191-204.

Tobin R. L. and Friesz T. 1988. Sensitivity Analysis for Equilibrium Network Flow. *Transportation Science*, **22**(4). 242-250.

Transportation Research Board. 1985. *Highway Capacity Manual*, Special Report 209, TRB, Washington DC.

Van Aerde M. 1992. INTEGRATION : A Dynamic Traffic Simulation Assignment Model. *Presented at the IVHS Dynamic Traffic Assignment and Simulation Workshop*, Federal Highway Administration Mclean, Virginia.

van Vuren T. and Van Vliet D. 1992. *Route Choice and Signal Control -- The Potential for Integrated Route Guidance*, Avebury, Great Britain.

Van Vliet D. 1982. SATURN: A Modern Assignment Model. *Traffic Engineering Control*, **23**, 578-581.

Van Vliet D. 1990. *Fundamental Requirements of Full-scale Dynamic Route Guidance: Recent Developments to SATURN*. Technical Note 1, Institute for Transport Studies, University of Leeds.

Vaughan R, Hurdle V. F. and Hauer E. 1984. A Traffic Flow Model with Time Dependent O-D Patterns. *Proceedings of the 9th International Symposium on Transportation and Traffic Theory*, Edited by Vollmuller, J. and Hamerslag, R., VNU Science Press, Utretch, 155-178.

Vaughan R. and Hurdle V. F. 1992. A Theory of Traffic Flow for Congested Conditions on Urban Arterial Streets I: Theoretical Development. *Transportation Research*,

**26B**(5), 397-415.

Vythoulkas P. C. 1990. Two Models for Predicting Dynamic Stochastic Equilibria in Urban Transportation Networks. *Proceedings of 11th International Symposium on Transportation and Traffic Theory*, Elsevier Science, Amsterdam, 253-272.

Wang H. F. and Liao H. L. 1994. Variational Inequality with Fuzzy Functions, Working Paper, National Tsing-Hua University, Hsinchu, Taiwan.

Wang H. F. and Liao H. L. 1995. Variational Inequality with Fuzzy Function, Submitted to *Fuzzy Mathematics*.

Wang H. F. and Liao H. L. 1997. Resolution of the Variational Problems, Submitted to *the Journal of Optimization Theory and Applications* (First Revision).

Wardrop J. G. 1952. Some Theoretical Aspects of Road Traffic Research. *Proceedings of the Institution of Civil Engineers, Part II*, **1**, 325-378.

Webster F. V. 1958. Traffic Signal Settings. *Road Research Laboratory Technical Paper*, **39**, HMSO, Landon.

Weintraub A. C., palmas C. and Jaime G. 1985. Accelerating Convergence of the Frank-Wolfe Algorithms, *Transportation Research*, **19B**, 113-122.

Wicks D. A. 1977. *INTRAS-A Microscopic Corridor Simulation Model*, Vol. 1, FHWA, U.S. Department of Transportation.

Wicks D. A., Lieberman E. B. and KLD Associates 1977. *Development and Testing of INTRAS-A Microscopic Corridor Simulation Model*, Vol. 1: *Program Design, Parameter Calibration, and Freeway Dynamics Component Development*. Report FHWA-RD-80-106, FHWA, U.S. Department of Transportation.

Wie B. W., Friesz T. L. and Tobin R. L. 1990. Dynamic User Optimal Traffic Assignment on Congested Multidestination Networks. *Transportation Research*, **24B**, 443-451.

Wie B. W., Tobin R. L. and Friesz T. L. 1994. The Augmented Lagrangian Method for Solving Dynamic Network Traffic Assignment Model in Discrete Time. *Transportation Science*, **28**, 204-220.

Yang H. and Yagar S. 1994. Traffic Assignment and Traffic Control in General Freeway-Arterial Corridor Systems. *Transportation Research*, **28B**(6), 463-486.

Yang H. and Yagar S. 1995. Traffic Assignment and Signal Control in Saturated Road Networks. *Transportation Research*, **29A**(2), 125-139.

Zadeh L. A. 1978. Fuzzy Sets as a Basis for a Theory of Possibility. *Fuzzy Sets and Systems* **1**, 3-28.

Ziliaskopoulos A. K. and Mahmassani H. S. 1993?. Time-Dependent, Shortest-Path Algorithm for Real-Time Intelligent Vehicle Highway System Applications. *Transportation Research Record* **1408**. 94-100.

# List of Figures

# List of Tables

# Subject Index

# Author Index

# Summary of Notation

This is a summary of symbols used in this book. Other symbols are defined in this book, as needed. A vector is assumed to be a column vector, unless noted otherwise.

| | |
|---|---|
| $a$ | link designation |
| $A(j)$ | set of links whose tail node is $j$ |
| $A^{-1}$ | the inverse of the matrix $A$ |
| $B(j)$ | set of links whose head node is $j$ |
| $c_a(t)$ | travel time for link $a$ during time interval $t$ |
| $c_{a_0}(t)$ | free flow travel time for link $a$ during interval $t$ |
| $c_{ma}(t)$ | travel time for link $a$ by mode $m$ during time interval $t$ |
| $\hat{c}_a(t)$ | marginal travel time for link $a$ during time interval $t$ |
| $\widehat{c}_a(t)$ | perceived travel time for link $a$ during time interval $t$ |
| $\breve{c}_a(t)$ | capacitated travel time for link $a$ during time interval $t$ |
| $\tilde{c}_a(t)$ | fuzzy travel times for link $a$ during time interval $t$ |
| $\tilde{c}_{a,\alpha}(t)$ | interval-valued travel times at $\alpha$-cut level for link $a$ during time interval $t$ |
| $\overline{\tilde{c}}_{a_0}(t)$ | representative believed free flow travel time for link $a$ during interval $t$ |
| $\overline{\tilde{c}}_a(t)$ | representative believed travel time for link $a$ during interval $t$ |
| $\overline{\tilde{c}}_{a,\alpha}(t)$ | representative believed travel time for link $a$ during time interval $t$ at $\alpha$-cut level ($\geq \alpha$) |
| $\tilde{c}_a^L(t)$ | lower bound of the possibilistic distribution for link $a$ during time interval $t$ |
| $\tilde{c}_a^M(t)$ | main value of possibilistic distribution for link $a$ during time interval $t$ |
| $\tilde{c}_a^U(t)$ | upper bound of possibilistic distribution for link $a$ during time interval $t$ |

| | |
|---|---|
| $\tilde{c}^{L}_{a,\alpha}(t)$ | lower bound of possibilistic distribution for link *a* during time interval *t* at $\alpha$-cut level ($\geq \alpha$) |
| $\tilde{c}^{U}_{a,\alpha}(t)$ | upper bound of possibilistic distribution for link *a* during time interval *t* at $\alpha$-cut level ($\geq \alpha$) |
| $\bar{\tilde{c}}^{rs}_{p}(k)$ | representative believed travel time for route *p* between O-D pair *rs* during interval *k* |
| $\bar{\tilde{c}}^{rs}_{p,\alpha}(k)$ | representative travel time at $\alpha$-cut level for route *p* between O-D pair *rs* during time interval *k* |
| $c^{rs}_{p}(k)$ | travel time for route *p* between O-D pair *rs* during time interval *k* |
| $c^{rs}_{mp}(k)$ | travel time for route *p* by mode *m* between O-D pair *rs* during time interval *k* |
| $\hat{c}^{rs}_{p}(k)$ | marginal travel time for route *p* between O-D pair *rs* during time interval *k* |
| $\tilde{c}^{rs}_{p}(k)$ | perceived travel time for route *p* between O-D pair *rs* during time interval *k* |
| $\tilde{c}^{rs}_{p}(k)$ | capacitated travel time for route *p* between O-D pair *rs* during time interval *k* |
| $\tilde{c}^{rs}_{p}(k)$ | fuzzy travel times for route *p* between O-D pair *rs* during time interval *k* |
| $\tilde{c}^{rs}_{p,\alpha}(k)$ | interval-valued travel times at $\alpha$-cut level for route *p* between O-D pair *rs* during time interval *k* |
| $\bar{c}^{rs}_{p,\alpha}(k)$ | upper bound of interval-valued route travel times $\tilde{c}^{rs}_{p,\alpha}(k)$ |
| $\underline{c}^{rs}_{p,\alpha}(k)$ | lower bound of interval-valued route travel times $\tilde{c}^{rs}_{p,\alpha}(k)$ |
| $\mathbf{c}$ | vector of link travel times |
| $\mathbf{c}^{T}$ | transpose of a vector $\mathbf{c}$ |
| $CAP_{a}(t)$ | capacity for link *a* during time interval *t* |
| $C^{I}(t)$ | cycle length for intersection *I* during time interval *t* |
| $D^{-1}_{rs}(\bullet)$ | inverse of demand function between O-D pair *rs* |
| $D^{-1}_{rsk}(\bullet)$ | inverse of demand function between O-D pair *rs* during time interval *k* |
| $\mathbf{D}^{-1}$ | vector of the inverse of demand function |
| $\mathbf{d}$ | vector of search direction |
| $d$ | traffic delay or search direction of descent methods |
| $e^{rs}(k)$ | excess demand between O-D pair *rs* during time interval *k* |

| | |
|---|---|
| $e^{rs}$ | excess demand between O-D pair *rs* regardless of departure time |
| $e_{a\max}$ | maximal traffic density for link *a* |
| $f_a$ | flow on link *a* |
| $F$ | travel time function of link flows |
| $g_I^m(t)$ | green time associated with phase *m* at intersection *I* during time interval *t* |
| $g'^m_I(t)$ | green time associated with phase *m* at intersection *I* during time interval *t* (subproblem) |
| $\underline{g}_I^m$ | lower limit for the green time associated with phase *m* at intersection *I* |
| $h_p^{rs}$ | departure flow rate on route *p* from origin *r* toward destination *s* |
| $h_p^{rs}(k)$ | departure flow rate on route *p* from origin *r* toward destination *s* during time interval *k* |
| $h_{mp}^{rs}(k)$ | departure flow rate of class *m* vehicles on route *p* from origin *r* toward destination *s* during time interval *k* |
| $H_a^m(t)$ | degree of saturation over link *a* associated with phase *m* during time interval *t* |
| $I$ | intersection designation |
| $j$ | node designation |
| $k$ | time interval designation which usually denotes the departure time interval for a route |
| $l_I^m$ | loss time associated with phase *m* at intersection *I* |
| $L$ | *Lipschitz* constant |
| $m$ | phase or mode designation |
| $M^s(k)$ | attractiveness associated with destination *s* during time interval *k* |
| $M^s$ | attractiveness associated with destination *s* regardless of departure time |
| $o_a(t)$ | exit flow from link *a* during time interval *t* (subproblem variable) |
| $p$ | route designation |
| $p_a(t)$ | inflow rate into link *a* during time interval *t* (subproblem variable) |
| $p_{apk}^{rs}(t)$ | part of inflow rate on link *a* during time interval *t* that is departing origin *r* toward destination *s* over route *p* during time interval *k* (subproblem variable) |
| $\mathbf{p}$ | vector of link inflow rates (subproblem variable) |
| $q^{rs}$ | departure flow rate from origin *r* toward destination *s* |

| | |
|---|---|
| $q^{rs}(k)$ | departure flow rate from origin $r$ toward destination $s$ during time interval $k$ |
| $q^{r}(k)$ | departure flow rate from origin $r$ during time interval $k$ |
| $q_m^{rs}(k)$ | departure flow rate of class $m$ vehicles from origin $r$ toward destination $s$ during time interval $k$ |
| $q_m^{r}(k)$ | departure flow rate of class $m$ vehicles from origin $r$ during time interval $k$ |
| $q^{r}$ | departure flow rate from origin $r$ regardless of departure time |
| $\bar{q}^{rs}(k)$ | fixed departure flow rate from origin $r$ toward destination $s$ during time interval $k$ |
| $\bar{q}^{rs}$ | fixed departure flow rate between O-D pair $rs$ regardless of departure time |
| $\bar{q}^{r}$ | fixed departure flow rate from origin $r$ regardless of departure time |
| $\bar{q}_{\max}^{rs}$ | maximal departure flow rate between O-D pair $rs$ |
| $\bar{q}_{\max}^{r}$ | maximal departure flow rate from origin $r$ |
| $\mathbf{q}$ | vector of O-D demands |
| Q | number of queuing vehicles |
| $R[\cdot]$ | operator for representative link travel times |
| R | the real line |
| $R^n$ | Euclidean $n$-dimensional space |
| $R_+^n$ | Euclidean $n$-dimensional space |
| $r$ | origin designation |
| $s$ | destination designation |
| $S_a^m$ | saturation flow rate for link $a$ associated with phase $m$ |
| $t$ | time interval designation which usually denotes the link entering time interval |
| $T$ | analysis period |
| $\lvert T \rvert$ | total number of time intervals |
| $\mathbf{u}$ | vector of link inflow rates |
| $u_a(t)$ | inflow rate into link $a$ during time interval $t$ |
| $u_{ma}(t)$ | inflow rate of class $m$ vehicles into link $a$ during time interval $t$ |
| $u_a^{rs}(t)$ | inflow rate into link $a$ that is associated with O-D pair $rs$ during time interval $t$ |
| $u_{ap}(t)$ | inflow rate into link $a$ on route $p$ during time interval $t$ |

| | |
|---|---|
| $u_{ap}^{rs}(t)$ | inflow rate into link $a$ on route $p$ between O-D pair $rs$ during time interval $t$ |
| $u_{apk}^{rs}(t)$ | part of inflow rate for link $a$ during time interval $t$ that is departing origin $r$ over route $p$ toward destination $s$ during time interval $k$ |
| $v_a(t)$ | exit flow rate from link $a$ during time interval $t$ |
| $v_{a\max}$ | exit flow capacity for link $a$ |
| $v_a^{rs}(t)$ | exit flow rate from link $a$ associated with O-D pair $rs$ during time interval $t$ |
| $v_{ap}(t)$ | exit flow rate from link $a$ on route $p$ during time interval $t$ |
| $v_{ap}^{rs}(t)$ | exit flow rate from link $a$ on route $p$ between O-D pair $rs$ during time interval $t$ |
| $v_{apk}^{rs}(t)$ | part of exit flow rate from link $a$ during time interval $t$ that is departing origin $r$ over route $p$ toward destination $s$ during time interval $k$ |
| $w_p^{rs}(k)$ | departure flow rate from origin $r$ toward destination $s$ on route $p$ during time interval $k$ (subproblem) |
| $w_{mp}^{rs}(k)$ | departure flow rate of class $m$ vehicles from origin $r$ toward destination $s$ on route $p$ during time interval $k$ (subproblem) |
| $x_a(t)$ | number of vehicles on link $a$ at the beginning of time interval $t$ |
| $x_a^{rs}(t)$ | number of vehicles on link $a$ at the beginning of time interval $t$ that is associated with O-D pair $rs$ |
| $x_{ap}^{rs}(t)$ | number of vehicles on link $a$ at the beginning of time interval $t$ that is departing from origin $r$ over route $p$ toward destination $s$ |
| $y_a(t)$ | number of vehicles on link $a$ at the beginning of time interval $t$ (subproblem variable) |
| $y_a^{rs}(t)$ | number of vehicles on link $a$ at the beginning of time interval $t$ that is associated with O-D pair $rs$ (subproblem variable) |
| $y_{ap}^{rs}(t)$ | number of vehicles on link $a$ at the beginning of time interval $t$ that is departing from origin $r$ over route $p$ toward destination $s$ (subproblem variable) |
| $z_\alpha$ | objective value at $\alpha$ -cut level |
| $\alpha$ | $\alpha$ -cut of the membership function of a fuzzy variable; minimum acceptance level |
| $\beta$ | dual variable associated with link capacity constraint |
| $\gamma$ | weight |
| $\Delta_1, \Delta_2$ | small amount units |

| | |
|---|---|
| $\delta_{apk}^{rs}(t)$ | 1, if inflow rate on link $a$ during time interval $t$ departs from origin $r$ over route $p$ toward destination $s$ during time interval $k$, otherwise, 0 |
| $\delta_{mapk}^{rs}(t)$ | 1, if inflow rate of class $m$ vehicles on link $a$ during time interval $t$ departs from origin $r$ over route $p$ toward destination $s$ during time interval $k$, otherwise, 0 |
| $\bar{\delta}_{apk}^{rs}(t)$ | estimated indicator variable that determines the presence of inflow rate $u_{apk}^{rs}(t)$ |
| $\delta_{apk}^{rs,B}(t)$ | 1, if inflow rate on link a during interval t departs origin r over route p toward destination s during interval k; otherwise, 0. (The realization of the indicator variable $\delta_{apk}^{rs,B}(t)$ is based on the believed actual link travel time $\tau_a^B(t)$). |
| $\delta_{a_1}^t(i)$ | 1, if inflow rate on link $a$ during time interval $i$ exits the link during time interval $t$, i.e., $t = i + \tau_a(i)$; otherwise, 0 |
| $\delta_{a_2}^t(i)$ | 1, if inflow rate on link $a$ during time interval $i$ ($i<t$) remains on the link at the beginning of time interval $t$, otherwise, 0 |
| $\varepsilon$ | random error or convergence criterion |
| $\theta$ | parameter of Logit model |
| $\lambda$ | move size |
| $\mu$ | membership function of fuzzy variables |
| $\xi$ | penalty parameter |
| $\Pi$ | possibility distribution |
| $\Pi_{c_a(t)}$ | possibilistic distribution of travel time for link $a$ during time interval $t$ |
| $\Pi_{c_a(t)}(\chi)$ | grade membership of possibilistic distribution of travel time for link $a$ during time interval $t$ |
| $\pi^{rs}(k)$ | minimal route travel time between O-D pair $rs$ during time interval $k$ |
| $\pi_m^{rs}(k)$ | minimal route travel time by mode $m$ between O-D pair $rs$ during time interval $k$ |
| $\pi_i^{rs}(k)$ | minimal subroute travel time between origin $r$ and intermediate node $i$ toward destination $s$ during time interval $k$ |
| $\pi^{rs}$ | minimal route travel time between O-D pair $rs$ regardless of departure time |
| $\pi_m^{rs}$ | minimal route travel time by mode $m$ between O-D pair $rs$ regardless of departure time |
| $\pi_i^{rs}$ | minimal subroute travel time between origin $r$ and intermediate node $i$ toward destination $s$ regardless of departure time |

| | |
|---|---|
| $\hat{\pi}^{rs}(k)$ | minimal marginal route travel time between O-D pair *rs* during time interval *k* |
| $\widehat{\pi}^{rs}(k)$ | minimal perceived route travel time between O-D pair *rs* during time interval *k* |
| $\breve{\pi}^{rs}(k)$ | minimal capacitated route travel time between O-D pair *rs* during time interval *k* |
| $\widetilde{\pi}^{rs}(k)$ | minimal fuzzy route travel times between O-D pair *rs* during time interval *k* |
| $\widetilde{\pi}^{rs}$ | minimal fuzzy route travel times between O-D pair *rs* regardless of departure time |
| $\overline{\widetilde{\pi}}_{\alpha}^{rs}(k)$ | minimal representative believed travel time at $\alpha$ -cut level between O-D pair *rs* during interval *k* |
| $\widetilde{\pi}_{\alpha}^{rs}(k)$ | minimal interval-valued route travel times at $\alpha$ -cut level between O-D pair *rs* during time interval *k* |
| $\widetilde{\pi}_{\alpha}^{rs}$ | minimal interval-valued route travel times at $\alpha$ -cut level between O-D pair *rs* regardless of departure time |
| $\rho$ | parameter of projection type methods |
| $\sigma$ | perturbation parameter |
| $\tau_a(t)$ | actual travel time for link *a* during time interval *t* |
| $\tau_a^B(t)$ | believed actual link travel time for link *a* during interval *t* |
| $\dot{\tau}_a(t)$ | changing rate of actual travel time $\tau_a(t)$ |
| $\overline{\tau}_a(t)$ | estimated actual travel time for link *a* during time interval *t* |
| $\varphi^{rs}(k)$ | auto preference for flows departing origin *r* toward destination *s* during time interval *k* |
| $\varphi^{rs}$ | auto preference for flows departing origin *r* toward destination *s* regardless of departure time |
| $\widetilde{\Psi}_{\alpha}$ | vector of fuzzy mapping |
| $\Omega$ | feasible region |
| $\overline{\Omega}$ | feasible region associated with estimated link actual travel times $\overline{\tau}_a(t)$ |
| $\Omega_a$ | feasible region specific for link inflows |
| $\Omega_{\alpha}$ | fuzzy feasible region at $\alpha$ -cut level |
| $\overline{\Omega}_{\alpha}$ | subset of $\Omega_{\alpha}$ with $\delta_{apk}^{rs}(t) = \overline{\delta}_{apk}^{rs}(t)$ |
| $\Omega_{\alpha}^{*}$ | subset of $\Omega_{\alpha}$ with $\delta_{apk}^{rs}(t) = \delta_{apk}^{rs*}(t)$ |
| * | equilibrium condition |

| | |
|---|---|
| $\arg\min_{x \in k}$ | the set of $x \in k$ attaining the minimum |
| $\nabla z$ | gradient of $z$: $R^n \mapsto R$ |
| $\mapsto$ | maps to |
| $\to$ | tends to |
| $\forall$ | for all |
| $\in$ | an element of |
| $\subset$ | subset of |
| $\Rightarrow$ | implies to |
| $\exists$ | there exists |
| $\infty$ | infinity |